Redox Metabolism and Longevity Relationships in Animals and Plants

EXPERIMENTAL BIOLOGY REVIEWS

Environmental Stress and Gene Regulation
Sex Determination in Plants
Plant Carbohydrate Biochemistry
Programmed Cell Death in Animals and Plants
Biomechanics in Animal Behaviour
Cell and Molecular Biology of Wood Formation
Molecular Mechanisms of Metabolic Arrest
Environment and Animal Development: Genes, Life Histories and Plasticity
Brain Stem Cells
Endocrine Interactions of Insect Parasites and Pathogens
Vertebrate Biomechanics and Evolution
Osmoregulation and Drinking in Vertebrates
Host–Parasite Interactions
The Nuclear Envelope
The Carbon Balance of Forest Biomes
Comparative Genomics and Proteomics in Drug Discovery
The Eukaryotic Cell Cycle
Drosophila: A Toolbox for the Study of Neurodegenerative Disease
Practical Systems Biology
Insect Taste

Redox Metabolism and Longevity Relationships in Animals and Plants

Edited by

CHRISTINE H. FOYER
School of Agriculture, Food and Rural Development,
University of Newcastle upon Tyne, Newcastle upon Tyne, UK

RICHARD FARAGHER
School of Pharmacy and Biomolecular Sciences,
University of Brighton, Brighton, UK

PAUL J. THORNALLEY
Warwick Medical School, University of Warwick, Coventry, UK

Taylor & Francis
Taylor & Francis Group
LONDON AND NEW YORK

Published by:
Taylor & Francis Group
In UK: 2 Park Square, Milton Park
Abingdon, Oxon OX14 4RN
In US: 270 Madison Avenue
New York, NY 10016

Transferred to Digital Printing 2009

First published 2009

ISBN: 9780415419543

A catalog record for this book is available from the British Library.

Library of Congress Cataloging-in-Publication Data

Redox metabolism and longevity relationships in animals and plants / edited by Christine H. Foyer, Richard Faragher, Paul J. Thornalley. - - 1st ed.
p. cm. - - (SEB experimental biology series; 62)
Includes index.
ISBN 978-0-415-41954-3
1. Animals- -Longevity. 2. Plants- -Longevity. 3. Oxidation-reduction reaction. 4. Energy metabolism. I. Foyer, Christine H. II. Faragher, Richard. III. Thornalley, Paul J.
QP85.R44 2008
612.3'9--dc22

2008038761

Editor: Elizabeth Owen
Editorial Assistant: Sarah E. Holland
Production Editor: Helen Powis
Typeset by: Phoenix Photosetting, Chatham, Kent, UK

Taylor & Francis Group, an Informa business Visit our web site at http://www.garlandscience.com

Contents

Contributors

Antonysunil Adaikalakoteswari, Protein Damage and Systems Biology Research Group, Clinical Sciences Research Institute, Warwick Medical School, University of Warwick, Coventry, UK

Paul C. Bethke, USDA ARS, Department of Horticulture, University of Wisconsin, Madison, WI, USA

Christine Chang, INRA, Sophia-Antipolis, France

João Pedro de Magalhães, Department of Genetics, Harvard Medical School, Boston, MA, USA

Radhika Desikan, Division of Biology, Imperial College London, London, UK

Richard Faragher, School of Pharmacy and Biomolecular Sciences, University of Brighton, Brighton, UK

Christine H. Foyer, School of Agriculture, Food and Rural Development, Agriculture Building, Newcastle University, Newcastle upon Tyne, UK

David Gems, Department of Biology, University College London, London, UK

John Hancock, Centre for Research in Plant Science, Genomics Research Institute, University of the West of England, Bristol, UK

Christopher J. Hutchison, School of Biological and Biomedical Sciences, Integrated Cell Biology Laboratories, Durham University, Durham, UK

Barbara Karpinska, Department of Life Sciences, Södertörn University College, Sweden

Stanisław Karpiński, Department of Genetics, Breeding and Biotechnologies, University of Life Sciences, Warszawa, Poland

Karl Kunert, Forestry and Agricultural Biotechnology Institute, Botany Department, University of Pretoria, South Africa

Kelly J. Lamb, University of Glasgow, Department of Surgery, Glasgow, UK

Per Mühlenbock, Department of Molecular Genetics, University of Ghent, Belgium

Ananda Mustafiz, Department of Plant Molecular Biology, International Centre for Genetic Engineering and Biotechnology, Aruna Asaf Ali Marg, New Delhi, India

Steven J. Neill, Centre for Research in Plant Science, Genomics Research Institute, University of the West of England, Bristol, UK

Vanja Pekovic, School of Biological and Biomedical Sciences, Intergrated Cell Biology Laboratories, Durham University, Durham, UK

Anneke Prins, Forestry and Agricultural Biotechnology Institute, Botany Department, University of Pretoria, South Africa and School of Agriculture, Food and Rural Development, Agriculture Building, Newcastle University, Newcastle upon Tyne, UK

Naila Rabbani, Protein Damage and Systems Biology Research Group, Clinical Sciences Research Institute, Warwick Medical School, University of Warwick, Coventry, UK

Paul G. Shiels, Department of Surgery, University of Glasgow, Glasgow, UK

Sneh L. Singla-Pareek, Department of Plant Molecular Biology, International Centre for Genetic Engineering and Biotechnology, Aruna Asaf Ali Marg, New Delhi, India

S. K. Sopory, Department of Plant Molecular Biology, International Centre for Genetic Engineering and Biotechnology, Aruna Asaf Ali Marg, New Delhi, India

Paul J. Thornalley, Protein Damage and Systems Biology Research Group, Clinical Sciences Research Institute, Warwick Medical School, University of Warwick, Coventry, UK

Ian Wilson, Centre for Research in Plant Science, Genomics Research Institute, University of the West of England, Bristol, UK

Mingzhan Xue, Protein Damage and Systems Biology Research Group, Clinical Sciences Research Institute, Warwick Medical School, University of Warwick, Coventry, UK

Sudesh K. Yadav, Department of Plant Molecular Biology, International Centre for Genetic Engineering and Biotechnology, Aruna Asaf Ali Marg, New Delhi, India

Ulrike Zentgraf, ZMBP, General Genetics, University of Tübingen, Tübingen, Germany

Abbreviations

AASA	2-aminoadipic semialdehyde
α-TQ	α-tocopherolquinone
ABA	abscisic acid
ABI4	ABA-insensitive 4
ACC	1-aminocyclopropane-1-carboxylic acid
AD- and AR-EDMD	autosomal dominant and recessive forms of Emery-Dreyfuss muscular dystrophy
AF	atrial fibrillation
AGPase	ADP-glucose pyrophosphorylase
AKRd	aldoketo reductases
ALR	aldose/aldehyde reductase
ALT	alternative lengthening of telomeres
AOPPs	Advanced oxidation protein products
AP	activating protein
AP	Apetala 2
APC	Adenomatous Polyposis Coli
APX	ascorbate peroxidase
ARE	antioxidant response elements
AS	arthropathy syndrome
ASH	ascorbate
ATM	Ataxia Telangiectasia Mutated protein
ATR	ATM and Rad 3-related
AWS	Atypical Werner's syndrome
BAP	Benzylaminopurine
BMP	bone morphogenic protein
bZIP	basic leucine zippers
CaBPs	calcium-binding proteins
CaM	calmodulin
CAT	Catalase
CBF	C-repeat binding protein
CCS	copper chaperone of SOD protein
CDK	cyclin-dependent kinase
CEL	N_ε-carboxyethyl-lysine
CGCs	cerebral giant cells
ChIP	chromatin immunoprecipitation
CMA	N_ω-carboxymethylarginine
CMC	N_ε-carboxymethyl-cysteine
CML	N_ε-carboxymethyl-lysine
CMT2B1	Charcot-Marie-Tooth syndrome type 2B1
CoQ	ubiquinone

CPs	cysteine proteases
c-PTIO	2-(4-carboxyphenyl)-4,4,5,5-tetramethylimidazoline-1-oxyl-3-oxide
CR	caloric restriction
CRM	exportin chromosome region maintenance
C/R ratio	calorimetric/respirometric ratio
CRY	cryptochrome
DBMIB	2,5-dibromo-3-methyl-6-isopropyl-p-benzoquinone
DCM-CD	dilated cardiomyopathy with conduction system disease
-CF	DCM with early-onset cardiac fibrosis
-VA	DCM with apical left ventricular aneurysm
DCMU	3-(3,4-dichlorophenyl)-1,1-dimethyl urea
3DG	3-deoxyglucosone
3DG-H	N_δ-(5-hydro-5-(2,3,4-trihydroxybutyl)-4-imidazolon-2-yl)-ornithine
DHS	dropped head syndrome
DMEM	Dulbecco's modified Eagle's medium
DMQ9	demethoxy-ubiquinone-9
DNA-PK*cs*	DNA-dependent protein kinase catalytic subunit
DOD	2-deoxy-D-glucose
DR	dietary restriction
DSBs	DNA double-strand breaks
E	Enzyme
EEE	excess excitation energy
EDMD	Emery-Dreyfuss muscular dystrophy
EDS1	enhanced disease susceptibility 1
EIN2	ethylene insensitive 2
EL	excess light
ER	endoplasmic reticular/reticulum
ESI	electrospray ionization
ESTs	expression sequence tags
ETC	electron transport chain
FAD	flavin adenine dinucleotide
FCCP	carbonyl cyanide p-trifluoromethoxyphenylhydrazone
FL	N_ε-fructosyl-lysine
FPLD	Dunnigan's familial partial lipodystrophy
FTI	farnesyl transferase inhibitor
γ-ECS	γ-glutamyl cysteine synthetase
GA	gibberellins
GAPDH	glyceraldehyde 3-phosphate dehydrogenase
GBF1	G-box binding factor 1
GC-MS	gas chromatography with mass spectrometric detection
GFP	green fluorescent protein
G-H1	N_δ-(5-hydro-4-imidazolon-2-yl)-ornithine
GID1	gibberellin-insensitive dwarf 1
Glo1	glyoxalase 1
GPX	glutathione peroxidase

GSA	glutamic semialdehyde
GSH	reduced glutathione
GSSG	glutathione disulphide (oxidized glutathione)
GSTs	glutathione S-transferases
HACA	6-hydroxy-2-aminocaproic acid
HAVA	5-hydroxy-2-aminovaleric acid
HMG-CoA reductase	3-hydroxy-3-methylglutaryl-Coenzyme A reductase
HGPS	Hutchinson-Gilford progeria syndrome
3-NHA	3-nitro-4-hydroxyphenylacetic acid
HNE	4-hydroxynon-2-enal
H_2O_2	hydrogen peroxide
HPLC	High performance liquid chromatography
HR	hypersensitive disease response
HSF-1	heat shock factor 1
HSP	Heat shock protein
hTERT-BMCs	telomerase-immortalized bone marrow cells
HXK1	Hexokinase 1
IAA	Auxin
IAF	5′-iodoacetamide fluorescein
IF	intermediate filament (proteins)
IGF	Insulin-like growth factor
IIS	insulin/IGF-1 signalling
INM	inner nuclear membrane
JA	jasmonic acid
JNK	Jun N-terminal kinase
KFA	kynurenine formamidase
KO	knockout
Kyn	kynurenine
LAP2α	lamina-associated polypeptide 2α
LC-MS	liquid chromatography with mass spectrometric detection
LFA	Lethal foetal akinesis
LGMD1B	limb girdle muscular dystrophy with atrioventricular conduction disturbances type 1B
LL	Low light
LMNA	lamin A/C
LMNA	lamin A gene
LMNA	lamin B gene
LSD1	lesion simulating disease 1
MAD	mandibulo-acral dysplasia
MALDI	matrix-assisted laser desorption/ionization
MAPKs	mitogen-activated protein kinases
MAPKKKs	MAPK kinase kinases
Met	methionine
MetSO	methionine sulphoxide
MG	methylglyoxal
MG-H1	N_δ-(5-hydro-5-methyl-4-imidazolon-2-yl)-ornithine

Msr	methionine sulphoxide reductase
mtDNA	mitochondrial DNA
NAA	naphthaleneacetic acid
NAC	*N*-acetyl-cysteine
$NADP^+$	Nicotinamide adenine dinucleotide phosphate
NADPH	Nicotinamide adenine dinucleotide phosphate (reduced form)
NCEs	new chemical entities
NE	nuclear envelope
NF	nuclear factor
NFK	N-formyl-kynurenine
NHEJ	non-homologous end-joining pathway
NahG	bacterial salicylate hydroxylase
NLS	nuclear localization sequence
NNT	nicotinamide nucleotide transhydrogenase
NO	nitric oxide
NPCs	nuclear pore complexes
NPQ	non-photochemical quenching
NPR1	Non-Expressor of Pathogenesis Related Genes 1
NR	nitrate reductase
NT	nuclear transfer
3NT	3-nitrotyrosine
ONM	Outer nuclear membrane
OXI1 kinase	Oxidative signal-inducible 1 kinase
PAD4	phytoalexin deficient 4
PAGE	Polyacrylamide gel electrophoresis
PBL	peripheral blood leucocyte
PCD	programmed cell death
3PGA	3-phosphoglycerate
PGlyco	phosphoglycolate
PHOT	phototropin
PHY	phytochrome
PIF3	phytochrome-interacting transcription factor 3
PM	peripheral membrane
PMSR	protein methionine sulphoxide reductase
POS	polycystic ovaries with type-A insulin resistance syndrome
PP2C	protein phosphatase 2C
PR	Pathogenesis-related
PSI/PSII	photosystems I and II
PUFAs	polyunsaturated fatty acids
QM-DCM	quadricipital myopathy with dilated cardiomyopathy
QTLs	Quantitative trait loci
Rbcs	RuBisCO small subunit
Rboh	Respiratory burst oxidase homologue
RD	restrictive dermopathy
Rel	reticuloendotheliosis viral oncogene
RGR	relative growth rate
RNAi	RNA interference

RNS	reactive nitrogen species
ROS	reactive oxygen species
R-PCD	runaway programmed cell death
RuBisCO	ribulose bisphosphate carboxylase/oxygenase
RuBP	ribulose-1,5-bisphosphate
SA	Salicylic acid
SAA	Systemic acquired acclimation
SAG	Senescence-associated gene
SAGs	senescence-associated genes
SAPK	stress-activated protein kinase
SAR	Systemic acquired resistance
SCF	Skp, Cullin, F-box-containing complex
SDH	succinate dehydrogenase
SIPS	stress-induced premature senescence
SNAP	S-nitroso-*N*-acetylpenicillamine
SNF1	sucrose non-fermenting-1
SN	sodium nitroprusside
SNP	single-nucleotide polymorphism
SnRKs	SNF1-related protein kinases
SOD	superoxide dismutase
SR	sarcoplasmic reticulum
SREBP1	sterol-regulatory element-binding protein
STASIS	stress- and stimulation-induced senescence
TCA	Tricarboxylic acid cycle
TCF/LEF	T cell factor/lymphoid enhancer factor
TDO	tryptophan 2,3-dioxygenase
TGF	tumour growth factor
TOF	train of four
trp	tryptophan
TRX	thioredoxin
UCPs	uncoupling proteins
UPLC	ultra-high-performance liquid chromatography
VSP	vegetative storage protein
Wnt	Wingless/INT
WRN	Werner's protein
WS	Werner's syndrome
WT	wild type

Preface

The massive social and economic changes that occurred during the last century not only led to the virtual eradication of nutrient deficiency diseases in the developed world, but they also focused attention on the problems associated with the appearance of a relatively new phenomenon, an increasingly ageing population. In the last two decades, scientists studying the animal and plant kingdoms have become increasingly occupied with gaining a deeper understanding of the genetic basis for ageing and the impact of the environment on this process. Since, the maintenance of youth, youthfulness or at least healthy ageing in animals is a goal aspired to by many, researchers in this area have become increasingly engaged in developing innovative strategies for delaying the ageing process. Similarly, plant stay-green phenotypes are highly sought after because of their enhanced performance in the field, particularly during periods of exposure to environmental stress. This volume addresses the recent issues that have emerged in ageing research, outlined in the context of the individual personal perspectives and insights of key researchers working in the field.

Current knowledge would suggest that in all probability, the mythical Fountain of Youth resides in the ability of animals and plants to recycle damaged cells. While the complex and multifactorial aetiology of the ageing process remains far from understood, a consensus of view is emerging regarding the nature of the ageing process as it operates in the few model animal species that have been subjected to intensive study. The original simple concept defined in the 'oxidant theory of ageing', of a clear causal relationship between oxidative damage and animal ageing, has not been substantiated and is giving way to a much more complex and intellectually stimulating picture of ageing as a failure to recycle damaged cells and macromolecules. All aerobic organisms exploit the potential of oxygen chemistry and effectively deal with the ever-present generation of reactive oxygen species (ROS) and reactive nitrogen species (RNS). A late addition to this grouping of reactive metabolites is sugar-derived dicarbonyls involved in physiological glycation. Within the context of the sugar economy of the cell, redox metabolism impacts directly on genome and proteome, which are modified by processes such as oxidation, nitration and glycation. The field of redox biology has witnessed a reappraisal of the function of ROS and antioxidants in recent years. Once considered only as 'molecular hoodlums' to be suppressed or policed by the antioxidant system, ROS are now considered to be dynamic information-rich signalling molecules in some instances and mediators of oxidative damage in others. Similarly, taking low-molecular-weight antioxidants such as vitamins C and E as dietary supplements is no longer considered to be beneficial in retarding the ageing process. This volume reviews current concepts concerning cellular redox homeostasis and ageing in animals and plants, relationships to programmed cell death, the production of oxidants and dicarbonyls, the ways that different organisms perceive and respond to oxidative, nitration and glycation challenges, and how this might be intricately connected to ageing and lifespan. The topics to be covered in this volume include the science of ageing, the ecology of ageing, the role of mitochondria and energy metabolism, the free radical theory of ageing, redox homeostasis and antioxidant signalling, reactive oxygen and reactive nitrogen species, senescence and programmed cell death, glycation, quantifying protein and nucleotide modifications during ageing, redox proteomics in ageing in animals and plants, genomics of ageing and caloric restriction, bioinformatics tools in human ageing genomics and caloric restriction mimetics. The chapter by Richard Faragher provides an overview of ageing in animals, while the chapter presented by Paul Thornalley and co-investigators provides a critical appraisal of methods used to assess damage to the proteome in ageing, and considers the emerging evidence for a decline in enzymatic defences against oxidation, nitration and glycation with age as an explanation for increased proteome damage of the ageing phenotype. The protective effects of glyoxalase 1 in plants exposed to stress are considered in the chapter by Sudir Sopory. An anti-ageing effect of over-expression of glyoxalase 1 in *Caenorhabditis elegans* (see cover) was recently reported by Thornalley and collaborators. While concepts of lifespan are rarely considered in plants, the possible roles of oxidants in leaf senescence are discussed by Ulrike Zentgraf and in programmed cell death in seeds by Paul Bethke. The roles of ROS as signalling molecules in plants are discussed in detail by John Hancock, who considers the paradox between their functions in the stimulation of growth and development and as triggers for programmed cell death. The role of cellular redox metabolism in the developmental responses of plants to enrichment in atmospheric carbon dioxide are described by Christine Foyer and colleagues, while the chapter from Stanisław Karpiński and his team considers the interfacing pathways of light signalling and plant/pathogen responses in programmed cell death.

The chapter by João Pedro de Magalhães considers informatics and data approaches, providing a database of genes related to human ageing (GenAge) and related ageing research informatics (http://genomics.senescence.info/index.html). The battery of protective genes that are involved with detoxification of xenobiotics and the enzymatic defences against protein damage, and that are associated with longevity in *C. elegans*, is reviewed in the chapter by David Gems. The latest research on increased longevity through dietary restriction and therapeutic agents with related mimetic activity is discussed by Stephen Spindler, who demonstrates that the calorie-restricted state in mammals is associated with significant tissue re-sorption and the up-regulation of enzymes involved in phase 1 and phase 2 drug metabolism – the same pathways associated with extended lifespan in the simple metazoan organisms. Finally, Lamb and Shiels provide an updated review on the role of telomere biology in ageing at the cellular level while Pekovic and Hutchison put forward a cellular stress hypothesis to explain the links between A-type lamins, ageing and premature ageing diseases.

These chapters bring together current knowledge on cell senescence and programmed cell death in animals and plants with the underpinning processes of tissue turnover and molecular recycling. While Ponce de Leon's never-ending search for the mythical Fountain of Youth proved to be fruitless, the contents of this volume demonstrate that we are now in a better position to understand the respective roles of accumulating damage to protein and nucleotides, and thus develop nutritional and prospective interventions to slow the ageing process.

June 2008. Christine Foyer, Richard Faragher & Paul Thornalley

1

What can we learn from the cross-species biology of ageing?

Richard G.A. Faragher

1 Introduction: What is ageing and why should we care?

In 1967, the Society for Experimental Biology played host to a symposium on the biology of ageing which included studies of the comparative biology of ageing in plants, insects, protozoa and mammals (Woolhouse *et al.*, 1967). Scientists lacking an all-consuming interest in the field were less than impressed. One participant in particular was heard to remark that 'a professional biochemical team could have all this sorted out in a fortnight' (Woolhouse *et al.*, 1967). Four decades on, this chapter is written with much the same intent as that original SEB volume. It is a newscast-like account of the longest fortnight in scientific history (or perhaps 10 days since many areas in ageing research remain opaque) written for the student who is embarking on a career as a gerontologist or the specialist who has, perhaps by chance, blundered into the minefield of confused terminology and muddled debate that until recently characterized ageing research.

The thing which is most striking as a contemporary gerontologist looking back at the literature from the beginnings of the field is the complete absence of any organizing theoretical principles which would allow investigators to gain understanding (rather than simply experimental results) from the systems they chose to study. There seems to have been little consensus on what the ageing process actually was and this rendered discussion of any subsequent questions turbid to say the least. Some years before, Strehler (Strehler and Mildvan, 1960) had set out four principle criteria that defined the ageing process (or at least distinguished it from maturation and development) but there is little evidence from the 1967 SEB symposium that these were enthusiastically embraced by his contemporaries. Strehler defined ageing as a process which was:

- Universal (i.e. all members of a population of organisms will show it, a distinction from infectious disease);
- Progressive (the process was continual and incremental rather than sudden as in the case of 'programmed' organismal death);
- Intrinsic (distinguishing ageing from death due to outside events);
- Degenerative (this captures the idea that ageing is associated with both increasing chances of mortality but also an increasing level of morbidity).

Applying these criteria helps distinguish ageing from other pathological processes at work within organisms. However, they do not lay sufficient stress on one key aspect of the physiology of ageing, the increased frailty of old organisms compared to their young counterparts. Put simply, aged organisms often fail to survive physiological stresses which young organisms are able to weather effectively. For example, the budding yeast *Saccharomyces cerevisiae* reproduces asexually but shows an ageing process marked by the eventual cessation of reproductive capacity in older mother cells (Powell *et al.*, 2003). If young and old cultures of yeast are exposed to an ageing that induces physiological stress (such as UV radiation), old yeast cells are markedly more prone to die (Kale and Jazwinski, 1996). In the nematode *Caenorhabditis elegans*, multiple mutations or interventions that lengthen the life of the animal also impart a stress resistance phenotype compared to wild-type controls (Gill *et al.*, 2003). Many other examples of similar phenomena across the biosphere could be listed.

Understanding and dealing with human frailty, along with human morbidity, are both the major challenges facing modern gerontology and its primary justification as more than a disinterested search for the truth concerning our condition. Morbidity, the time spent sick before either death or recovery occurs, is prolonged, painful, expensive and undesirable for all concerned (it has been calculated that a 1% reduction in morbidity will result in savings of billions of pounds per year in long-term care costs by the middle third of this century). In contrast, death as a result of frailty can be both abrupt and apparently cheap (as in the case of death by infection in elderly humans following hip fracture); however, the fact that it falls on members of the population who are healthy, active and socially engaged means that it also has significant financial, as well as emotional, costs.

To venture upon a mechanistic analogy, ageing is the study of the biology of worn parts and the consequences thereof. But, before discussing the mechanisms of wear in more detail, it is necessary to consider briefly a question which has troubled many thinkers in human history: 'why does ageing happen at all?'

2 Why does ageing happen?

Ageing is not universal across all species, or even among all metazoans, but it is extremely common and thus presumably provides some form of selective advantage to organisms that show it compared to those that do not. The theories I am about to present concerning ageing have been formulated with regard primarily to species that show a germ-line to soma distinction (the soma is the bits of you that are reading this chapter). Although a soma could, in principle, be either ageing or non-ageing the germ-line must, by definition, be immortal in order to allow the organism to contribute any offspring at all to future generations (an observation that appears to have been first made by the 19th-century evolutionary biologist August Weismann). These theories thus have only a tangential bearing on organisms that can reproduce clonally for extended periods (e.g. some plants). Nonetheless, at least some asexual metazoans do display an individual ageing process (Martínez and Levinton, 1992) and clonal lineages of these organisms can suffer a reduction of average fitness through time by the accumulation of slightly deleterious mutation which cannot be repaired without sexual recombination (a mechanism known as Muller's Ratchet). Muller's Ratchet should, in theory, drive purely asexual replicators to extinction unless the population is extremely large.

However, many metazoan species that replicate clonally retain the ability to initiate sexual reproduction at specific time points (often generating germ-line from undifferentiated stem cells). Such organisms can thus be seen as opting in and out of the germ-line/soma distinction and when they are 'in' the evolutionary forces described below most definitely bite. The best current example of this is probably the cnidarian *Hydra*, which has been shown to be non-ageing in its asexual form (Martinez, 1998) but after sexual reproduction the population shows an exponential increase in mortality rate associated with functional deficits in cellularity, contractile movement and food capture in individual organisms (Yoshida *et al.*, 2006).

Modern explanations for 'why ageing happens' are all based on the evolutionary truism that the force of natural selection declines with age. This means that, even in a population of immortal organisms, there are always far fewer chronologically old ones than young ones around (because the longer a given organism has been around the more likely it is to have been eaten, met with an accident, etc.). Thus, even though the reproductive ability of 'old' and 'new' non-ageing organisms is the same, the 'old' organisms contribute fewer offspring to the next generation than the 'new' organisms *simply because there are fewer of them*. Thus, any mutation that favours early life fecundity will be selected for even if it results in deleterious effects later on in the lifetime (a type of gene action termed antagonistic pleiotropy; Williams, 1957). This view of ageing argues against the operation of a 'programme' controlling the ageing of individuals (i.e. the existence of a genetic pathway or process that causes the organism to age but does nothing else). Rather, it suggests that ageing will result from an accumulation of unrepaired faults, which are generated at different rates in different tissues.

Antagonistic pleiotropy and related theories which conceptualize ageing as the by-product of selection for early-life fecundity (such as the 'disposable soma' theory of Thomas Kirkwood, which considers the problem in terms of resource allocation between somatic repair and reproduction) have proved highly satisfactory in explaining why ageing happens (Kirkwood, 2005). According to antagonistic pleiotropy, optimal lifespan is determined by selection pressure for maximum reproductive success. This is itself determined by the level of environmental risk (e.g. the chance of predation) to which members of the population are exposed as a consequence of the ecological niche which they occupy. The huge range of lifespans seen among organisms on this planet can thus be conceived of as points lying on a continuum between two opposing evolutionary strategies. Organisms at very high levels of risk tend to display 'prodigal' (or *r* selected) life history strategies marked by single reproductive events which generate many small and rapidly maturing young which then disperse without parental care. In contrast, species following 'prudent' (or *K*-selected) strategies tend to reproduce steadily and produce a few large young that receive intensive parental care and mature slowly. Clearly, prudence is only an option for a species whose members have a chance of making it to the end of the week in one piece. These models allow an organism to be non-ageing only under two conditions: (i) the lack of a germ-line soma distinction (exemplified by non-sexually differentiated *Hydra* discussed above) or (ii) if the efficiency of the organism at producing offspring per unit energy increases over time (thus counteracting the beneficial effect of mutations favouring early fecundity).

Fecundity-based theories have also provided a powerful conceptual framework for the manipulation of lifespan itself. This is best exemplified by the work of Charlesworth, Rose and colleagues (Rose and Charlesworth, 1981). These researchers were able to extend significantly the overall longevity of female *Drosophila* by the simple expedient of selecting those females that showed the highest fecundity in the late phase of the lifespan (typically 20–25 days) as the parents of the F_1. If performed over many generations this procedure produces increased longevity, increased late-life fecundity and generally enhanced resistance to physiological stress compared to age-equivalent wild-type flies (Nghiem *et al.*, 2000). Essentially similar findings have been reported by Austad, comparing populations of opossums living in relatively high and low predation environments. These studies are consistent with the hypothesis that selection for early-life fecundity is the primary reason for organismal ageing.

Despite these successes, care should be taken always to view antagonistic pleiotropy and disposable soma as theories or hypotheses (q.v.) not immutable laws delivered from on high. The number of species in which their predictions have been tested is extremely limited and the majority of these have been organisms maintained under laboratory conditions for many generations. It has even been suggested that these hypotheses may have actually hindered progress in evolutionary studies of ageing research (Promislow and Pletcher, 2002). This question will be revisited at the end of this chapter.

3 How does ageing happen?

Johnson once famously described patriotism as the 'last refuge of the scoundrel' and in a similar vein gerontology could sometimes be described as the last refuge of the theorist. At the last count, there were approximately 300 distinct 'theories of ageing' (Medvedev, 1990). Some gerontologists (often new students) see this as evidence of a flourishing and vigorous field. However, like many of my colleagues, I do not see this morass in a positive light; after all, no-one in the mitochondrial field now hankers for the good old days when the chemiosmotic hypothesis stood alongside rival theories (including the production of the elusive high energy '~' and conformational strain) as an attempt to explain how ATP was produced. The recognition that proton-motive force generates ATP allowed the field to move forward. Thus a field with a few well-tested ideas is a field which is going places. A field which lovingly retains many theories (some of them discredited antique curios fit only for the historian of science) probably deserves the jibe made about archaeology (sometimes called by wags the 'science of rubbish'): 'As fast as the rubbish is dug up; it is written down'. I would advise anyone new to the field of ageing to treat many theories of 'how we age' as though they carried a health warning.

To be fit for purpose as a research tool any scientific hypothesis must satisfy the following criteria (best described by Karl Popper). To avoid the fundamental philosophical problem posed by induction (the commonplace but unprovable assumption that objects outside the study will behave in the same way as objects within the study) any theory must be capable of being refuted or falsified (i.e. it should make predictions that are incompatible with certain possible results of observation). Thus all good scientific theories should be very precise (to maximize the number of potentially incompatible observations that could be made) and every genuine test of a hypothesis should be an attempt to falsify it (because it is easy to obtain verifications of a theory if you do nothing but look for them). In addition, every theory formulated to explain ageing needs to meet three common-sense criteria.

- In order to cause organismal ageing the postulated mechanism must be present *in vivo*.
- If the mechanism is present *in vivo* it must be capable of exerting degenerative effects.
- Increasing the rate at which the mechanism operates should increase the rate of ageing and (perhaps more convincingly) slowing the rate at which the mechanism works should slow the rate of ageing.

Looked at from this point of view very many theories of ageing (e.g. the original formulation of the somatic mutation hypothesis) either collapse due to insufficient precision (they are simply untestable) or have already been tested and refuted. The classic example of the latter is probably Orgel's error-catastrophe theory (Orgel, 1963). This theory proposed that insertion of the wrong tRNA into a growing polypeptide chain during translation could give rise to long-lived molecules (such as DNA or RNA polymerase) which were themselves error-prone. This was postulated to drive an exponential propagation of 'errors' through the cell leading to cell death. The most elegant refutation of this is probably a study in which senescent cells (q.v.) were infected with a virus that used the host replication machinery in order to reproduce. If 'error-catastrophe' had occurred, this machinery should have been dysfunctional[1]. In fact, it gave rise to yields of virus essentially identical to those found in non-senescent cells from the same donors (Danner *et al.*, 1978). Unfortunately, well-designed experiments of this type have been uncommon in ageing research until quite recently.

Attempts to arrive at a mature understanding of the cross-species biology of ageing also run the risk of foundering on an unusual problem in biology. Cross-species studies have been hugely successful in elucidating the operation of a very wide variety of fundamental processes (e.g. DNA replication and repair) because the molecular machinery necessary to carry out the process under study is often tightly conserved (i.e. it is a safe bet that a DNA polymerase from *E. coli* will be substantially similar to that found in *Homo sapiens* because both species ultimately derive from a common ancestor and the functional requirements of replicating human and bacterial DNA are very similar indeed). There has thus been an expectation in some quarters of close similarities between species when the mechanistic causes of ageing are the subject of investigation (i.e. the study of ageing in yeast will inform on human ageing). To an extent this expectation has been justified. A series of mutants in genes affecting the insulin-IGF axis have been shown to lengthen lifespan in very different species by mechanisms that have yet to be fully clarified (Partridge *et al.*, 2005). Unfortunately, there is no logical reason that this must always be the case.

Ageing is the unprogrammed result of selection for early reproductive success and thus there is no *a priori* reason why the deleterious changes that result from that selection pressure have to be universal across species (by the same token there is also nothing to exclude similar changes if they impact a common pathway). In fact, there

[1] In retrospect, it probably wasn't worthwhile carrying out this experiment at all since error catastrophe predicted cell death and senescent cells were known by this time not to do this at appreciably greater frequencies than their growing counterparts. However, in fairness the theory *was* concerned with cells in culture and *was* being invoked to explain senescence at the time (e.g. Holliday and Tarrant, *Nature* **238:** 26–30).

is evidence that the primary driver of organismal ageing can sometimes be quite different between species. For example, a major cause of ageing and death in female *D. melanogaster* is the toxic effect of compounds present in the seminal fluid products secreted from the main cells of the male fruit fly accessory gland (Barnes *et al.*, 2008). Seminal fluid toxicity is not seriously advanced as a primary cause of mammalian ageing. Similarly, replicative senescence (the loss of divisional capacity in the mitotic tissue compartments of the soma) is not seriously advanced as an ageing mechanism for species such as *C. elegans*, that have a completely post-mitotic soma. In summary, an ageing mechanism which is 'universal' is not necessarily 'primary' in all species and a mechanism which is 'primary' in one species is not necessarily 'universal' in all the others. Picking one's way through this morass requires clear thought and by way of illustration I give two theories which stand out from the confusion. One is a much-touted theory which (in its original form) probably captures everything that used to be wrong with studies of 'how ageing happens' but which is now being reformulated in a much more testable manner. The other is a much-maligned theory that does seem a candidate for a plausible ageing mechanism in some species at some times.

3.1 *Who's afraid of 'free' radicals?*

The free radical theory of ageing proposes that damage caused by free radical reactions is a key contributor to both ageing and age-associated disease. It was first proposed by Denham Harman in the 1950s (Harman, 1956) and is probably the only theory of ageing that the general public are familiar with. Perhaps no single theory in gerontology has attracted quite as much interest among researchers.

The free radical theory passes two of the three common-sense criteria I gave above. Free radical reactions certainly take place in real bodies (a fact not formally established at the time the theory was proposed) and both radicals and reactive oxygen species clearly have the potential to exert degenerative effects. However, the evidence that slowing the rate at which oxidative damage occurs slows the rate of ageing is far more problematic. Much of the data on which the oxidative damage theory of ageing is based is correlative and is based on the examination of antioxidant defence levels in different species to determine if the ability to defend against oxidative stress correlates with species-specific lifespan (in general the correlation is rather good but see Section 4.0). Similarly, long- and short-lived strains of the same organism have been compared for antioxidant defence levels (longer-lived animals have more; see Dudas and Arking, 1995). However, when interventional tests of the theory have been attempted (typically by feeding ageing rodents dietary antioxidants such as vitamin E to try and extend their lifespans) the results have been far less successful. One particularly well-conceived series of studies may stand for the whole. A class of molecules known as the salens were shown some years ago to act as superoxide dismutase and catalase mimics *in vivo* and *in vitro*. In particular these molecules were shown to be able to somewhat extend the lifespan of mice carrying a loss-of-function mutation in mitochondrial superoxide dismutase (these animals normally have a lifespan of only a few days, the salen was able to increase this to some weeks; Melov *et al.*, 2001). An initial report also appeared to show a clear increase in normal nematode lifespan when treated with the same compounds (Melov *et al.*, 2000). However, a subsequent series of nematode studies: (i) determined the dose of salen needed to give a protective effect against paraquat-mediated killing (to ensure physiological efficacy) and (ii) measured the effect of continuous supplementation at that dose on the mean of maximum lifespans of a population of nematodes. No extension of lifespan could be observed (Keaney and Gems, 2003; Keaney *et al.*, 2004). This is quite a typical pattern in which a study seeking to test the free radical theory reports a positive effect and others (often in less commonly read journals) fail to reproduce it. Explanations for this failure to produce a consistent lifespan extension effect by interdiction of a pathway that is proposed to be of primary (and perhaps universal) importance generally include some form of the notion that the effective dose of antioxidants delivered is not high enough. Whilst there can be a certain amount of truth in these statements it is hard not to consider them a variant of the 'conventionalist twist'[2].

In at least one species (*C. elegans*) the classical formulation of the free radical theory of ageing appears to have been refuted. Knocking out the manganese-containing forms of the antioxidant enzyme superoxide dismutase has no apparent effects on lifespan (although loss of major Cu/Zn SOD, *sod-1*, does shorten life). Over-expression of *sod-1* does not produce any increase in longevity in this organism. These data are simply not consistent with a simple relationship between free radical (or oxidative) damage and organismal ageing. What is going on?

One clue is provided by microarray analysis of genes regulated by insulin/insulin-like growth factor-1 (IGF-1) signalling in this species. Mutants in this pathway significantly increase lifespan and that increased lifespan is associated with up-regulated antioxidant defences (including superoxide dismutase levels). However, many other enzymes are also up-regulated which have no role in dealing with oxidative stress but instead are components of either the phase 1 or phase 2 detoxification system used for xenobiotic metabolism. Thus it appears that long life is associated with increased activity in pathways that are primarily designed to detoxify a broad spectrum of damaged macromolecules. Molecules damaged by reactive oxygen species (ROS) are only one feedstock going into this recycling plant. This 'green theory' of ageing (Gems and McElwee, 2005) represents a considerable conceptual advance on anything which has gone before and is consistent with a number of other studies (not least the observation that over-expression of glyoxalase I increases lifespan in *C. elegans*; see Morcos *et al.*, 2008).

3.2 *Replicative senescence causes ageing: Definitely maybe*

Replicative senescence is a permanent block to further replication in cells from the mitotic tissue compartments of a metazoan organism. First reported in the 1960s by Leonard Hayflick and Paul Moorhead (reviewed in Faragher and Kipling, 1998) the phenomenon has been extensively studied in cells from mammals but much less so in cells from lower vertebrates (although it is known to occur in amphibians and may occur in molluscs). Senescent cells are normally unable to traverse the cell cycle as the result of an active block to cell-cycle traverse. This block usually occurs in the G_1

---

[2] According to Popper, conventionalist twists are the introduction of *ad hoc* assumptions into a theory by proponents when the original form of the theory appears to have been falsified by experiment. This rescues the theory from refutation at the price of lowering (or perhaps destroying) its scientific merit. Not all twists do this. It is perfectly acceptable to modify a theory provided the 'twists' themselves give rise to falsifiable predictions.

phase of the cell cycle and is mediated by one or more cyclin-dependent kinase inhibitors. In most circumstances the result is a living, viable and metabolically active, but reproductively sterile cell. The primary physiological role of entry into senescence is almost certainly to act as a longevity assurance mechanism by preventing the unconstrained proliferation of clones of cells that may have accumulated pro-carcinogenic mutations. Senescent cells are responsible for the decline in growth potential of *in vitro* cell populations that have undergone substantive turnover. They display many biochemical features that are distinct from their growing counterparts as a result of highly selective changes in gene expression. The expression of some genes goes up (as a consequence of both increased transcription and mRNA stabilization), the expression of others goes down and the expression of still more is unaffected. The overall effect is to produce a cell that is locked into a pro-inflammatory phenotype by comparison to cells of the same time that are still capable of division. It has been proposed that progressive tissue turnover during life leads to ever-increasing numbers of cells entering the senescent state.

For many years replicative senescence was a remarkably poor candidate as an *in vivo* ageing mechanism (measured against the three common-sense criteria above). From the 1960s to the 1990s, there was very little evidence that senescent cells were present in bodies. There was even less evidence that senescent cells were capable of exerting degenerative effects. The best evidence that there might be correlations between senescence and organismal ageing were probably: (i) the observation that the proliferative capacity of fibroblasts co-varied in a roughly linear fashion with maximum lifespan in eight different mammalian species (Röhme, 1981) and (ii) reports that fibroblasts derived from humans with progeroid (accelerated ageing) syndromes, such as Werner's syndrome and progeria, showed a greatly attenuated capacity to proliferate in culture (Kipling *et al.*, 2004). This was hardly impressive, but the lack of data represented a lack of interest in these questions. The majority of workers in the field were far more concerned with dissecting the molecular mechanisms that led to the initiation and establishment of the senescent state *in vitro*.

However, the situation is now very different. A landmark paper in the mid-1990s used a modified catalytic histochemical assay for β-galactosidase to demonstrate that senescent cells were present *in vivo* and that they increased in an age-dependent manner (Dimri *et al.*, 1995). Techniques for the visualization of senescent cells have improved considerably since this paper was published (although it is still far from easy) and a recent study using multiple markers has concluded that upwards of 15% of cells in the skin of old baboons are in fact senescent (Herbig *et al.*, 2006). Thus, there is limited but solid evidence consistent with senescent cells being present in real bodies.

When incorporated into reconstituted human skin equivalents, human dermal populations made senescent *in vitro* are known to increase dermal fragility and subepidermal blistering in a cell-number-dependent manner demonstrating that the presence of such cells could exert degenerative effects in *ex-vivo* equivalents (Funk *et al.*, 2000). However, a still more striking demonstration that senescent cells can produce life-threatening pathology was provided by experimental induction of senescence in living rat carotid arteries (Minamino *et al.*, 2004). This produced severe vascular inflammation and changes consistent with the development of atheroma. This study is probably the best current evidence that senescent cells are deleterious to living organisms.

Since the early 1990s, it has been known that the short replicative lifespan of Werner's syndrome is due to a three- to fivefold increase in the rate at which cells exit the cell cycle and become senescent *in vitro* (Faragher *et al.*, 1993). Although this strongly suggests that a similar process drives the accumulation of senescent cells in Werner's patients the best evidence that senescent cells cause accelerated ageing comes from work on two strains of mice made null for the Werner's syndrome gene (*wrn*). Mice singly deficient for *wrn* show little by way of any obvious premature ageing phenotype despite the fact that they show other cellular phenotypes of human WS. However, the differences in replicative lifespan between wild-type and $wrn^{-/-}$ mouse fibroblasts were small. However, a second knock out strain (a $wrn^{-/-}$ $terc^{-/-}$ double null) developed age-dependent pathologies that closely paralleled those seen in Werner's syndrome humans. These include grey hair, osteoporosis, type-II diabetes, cataracts, an elevated frequency of non-epithelial malignancies and premature death. Fibroblasts from the $wrn^{-/-}$ $terc^{-/-}$ animals showed accelerated replicative senescence *in vitro* (Kipling *et al.*, 2004). The conclusion that senescent cells are driving the 'ageing' changes is very hard to avoid.

4 Strengths and weaknesses of a cross-species approach

4.1 *Short-lived ageing models*

In a sense, the majority of studies of the ageing process are cross-species studies by inference. After all, there is limited utility in learning about the mechanisms of ageing in *C. elegans* or *Drosophila melanogaster* unless the data gained are eventually of value in advancing our understanding of human ageing. However, we may perhaps distinguish between three styles of cross-species study.

- The study of short-lived organisms that show a defined ageing process.
- The study of ultra-long-lived species that still show ageing.
- Comparative studies across a wide range of different species intended to test a mechanistic or evolutionary hypothesis.

The first of these styles of study is quite familiar. The use of organisms such as *C. elegans* and *Drosophila* has brought much needed rigour to biogerontology. The primary advantages of these species are that they reproduce quickly, have lifespans of only a couple of months and have excellent genetic tools which allow any 'ageing' phenotypes to be quickly identified and dissected. The chief disadvantage of these classic short-lived models is that they are a long way (phylogenetically speaking) from humans. To that end, there is a need for alternative short-lived ageing models that are closer to mammals (Gerhard, 2007). A particularly promising candidate is the zebrafish (*Danio rerio*). This animal is small, prolific, easily reared and has excellent molecular genetics resources available due to its use as a developmental model (forward and reverse genetics can be performed in zebrafish and there are a variety of expression constructs available which permit the construction of transgenic animals). Some physiological parameters are perturbed in the animal (e.g. circadian rhythm) that are of direct relevance to human ageing and thus potential avenues of research for groups seeking to improve quality of life for the elderly (Zhdanova *et al.*, 2008). The chief disadvantage of the animal is, ironically, that it lives rather too long. Zebrafish have a maximum lifespan of about 5–6 years (and a mean lifespan of about 3.5 years). This is longer than that of the laboratory mouse, *Mus musculus*, which has a maximum lifespan of about 4 years. Promising alternative short-lived vertebrates are the killifish (*Nothobranchius furzeri*) and a closely related species, the Japanese medaka (*Oryzias latipes*). The medaka lives about as long as a mouse (2–4 years) but has a small genome (~800 Mb) that has already been cloned into BAC libraries. The killifish in contrast is much shorter-lived (just 3 months) and has the potential to represent a superb ageing resource. However, there is an almost complete lack of molecular or cellular tools for the study of the organism (although its genome seems to have a relatively high homology to that of medaka, which may facilitate molecular biology studies). In addition, *Nothobranchius* is potentially challenging to rear because the organism has a requirement for embryonic diapause (in the wild, killifish lay eggs in ponds which dry up during the summer). Establishing proper husbandry techniques for the animals is thus not trivial, but is likely to be rewarding for any group able to make the investment.

One other short-lived organism of great potential utility to ageing researchers is the pond snail, *Lymnaea stagnalis* (Patel *et al.*, 2006). Compared to higher organisms the snail has a relatively small central nervous system (~20 000 neurones) and has been extensively studied with regard to learning and memory (it can display both appetitive and aversive conditioning response) and as a result the neural circuits underlying a variety of physiological responses (including rhythmic feeding and respiration) are well understood. Recent studies have extended this analysis to the neurobiology of ageing in the animal and have found that the feeding response is significantly blunted with age (old animals bite less frequently following a stimulus and take longer to swallow food). This 'senility' is driven by changes in the connectivity of a pair of serotonergic neurones known as the cerebral giant cells (CGCs). Thus, although *Lymnaea* currently lacks a sequenced genome this is compensated for by its short lifespan (typically around 10–12 months) and the wealth of data available on cognition in the young animal.

4.2 *Long-lived ageing models*

If the advantage of short-lived organisms to ageing research is speed then the advantage of ultra-long-lived organisms is staying power; or more precisely, the fact that the organism in question is able to maintain normal physiological function for a much longer period than is possible for a human. Organisms which are documented as having much longer lifespans than humans are actually quite rare, and gerontologists seeking to work with them require some way of actually dating the organism to see how old it is (typically through some form of annual growth ring). Creatures known to live longer than humans include rockfish (~200 years), bowhead whales (at least one specimen dated to over 200) and sturgeon (~150 years of age). However, none of these organisms is suitable as a laboratory species for reasons which do not require elaboration (*Balaena mysticetus* weighs ~50 000 kg and is about 15 m long). The best candidate for an ultra-long-lived laboratory animal is almost certainly *Arctica icelandica* (the quahog or cyprine), a suspension-feeding bivalve mollusc which lives in deep water (below 25 m) in the shelf seas off the North European and North American continents. The species is dioecious (separate sexes), with larval development taking between 30 to 60 days and individuals become sexually mature within 15 years. Crucially, *Arctica* shells contain a growth record of the animal in the form of wide annual summer growth increments separated by narrow winter growth lines. This allows the organisms to be readily aged. Lifespans in excess of 200 years are common for *Arctica* and the oldest specimens have been dated at over four centuries. *Arctica* can readily be kept in tanks supplied with algae on which to filter feed and specimens of different ages can be obtained from discrete locations off the north-western coast of the United Kingdom. The genome size of the organism is unknown but a related species, *Ruditapes philippinarum*, has a genome of approximately 1900 Mb. Like the killifish, the disadvantage of working with *Arctica* is that gerontologists have to begin working essentially *de novo* without any of the molecular and cytological tools available taken for granted in standard laboratory species. The great advantage is the opportunity to begin to understand how an organism can retain functional muscle, gut and nerves about four times longer than we can.

4.3 *Hypothesis testing using comparative biology*

In contrast to the types of experimental approach outlined above, comparative studies across a range of species are generally conducted in order to test a single theoretical prediction. Quite a range of these studies has been carried out of which that by Valencak and Ruf (2007) is particularly noteworthy. These researchers set out to test the specific hypothesis that high levels of polyunsaturated fatty acids (PUFAs) might shorten organismal lifespan through a combination of elevated lipid peroxidation and an elevation of basal metabolic rate (sometimes called the 'membrane pacemaker hypothesis' of ageing). The team measured PUFAs in a dataset of 42 mammalian species and found no correlation between n3 polyunsaturate levels and lifespan, the degree of membrane unsaturation and maximum lifespan or between basal metabolic rate and lifespan. These data are not consistent with the membrane pacemaker theory and contradict earlier studies showing that a relationship exists between peroxidation and lifespan (e.g. Hulbert, 2005).

The Valencak study is a model of how to exploit the power of cross-species studies in order to conduct an informative and well-planned test of a mechanistic hypothesis for ageing (Valencak and Ruf, 2007). The team designed experiments that were capable of refuting the pacemaker hypothesis, which was a good hypothesis to test precisely because it made clearly falsifiable predictions. Crucially, the researchers were also aware of the potential for spurious correlations to distort their findings and did their utmost to correct for them. One major problem, which has dogged cross-species studies for many years, is the failure by some researchers to recognize that many traits (including lifespan) co-vary with body mass (i.e. big animals tend to be long lived). Valencak used body mass as a covariate in her analysis and combined it with a phylogenetic generalized least-square model to control for another spurious correlation (caused by close taxonomic relations between the species assayed). Many studies relating oxidative damage and lifespan have failed to do this with the result that quite a number of correlations (e.g. between increased resistance to oxidative stress and lifespan) have the potential to be spurious (Selman *et al.*, 2005; Speakman, 2005).

5 Unsolved questions in the cross-species biology of ageing

Returning to my starting point of the 1967 SEB symposium on the biology of ageing it should be apparent that the last 40 years have been a fairly busy time for gerontology generally. More gerontologists are now at least aware of the evolutionary theories that seek to explain why ageing happens than was the case in 1967. The application of standard short-lived laboratory organisms to ageing research (not least through the pioneering and largely unrecognized work of Klass (1983)) has produced tremendous insights into the malleability of lifespan and allowed the identification of genes which, contrary to expectations, seem to impact common pathways controlling lifespan. Experimental studies are much better designed than they were at the start of the discipline and several important gaps have been closed (such as the development of effective tools for the detection of senescent cells in tissue). What then is missing?

Earlier I alluded to the possibility that disposable soma provides such a good explanation for why ageing happens that the danger is emerging that we may be lulled into a false sense of security. Thinking that we know all we need to know about evolutionary theory we may be unpleasantly surprised when new observations fail to fit smoothly into standard versions of fecundity theory. A case in point concerns recent work conducted on ageing in the guppy (*Poecilia reticulate*). A series of studies have compared guppies in the same river system in Trinidad that live in high predation environments with those in low predation environmental niches (above barrier waterfalls). Classical resource allocation theory would predict that fish in low predation environments should display relatively longer lifespan and a reproductive strategy tending more towards the prudent end of the spectrum. In contrast, guppies in high predation environments should be driven towards prodigal strategies. However, it has been shown that guppies that evolved with predators mature at an earlier age but also show longer lifespans than their counterparts in environments with low extrinsic mortality (Reznik *et al*., 2006). This is difficult to reconcile with simple fecundity theory and demonstrates that other factors (e.g. foraging costs) may need to be factored into our current evolutionary models.

Little work has been done to date on relating the evolutionary drivers of ageing to those playing into other evolutionary strategies, such as kin selection. *In extremis*, this leads to the evolution of sterile social casts but also underlies the phenomenon of 'juvenile helpers' seen in some bird populations. Data from the study of the naked mole rat (which has a caste structure similar to that seen in social insects) has shown that workers have much longer lifespans than would be predicted for their body size and environment (Buffenstein, 2008). Lastly, we still know very little about true ageing (as opposed to senescence) in plants. The study of the nearest animal with a life history similar to members of this Kingdom (*Hydra*) has been very informative. The study of plants themselves may be predicted to be much more so.

To conclude, the study of cross-species ageing may yet yield mechanisms of ageing that are not just stranger than we imagine but are perhaps stranger than we can imagine at the present time. If SEB organizes another such symposium in 2048 I hope that my article will provide the same opportunity for reflection. However, I doubt I will be around to bore the author with personal anecdotes about organizing the meeting.

References

Barnes, A.I., Wigby, S., Boone, J.M., Partridge, L. and Chapman, T. (2008) Feeding, fecundity and lifespan in female *Drosophila melanogaster*. *Proc. Biol. Sci.* 22 April, epub.

Buffenstein, R. (2008) Negligible senescence in the longest living rodent, the naked mole-rat: insights from a successfully aging species. *J. Comp. Physiol. B* **178:** 439–445.

Danner, D.B., Schneider, E.L. and Pitha, J. (1978) Macromolecular synthesis in human diploid fibroblasts. A viral probe examining the effect of *in vivo* aging. *Exp. Cell Res.* **114:** 63–67.

Dimri, G., Lee, X., Basile, G., *et al.* (1995) A biomarker that identifies senescent human cells in culture and in aging skin *in vivo*. *Proc. Natl Acad. Sci. USA* **92:** 9362–9367.

Dudas, S.P. and Arking, R. (1995) A coordinate upregulation of antioxidant gene activities is associated with the delayed onset of senescence in a long-lived strain of Drosophila. *J. Gerontol. A Biol. Sci. Med. Sci.* **50:** B117–B127.

Faragher, R.G. and Kipling, D. (1998) How might replicative senescence contribute to human ageing? *Bioessays* **20:** 985–991.

Faragher, R.G., Kill, I.R., Hunter, J.A., Pope, F.M., Tannock, C. and Shall, S. (1993) The gene responsible for Werner syndrome may be a cell division 'counting' gene. *Proc. Natl Acad. Sci. USA* **90:** 12030–12034.

Funk, W.D., Wang, C.K., Shelton, D.N., Harley, C.B., Pagon, G.D. and Hoeffler, W.K. (2000) Telomerase expression restores dermal integrity to *in vitro*-aged fibroblasts in a reconstituted skin model. *Exp. Cell Res.* **258:** 270–278.

Gems, D. and McElwee, J.J. (2005) Broad spectrum detoxification: the major longevity assurance process regulated by insulin/IGF-1 signaling? *Mech. Ageing Dev.* **126:** 381–387.

Gerhard, G.S. (2007) Small laboratory fish as models for aging research. *Ageing Res. Rev.* **6:** 64–72.

Gill, M.S., Olsen, A., Sampayo, J.N. and Lithgow, G.J. (2003) An automated high-throughput assay for survival of the nematode *Caenorhabditis elegans*. *Free Radic. Biol. Med.* **35:** 558–565.

Harman, D. (1956) Aging: a theory based on free radical and radiation chemistry. *J. Gerontol.* **11:** 298–300.

Herbig, U., Ferreira, M., Condel, L., Carey, D. and Sedivy, J.M. (2006) Cellular senescence in aging primates. *Science* **311:** 1257.

Hulbert, A.J. (2005) On the importance of fatty acid composition of membranes for aging. *J. Theor. Biol.* **234:** 277–288.

Kale, S.P. and Jazwinski, S.M. (1996) Differential response to UV stress and DNA damage during the yeast replicative life span. *Dev. Genet.* **18:** 154–160.

Keaney, M. and Gems, D. (2003) No increase in lifespan in *Caenorhabditis elegans* upon treatment with the superoxide dismutase mimetic EUK-8. *Free Radic. Biol. Med.* **34:** 277–282.

Keaney, M., Matthijssens, F., Sharpe, M., *et al.* (2004) Superoxide dismutase mimetics elevate superoxide dismutase activity *in vivo* but do not retard aging in the nematode *Caenorhabditis elegans*. *Free Radic. Biol. Med.* **37:** 239–250.

Kipling, D., Davis, T., Ostler, E.L. and Faragher, R.G. (2004) What can progeroid syndromes tell us about human aging? *Science* **305:** 1426–1431.

Kirkwood, T.B. (2005) Understanding the odd science of aging. *Cell* **120:** 437–447.

Klass, M.R. (1983) A method for the isolation of longevity mutants in the nematode *Caenorhabditis elegans* and initial results. *Mech. Ageing Dev.* **22:** 279–286.

Martínez, D.E. (1998) Mortality patterns suggest lack of senescence in hydra. *Exp. Gerontol.* **33:** 217–225.

Martínez, D.E. and Levinton, J.S. (1992) Asexual metazoans undergo senescence. *Proc. Natl Acad. Sci. USA* **89:** 9920–9923.

Medvedev, Z.A. (1990) An attempt at a rational classification of theories of ageing. *Biol. Rev. Camb. Philos. Soc.* **65:** 375–398.

Melov, S., Ravenscroft, J., Malik, S., *et al.* (2000) Extension of life-span with superoxide dismutase/catalase mimetics. *Science* **289:** 1567–1569.

Melov, S., Doctrow, S.R., Schneider, J.A., *et al* (2001) Lifespan extension and rescue of spongiform encephalopathy in superoxide dismutase 2 nullizygous mice treated with superoxide dismutase-catalase mimetics. *Neuroscience* **21:** 8348–8353.

Minamino, T., Miyauchi, H., Yoshida, T., Tateno, K., Kunieda, T. and Komuro, I. (2004) Vascular cell senescence and vascular aging. *J. Mol. Cell Cardiol.* **36:** 175–183.

Morcos, M., Du, X., Pfisterer, F., *et al.* (2008) Glyoxalase-1 prevents mitochondrial protein modification and enhances lifespan in *Caenorhabditis elegans. Aging Cell* **7:** 260–269.

Nghiem, D., Gibbs, A.G., Rose, M.R. and Bradley, T.J. (2000) Postponed aging and desiccation resistance in *Drosophila melanogaster. Exp. Gerontol.* **35:** 957–969.

Orgel, L.E. (1963) The maintenance of the accuracy of protein synthesis and its relevance to ageing. *Proc. Natl Acad. Sci. USA* **49:** 517–521.

Partridge, L., Gems, D. and Withers, D.J. (2005) Sex and death: what is the connection? *Cell* **120:** 461–472.

Patel, B.A., Arundell, M., Allen, M.C., Gard, P., O'Hare, D., Parker, K. and Yeoman, M.S. (2006) Changes in the properties of the modulatory cerebral giant cells contribute to aging in the feeding system of *Lymnaea. Neurobiol. Aging* **27:** 1892–1901.

Popper, K. (1963) Conjectures and Refutations. Routledge, Abingdon.

Powell, C.D., Quain, D.E. and Smart, K.A. (2003) Chitin scar breaks in aged *Saccharomyces cerevisiae. Microbiology* **149:** 3129–3137.

Promislow, D.E. and Pletcher, S.D. (2002) Advice to an aging scientist. *Mech. Ageing Dev.* **123:** 841–850.

Reznick, D., Bryant, M. and Holmes, D. (2006) The evolution of senescence and post-reproductive lifespan in guppies (*Poecilia reticulata*). *PLoS Biol.* **4:** e7.

Röhme, D. (1981) Evidence for a relationship between longevity of mammalian species and life spans of normal fibroblasts *in vitro* and erythrocytes *in vivo*. *Proc. Natl Acad. Sci. USA* **78:** 5009–5013.

Rose, M.R. and Charlesworth, B. (1981) Genetics of life history in *Drosophila melanogaster*. II. Exploratory selection experiments. *Genetics* **97:** 187–196.

Selman, C., McLaren, J.S., Meyer, C., Duncan, J.S., Redman, P., Collins, A.R., Duthie, G.G. and Speakman, J.R. (2005) Life-long vitamin C supplementation in combination with cold exposure does not affect oxidative damage or lifespan in mice, but decreases expression of antioxidant protection genes. *Mech. Ageing Dev.* **127:** 897–904.

Speakman, J.R. (2005) Correlations between physiology and lifespan – two widely ignored problems with comparative studies. *Aging Cell* **4:** 167–175.

Strehler, B.L. and Mildvan, A.S. (1960) General theory of mortality and aging. *Science* **132:** 14–21.

Valencak, T.G. and Ruf, T. (2007) N-3 polyunsaturated fatty acids impair lifespan but have no role for metabolism. *Aging Cell* **6:** 15–25.

Williams, G.C. (1957) Pleiotropy, natural selection and the evolution of senescence. *Evolution* **11:** 398–411.

Woolhouse, H.W., ***et al.*** (1967) *Aspects of the Biology of Ageing*. Symposia of the Society for Experimental Biology XXI. Cambridge University Press, Cambridge.

Yoshida, K., Fujisawa, T., Hwang, J.S., Ikeo, K. and Gojobori, T. (2006) Degeneration after sexual differentiation in hydra and its relevance to the evolution of aging. *Gene* **385:** 64–70.

Zhdanova, I.V., Yu, L., Lopez-Patino, M., Shang, E., Kishi, S. and Guelin, E. (2008) Aging of the circadian system in zebrafish and the effects of melatonin on sleep and cognitive performance. *Brain Res. Bull.* **75:** 433–441.

2

Rebirth and death: Nitric oxide and reactive oxygen species in seeds

Paul C. Bethke

1 Introduction

1.1 *Seed germination and dormancy*

Seed germination and early seedling growth are critical phases in the plant life cycle (Bewley and Black, 1994). Young seedlings need to grow rapidly in order to effectively compete with neighbours, yet young seedlings have limited nutrient reserves to draw on as they resume growth. Root systems are rudimentary, and this limits the ability of young seedlings to explore the environment for water and minerals. Developing the photosynthetic capacity required for autotrophic growth is another draw on stored reserves. Young seedlings are also vulnerable to adverse environmental conditions such as insufficient water and extremes of temperature. In order to increase the likelihood of early survival, most plant seeds are dormant at maturity, that is, they will not germinate under environmental conditions that would promote germination of non-dormant seeds (Bewley, 1997; Bewley and Black, 1994; Finch-Savage and Leubner-Metzger, 2006). This is a highly adaptive property that allows plants to delay the growth of the next generation until the season or local environmental conditions are suitable. Dormancy loss may occur in dry storage after sufficient time has passed, a process referred to as after-ripening, or may require specific environmental signals. Examples of these are availability of soil nitrogen, exposure to light, a period of damp and cold, or exposure to compounds in smoke. These signals indicate times of the year or local environmental conditions that are likely to be favourable for seedling growth, and which, therefore, increase the likelihood of early seedling survival.

The embryo and nutrient tissues within dormant seeds have the potential for great longevity. In many cases, seeds remain viable for years or decades. Following dormancy loss and imbibition of water, combinations of plant hormones and signalling molecules promote germination, and the renewed growth and development of the quiescent embryo. In some seeds, these same hormones and signals initiate the programmed death of nutritive tissues.

This chapter will focus on the involvement of plant hormones, reactive oxygen species (ROS) and nitric oxide (NO) in seed germination and early seedling growth in two plant models, *Arabidopsis thaliana* (Arabidopsis) and *Hordeum vulgare* (barley). The data support the conclusion that NO functions to remove dormancy and promote early seedling growth. ROS production is an inevitable consequence of utilizing stored lipid reserves. In barley, ROS bring about the programmed death of specialized nutrient-storing cells. NO delays this programmed death, at least in part, by functioning as an antioxidant.

1.2 *The structure of Arabidopsis seeds and barley grains*

Each mature, dry Arabidopsis seed weighs approximately 20 μg and consists of three functional domains: the embryo, the aleurone layer and the seed coat (testa). These are illustrated in *Figure 1*. The embryo will develop into the young seedling; the Arabidopsis embryo exists within the seed as a rudimentary plant that consists of a single root linked to two cotyledons, or embryonic leaves. The aleurone layer is a living triploid tissue that surrounds the embryo and plays roles in dormancy maintenance and seedling nutrition. The aleurone layer contains a single cell type, the aleurone cell, which is terminally differentiated and rich in stores of lipid, protein and minerals. The seed coat encases both the embryo and aleurone layer, and at seed maturity it consists of dead, crushed cells. The barley grain, although a thousand times larger than the Arabidopsis seed, contains the same three functional domains, plus a starch-storing endosperm (*Figure 1*). The endosperm makes up the largest part of the grain. It is nonliving at grain maturity and contains nutrient reserves of starch, protein, minerals and nucleic acids in the corpses of cells that formed it.

1.3 *Abscisic acid and gibberellins are central to dormancy and germination*

Two plant hormones, abscisic acid (ABA) and gibberellins (GA) are central regulators of dormancy, germination and early seedling growth (Finch-Savage and Leubner-Metzger, 2006; Finkelstein *et al.*, 2002; Gubler *et al.*, 2005; Kermode, 2005; Peng and Harberd, 2002). ABA establishes and maintains seed dormancy. GA promotes germination, which is defined as the emergence of the embryonic root from the seed coat. GA also promotes utilization of endosperm reserves both in the aleurone cells and in the starchy endosperm, while ABA inhibits these processes. A diagram of the emerging model for

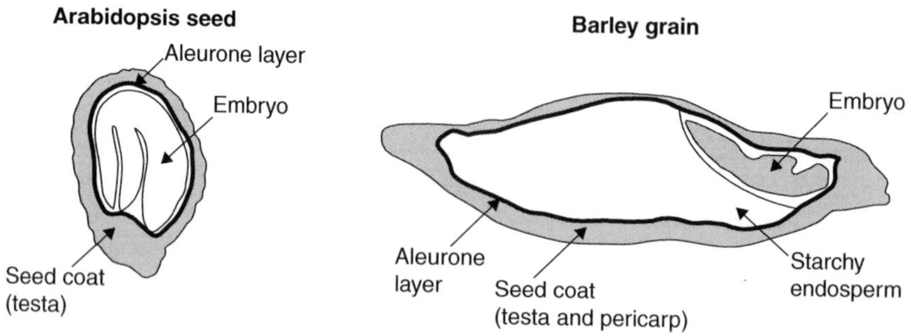

Figure 1. *Diagrams illustrating the arrangement of the embryo, aleurone layer and seed coat in an Arabidopsis seed and a barley grain. Barley grains also contain a large, starchy endosperm.*

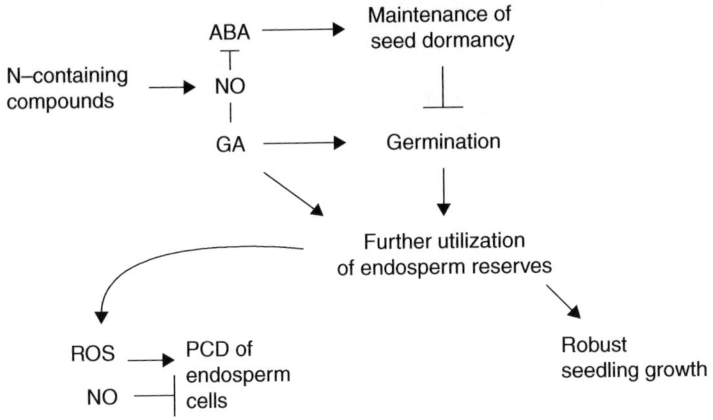

Figure 2. *A model for the involvement of nitric oxide (NO), reactive oxygen species (ROS), gibberellins (GA), and abscisic acid (ABA) in the transition from dormant seed to vigorously growing seedling (PCD, programmed cell death).*

the interaction between ABA, GA, NO and ROS during the transition from dormant seed to growing seedling is shown in *Figure 2*. Data supporting this model are presented below.

2 Nitric oxide lessens dormancy in Arabidopsis seeds and barley grains

Most Arabidopsis ecotypes have moderate to strong seed dormancy. Seeds stored at room temperature require from several months to over a year to lose dormancy (Koornneef *et al.*, 2000). When dormant Arabidopsis seeds of the C-24 ecotype were exposed to a continuous flow of NO gas for 48 h, however, approximately a quarter of the seeds lost dormancy and germinated within 5 days (Bethke *et al.*, 2007; Libourel *et al.*, 2006). Exposure of the same seeds to a continuous stream of air did not result in seed germination. Conversely, when a population of seeds that germinated to the 20% level in the absence of treatment were imbibed with the NO scavenger 2-(4-carboxyphenyl)-4,4,5,5-tetramethylimidazoline-1-oxyl-3-oxide (c-PTIO), dormancy was strengthened in a dose-dependent manner and the percentage of germinated seeds decreased (Bethke *et al.*, 2004b). The final germination percentages were 13% with 50 μM c-PTIO and 5% with 100 μM c-PTIO. The effect of c-PTIO was to strengthen dormancy rather than to cause an inhibition of germination. This was shown clearly in experiments where non-dormant Arabidopsis seeds were imbibed with c-PTIO. In these experiments treatment with c-PTIO at concentrations of up to 200 μM had virtually no effect on the rate of germination of the seeds, or on the final percentage of seeds germinated (Bethke *et al.*, 2006b). In each case, the final germination percentages were greater than 95%.

The dormancy observed in seed of cultivated barley is somewhat weaker than that in the Arabidopsis C-24 ecotype. Dormancy in barley seed is promoted by imbibition in the light. When dormant barley grains (cv. Proctor) were imbibed on moist filter paper for 5 days, germination in the dark was greater than 95% but values decreased in the light to only about 10% (Bethke *et al.*, 2004b). Germination in the light increased to almost 40% in the presence of a 100 μM concentration of the NO donor sodium nitroprusside (SNP) whereas imbibition with c-PTIO reduced germination in the light to less than 1%. As with Arabidopsis, this effect of c-PTIO was a strengthening of dormancy and not toxicity, as c-PTIO had no effect on the germination of dormant barley grain imbibed in the dark (to break dormancy), or on non-dormant grain imbibed in the light (Bethke *et al.*, 2004b).

3 The aleurone layer is a site of NO perception, response and synthesis

In order to determine the site of NO perception that was important for a loss of seed dormancy, microdissections were used to isolate embryos from briefly imbibed, dormant seeds. These embryos were still enveloped by the aleurone layer but they lacked the seed coat. Imbibition of seeds in the absence of the seed coat did not result in germination and dormancy was maintained for at least 28 days (Bethke *et al.*, 2007). Exposure of these seeds without the seed coats to NO gas, however, resulted in germination percentages approaching 100% (Bethke *et al.*, 2007). These data indicate that either the aleurone layer or the embryo responded to NO in a way that leads to a loss of dormancy. Isolated embryos lacking both the aleurone layer and the testa, however, showed no sign of dormancy. Embryos removed from dormant seeds of the C-24 ecotype, or more highly dormant Arabidopsis ecotypes such as Cape Verde (Cvi) or Kashmir2 (Kas2), grew and began to green as if they were non-dormant when placed on solid media (Bethke *et al.*, 2007). Furthermore, imbibition of isolated embryos with the NO scavenger c-PTIO had no effect on growth. These data suggest that early embryo growth was not dependent on NO, and that Arabidopsis seeds do not have true embryo dormancy.

Aleurone cells undergo morphological changes that are tightly linked to the dormancy status of the seed. In intact, imbibed, non-dormant seeds there is a progressive change in the number and size of the vacuoles within each aleurone cell, such that the average number of vacuoles per cell decreases to one or a few. As shown in *Figure 3A* and *B*, there are around 20 vacuoles per aleurone cell in Arabidopsis, and this number declines to around one at the time of germination (Bethke *et al.*, 2007). A similar change is observed in non-dormant barley grains (Bush *et al.*, 1986; Jones and Price, 1970). This process of rapid vacuolation does not occur in dormant seeds or grains (*Figure 3C*), and in dormant seeds of the Arabidopsis Cvi or Kas2 ecotypes, the number of vacuoles per cell may increase following imbibition (Bethke *et al.*, 2007). Thus, vacuolation of aleurone cells is a useful marker for the aleurone layer that indicates a loss of dormancy has occurred.

When aleurone layers from briefly imbibed, dormant Arabidopsis seeds were separated from the embryo and incubated on a solid medium, there was a progressive vacuolation of the aleurone cells (Bethke *et al.*, 2007). Isolated aleurone layers, therefore, are similar to isolated embryos in that they act as if they are not part of a dormant seed once they are removed from the seed. Imbibition of isolated aleurone layers with c-PTIO, however, inhibited this process of vacuolation and this indicated an involvement of NO in this process. Toxicity of c-PTIO to isolated aleurone layers was not a

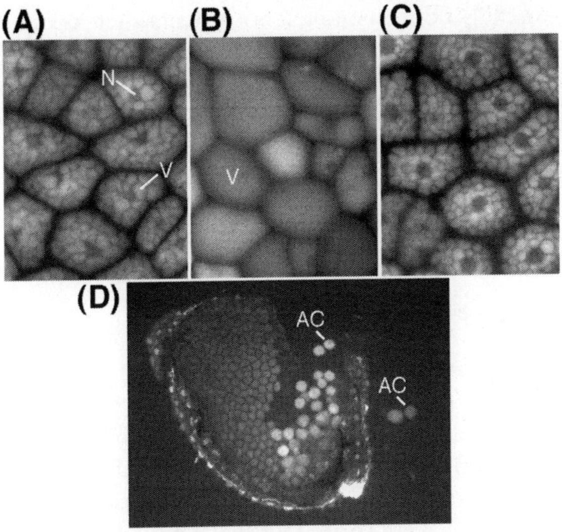

Figure 3. *Micrographs showing Arabidopsis aleurone cells in dormant and non-dormant seeds, and in seeds of the* Spy-1 *mutant. (A) Aleurone cells in a non-dormant seed 2 h after imbibition. Each cell contains many light coloured vacuoles (V) and a darker coloured nucleus (N). (B) Aleurone cells in a non-dormant seed shortly after germination. A single large vacuole nearly fills each cell. (C) Aleurone cells in a dormant seed 9 days after imbibition. The vacuolation that occurs in non-dormant seeds is prevented in dormant seeds and each cell contains numerous vacuoles. (D) Cells in the region of the aleurone layer adjacent to the root in* Spy-1 *mutant seeds vacuolate rapidly and enlarge. Individual aleurone cells (AC) are easily dislodged as a result of extensive wall weakening.*

cause of this inhibition, as imbibition of isolated aleurone layers with both c-PTIO and GA reversed the effect of c-PTIO alone, and aleurone cells became highly vacuolate (Bethke *et al.*, 2007). Taken together, these data indicate strongly that the aleurone layer is the site of perception for NO that leads to dormancy loss.

The inhibition of vacuolation observed in isolated aleurone layers by c-PTIO suggests that NO was also produced by the aleurone layer. Numerous sources of NO exist in plants, including NO synthase-like activities, nitrate reductase, a plasma membrane localized nitrite-NO reductase, and others (Crawford and Guo, 2005; Delledonne, 2005; Shapiro, 2005). Non-enzymatic NO synthesis, where NO is formed spontaneously from nitrite in an acidic environment, is an important source of NO in mammals (Gladwin *et al.*, 2005), and several lines of evidence indicate that it is an important source of NO in the aleurone layer of seeds as well (Bethke *et al.*, 2004a). Copious amounts of NO were produced within seconds when nitrite was added to samples containing barley aleurone layers and the medium in which they were incubated. This production of NO did not require aleurone cells. Nitrite-dependent NO production occurred when nitrite was added to the medium in which aleurone layers had been incubated. The rate of this reaction was dependent on the amount of added nitrite and was accelerated by native antioxidants such as the proanthocyanidins found in the seed coat, or exogenous phenolic antioxidants such as catechin. The observations that nitrite served as a substrate for NO production, and that NO reduced seed dormancy, fit well with the extensive research on seed physiology demonstrating that nitrite is a potent dormancy-breaking compound for seeds from a wide range of species (Hendricks and Taylorson, 1974).

4 GA synthesis and response act downstream of NO to promote germination

4.1 *Transcription of genes required for GA synthesis*

Several lines of experimental evidence suggest that NO acts upstream of GA in the processes that lead to germination. Quantitative RT-PCR was used to examine the accumulation of transcripts for GA biosynthetic genes in embryos and aleurone layers of Arabidopsis seeds (Bethke *et al*., 2007). *GA3ox1* and *GA3ox2* encode 3β-hydroxylases and catalyse the final step in the synthesis of biologically active GAs. When dormant Arabidopsis seeds were treated with a compound that initiated NO-dependent germination, in this case cyanide gas, *GA3ox1* and *GA3ox2* transcripts were not detectable in the aleurone layer at any time between 1 and 48 h after treatment. Embryos removed from treated seeds showed a significant accumulation of *GA3ox1* transcript that could be detected by the 24-h time-point. Both *GA3ox1* and *GA3ox2* transcripts accumulated to a much larger degree by 48 h after treatment. These increases in the abundance for GA biosynthetic gene transcripts were much less pronounced in the embryos from seeds treated with cyanide gas and imbibed with c-PTIO. Little accumulation of GA biosynthetic gene transcripts was found in embryos from seeds imbibed in water alone. These data suggest that the embryo synthesizes the GA required for germination, not the aleurone layer.

4.2 *Weakening of aleurone cell walls by secreted enzymes*

The above evidence shows that Arabidopsis aleurone cells become more vacuolate in seeds that are about to undergo germination and that this process can be prevented by the NO-scavenger c-PTIO but continues in the presence of both c-PTIO and GA. The process of vacuolation is one of many GA-dependent processes that have been characterized in barley aleurone cells. The aleurone layer in barley grain, however, is best known as a secretory tissue that synthesizes and secretes a wide range of hydrolytic enzymes (Fincher, 1989; Jones and Jacobsen, 1991). The most numerous of these are the α-amylases that break down the reserves of starch, and the proteases that hydrolyse the storage proteins in the dead starchy endosperm. Other secreted enzymes include cell-wall degrading enzymes that digest the thick aleurone cell walls. Wall hydrolysis makes additional carbohydrate available for uptake by the growing embryo. This digestion and presumed weakening of the cell walls surrounding the aleurone cells suggested a mechanism for germination in which the cell walls of the aleurone cells nearest to the embryonic root are weakened through the activity of enzymes secreted by the aleurone cells following perception of embryo-synthesized GA. Enzymatic wall weakening permits the root to overcome the restraint to germination imposed by the aleurone layer.

Three lines of evidence from Arabidopsis strongly support the hypothesis that enzymatic weakening of the cell wall permits the root to overcome the restraint to germination imposed by the aleurone layer. The first and most compelling line of evidence comes from experiments with the *Spy-1* mutant of Arabidopsis. *Spy-1* is a GA signalling mutant that has a lesion in a gene encoding an O-GluNAc transferase, and *Spy-1* mutant plants have phenotypes similar to wild-type plants sprayed repeatedly with GA. In aleurone layers of *Spy-1* mutant seeds, extreme wall weakening between the aleurone cells in the region near the embryonic root is observed (*Figure 3D*), to the extent that individual, greatly enlarged, living cells are readily dislodged from isolated aleurone layers (Bethke *et al.*, 2007). Cells in other regions of the aleurone layer are much less affected in the *Spy-1* mutant and resemble cells in wild-type seeds. Second, vacuolation of aleurone cells in imbibed, non-dormant Arabidopsis seeds was most rapid in those aleurone cells adjacent to the embryonic root, and least rapid in those cells adjacent to the cotyledons (Bethke *et al.*, 2007). That is, vacuolation, and presumed wall weakening, occurred fastest in that part of the aleurone layer that ruptures at germination. Finally, although imbibed dormant seeds with the seed coat removed remained dormant, physical damage to the aleurone layer resulted in rapid germination.

Taken together, these data are consistent with a model in which NO produced in the aleurone layer results in GA production in the embryo. Diffusion of GA from the embryo to the aleurone layer triggers the synthesis and secretion of cell-wall degrading enzymes. These enzymes weaken the walls between adjacent aleurone cells, and in doing so promote germination by reducing the physical force necessary for the root to penetrate through the aleurone layer. Many GA-dependent events in seeds are antagonized by ABA, and in Arabidopsis, NO not only stimulates the transcription of GA biosynthetic enzymes but may also decrease the sensitivity of seeds to ABA (Bethke *et al.*, 2006a).

5 The role of the aleurone layer after germination is species dependent

The embryo continues to grow and develop following germination, and becomes a self-sustaining, autotrophic seedling. For a critical period of time after germination, and to a degree that depends on the species, growth of the embryo is fuelled by nutrients that were in the seed. The post-germinative contribution of the aleurone layer to seedling growth is a direct consequence of the physical relationship between the germinated embryo and the aleurone layer. In Arabidopsis, for example, the seed coat and aleurone layer are pushed off the embryo as it unfolds and enlarges. Hence, the aleurone cells are of no consequence to the embryo after germination because they do not remain in contact with the embryo. In barley, on the other hand, the seed coats and aleurone layer remain tightly attached to the embryo. Because the aleurone layer remains in intimate contact with the embryo, the nutrients in it can be used to benefit the young seedling. This is accomplished, in the end, through the programmed death of the aleurone cells.

6 GA promotes and ABA inhibits programmed cell death of barley aleurone cells

Isolated barley aleurone layers have proved to be extremely useful in experiments focused on understanding the regulation of gene expression, hormone signalling and ion transport in the aleurone, as well as the cell biology and enzymology of the aleurone tissue in germination. A major reason for the usefulness of isolated aleurone layers is that control over cellular function can be achieved by incubating isolated aleurone layers with either GA or ABA. Layers incubated in ABA behave akin to those in the dry grain. Genes associated with early germination, such as the GA biosynthetic enzymes, are not transcribed in isolated barley aleurone layers. Similarly, nutrient hydrolysing amylases and proteases are not synthesized. Incubation of barley aleurone layers with GA, however, results in a rapid conversion of the cell from a quiescent storage cell to a dynamic secretary cell that rapidly hydrolyses its stored reserves of protein, lipid and potassium-storing phytate (Bethke *et al.*, 1998; Fincher, 1989; Jones and Jacobsen, 1991).

The amino acids derived from proteolysis of vacuolar storage proteins are used for the *de novo* synthesis of enzymes that will be secreted from the aleurone cells. These enzymes break down the carbohydrate, protein and other macromolecules in the starchy endosperm. The most abundant of these secreted enzymes are the α-amylases, which make up as much as 65% of the newly synthesized protein (Higgins *et al.*, 1982). Lipids, which are stored as triacylglycerides in oleosomes, are hydrolysed by lipases, and fatty acids are broken down via β-oxidation in the glyoxysome to support gluconeogenesis. These GA-dependent activities function to nourish the growing seedling, either directly, as in the export of minerals and sugar, or indirectly, as in the case of secreted enzymes. The final GA-dependent event for each aleurone cell is a programmed cell death (PCD); this also makes nutrients available to the growing embryo, since the contents of the dead aleurone cell become available for hydrolysis by the cocktail of digestive enzymes that the aleurone cells secreted while alive.

As illustrated in *Figure 4A*, cells in isolated aleurone layers remained viable for more than 48 h when incubated in media containing ABA (Fath *et al.*, 2001). When identical aleurone layers were incubated in media containing GA, however, a different picture emerged. For the first 24 h after GA treatment, all cells remained alive as judged by vital staining of intact aleurone layers. Between 24 and 48 h after GA treatment, however, there was a progressive increase in the number of dead cells, such that approximately half of the cells were dead at 36 h and 80–90% of the cells were dead by 48 h after treatment with GA (Fath *et al.*, 2001).

7 Programmed cell death results from damage by ROS

7.1 *Sensitivity to antioxidants and reactive oxygen species*

Several lines of evidence suggest that this hormone-dependent PCD results from increased oxidation and sensitivity to ROS. Firstly, incubating GA-treated aleurone layers with the antioxidant butylated hydroxytoluene slowed the rate of death, but did not slow the rate or extent of α-amylase secretion (Beligni *et al.*, 2002). Secondly, GA-treated aleurone layers became more sensitive to exogenous hydrogen peroxide with time after hormone treatment (*Figure 4B*; Bethke and Jones, 2001; Fath *et al.*, 2001). Cells in aleurone layers treated with either ABA or GA for 12 h remained viable when exposed to 1% H_2O_2 for 30 min. Eighteen hours after hormone treatment, however, about half of the cells in GA-treated aleurone layers died when exposed to 1% H_2O_2, while all cells in ABA-treated layers remained alive. By 24 h after hormone

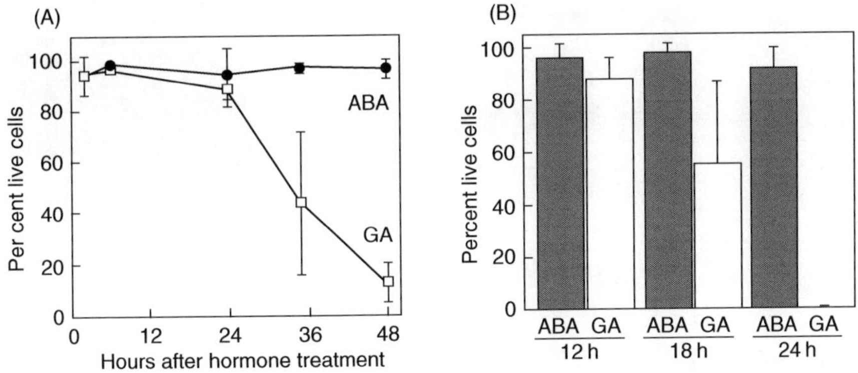

Figure 4. *Programmed cell death in barley aleurone cells is promoted by GA but inhibited by ABA. (A) A time-course showing the percentage of live cells in barley aleurone layers at various times after treatment with either GA or ABA. (B) The percentage of cells surviving exposure to 1% H_2O_2 for 1 h in aleurone layers 12 h, 18 h or 24 h after treatment with ABA or GA.*

treatment, no cells in GA-treated layers survived H_2O_2 treatment, but greater than 90% of cells in ABA-treated layers survived. A third line of evidence indicating that ROS are responsible for barley aleurone cell death comes from experiments with barley aleurone protoplasts. These are the individual living cells that are produced by treating aleurone layers with cell-wall digesting enzymes. Production of ROS within aleurone protoplasts was increased substantially under illumination with either bright blue (440–495) light or UV (370–390 nm) light. This process appears to result from the photoreduction of a flavin-containing oxidase, with a consequent production of ROS. ROS production within aleurone protoplasts was observed within seconds of illumination for both ABA-treated and GA-treated aleurone protoplasts. ABA-treated barley aleurone protoplasts remained viable when exposed to bright UV or blue light for up to 80 min. GA-treated protoplasts, especially highly vacuolate protoplasts that were approaching death, began to die within 20 min of illumination and 100% or 50% of cells were dead following 40 min exposure to UV or blue light respectively (Bethke and Jones, 2001). These experiments confirm that GA-treated aleurone cells are more sensitive to endogenously produced ROS than ABA-treated cells.

7.2 *Decreased capacity to metabolize reactive oxygen species*

The increased sensitivity of GA-treated aleurone cells to exogenous and endogenous ROS reflects a genetically programmed, GA-dependent down-regulation of enzymatic defences against ROS (Fath *et al.*, 2001). Northern analyses such as those in *Figure 5A* showed clearly that transcript amounts for superoxide dismutase (SOD), ascorbate peroxidase (APX) and catalase (CAT) were maintained with time in isolated barley aleurone layers treated with ABA. In GA-treated layers, however, SOD and APX transcripts were reduced 18 h after, and barely detectable 24 h after, hormone treatment. Catalase transcript was strongly down-regulated 6 h after hormone treatment and

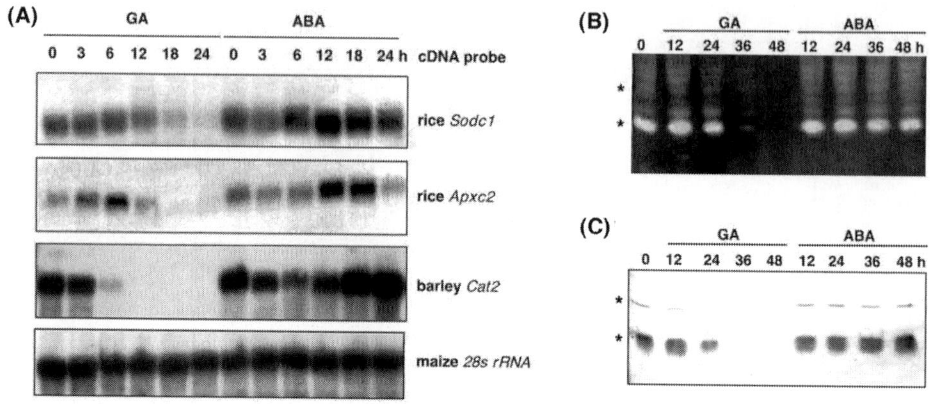

Figure 5. *Enzymatic defences against ROS are down-regulated in GA-treated barley aleurone cells. (A) Northern blots of barley aleurone RNA hybridized with probes to superoxide dismutase (rice sodc1), ascorbate peroxidase (rice apxc2), and catalase (barley cat2) genes. (B) A native gel showing ascorbate peroxidase activities in protein extracts from GA- and ABA-treated barley aleurone cells. At least six bands of activity are visible. (C) Western blot of barley aleurone extracts probed with an anti-ascorbate peroxidase antibody. The same bands indicated by asterisks in (C) are indicated by asterisks in (B).*

below the level of detection 12 h after GA treatment. It is important to note that these decreases in transcript amount occurred at a time when the cells remained alive and were actively producing and secreting enzymes. The amounts of the APX, SOD and CAT proteins and the activities of the enzymes (*Figures 5B and C*), followed the same pattern as transcript abundance, but changes occurred later in time (Fath *et al.*, 2001). Thus antioxidant proteins and activities were maintained in imbibed ABA-treated aleurone layers but they were strongly decreased in GA-treated layers. Native activity gels were used to assess the effects of GA on APX and SOD activities, as these are particularly informative with regard to the effects on the different isoforms of each enzyme. In all cases, enzyme activity was down-regulated by GA treatment. For SOD, at least three bands of activity could be detected and each was greatly reduced 24 h after GA treatment. For APX, at least six activity bands could be detected (*Figure 5B*), and a strong decrease in all of these bands was observed between 24 h and 36 h after GA treatment.

8 ROS production is a consequence of lipid breakdown

Plant cells generate ROS in many subcellular locations. The chloroplast electron transport chain is the largest source of ROS in photosynthetic tissues, while the respiratory electron transport chain in mitochondria also produces ROS in all cell types. In barley aleurone cells treated with GA, the largest single source of ROS is likely to be the glyoxysome, where fatty acyl chains derived from triacylglycerides are cleaved to produce acetyl-CoA through a series of enzymatic reactions. Acyl-CoA oxidase is one member of this group of enzymes and for each acetyl-CoA produced, acyl-CoA oxidase produces one molecule of H_2O_2. The potential for ROS production in this manner becomes obvious when one considers that 25–40% of the volume of a mature barley aleurone cell is stored lipid (Eastmond and Jones, 2005; Jones, 1969). Acyl-CoA oxidase activity is up-regulated by GA treatment of barley aleurone cells, as is lipid consumption (Eastmond and Jones, 2005). Hence, GA triggers a massive increase in ROS production while at the same time initiating a decrease in the capacity of the cells to metabolize ROS enzymatically. Within the span of a few days, this combination results in the death of the aleurone cells.

9 Nitric oxide acts as an antioxidant and delays programmed cell death in barley aleurone cells

Nitric oxide can act as either a reactive nitrogen species that increases oxidative stress, or as an antioxidant that decreases oxidative stress by reacting with ROS, particularly superoxide and lipid peroxyl radicals (Beligni and Lamattina, 1999a, 1999b, 2002). Pharmacological experiments with NO donors and NO scavengers have provided data that indicate that NO functions as a protective antioxidant in barley aleurone cells. For example, when barley aleurone layers were incubated with the NO donors SNP or S-nitroso-*N*-acetylpenicillamine (SNAP), the rate of PCD decreases such that approximately 50% of cells were alive 48 h after hormone treatment compared to 20% alive for controls (*Figure 6*; Beligni *et al.*, 2002). This effect of exogenous NO on PCD could be separated from early events related to GA perception and germination by adding SNP 18 h after GA treatment. Although only 20% of the cells were alive 48 h after adding GA alone, 45% of cells were alive at the same time when SNP was added at 18 h after GA treatment (*Figure 6A*). Complementary data were obtained using the

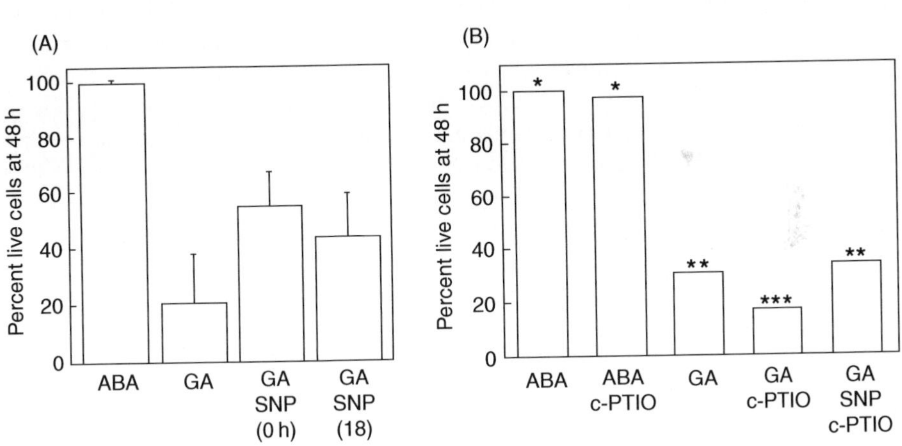

Figure 6. *Nitric oxide acts as an antioxidant in barley aleurone layers. (A) The NO-donor SNP added 0 h or 18 h after GA treatment of barley aleurone layers decreases the number of dead cells 48 h after GA treatment. (B) The NO scavenger c-PTIO accelerates the rate of GA-dependent death, and this effect of c-PTIO can be reversed by addition of SNP. The different number of asterisks indicate statistically different means calculated at P = 0.05 in a student's* t *test.*

NO scavenger c-PTIO (*Figure 6B*). Addition of c-PTIO increased the rate of GA-dependent PCD by a statistically significant amount, such that around 10% of cells were alive 48 h after treatment with GA and c-PTIO compared with around 25% of cells alive when treated with GA alone (Bethke *et al.*, 2004a). The effect of c-PTIO on death could be reversed by addition of SNP, and this further indicates that NO is involved in delaying death. Addition of c-PTIO to ABA-treated cells had no effect, and all cells remained alive. The potential advantage to the germinated seedling of NO delaying PCD was illustrated in experiments where NO donors were added to GA-treated aleurone layers and the activity of secreted α-amylase was determined. NO donors did not increase the rate at which α-amylase was secreted into the incubation medium during the first 24-h period after hormone addition. Between 24 and 48 h, however, during the period when cells treated with GA alone died, approximately 20% more α-amylase activity was found in the incubation medium surrounding aleurone layers treated with both an NO donor and GA compared to those incubated with GA alone (Beligni *et al.*, 2002).

10 Concluding remarks

Most seeds are required to enter a phase of metabolic and developmental quiescence in order to conserve stored nutrients until appropriate environmental cues lead to an end of dormancy and renewed growth. Once the process leading to germination has been initiated, rapid growth and development are desirable, since this allows roots to explore the soil for water and minerals, and leaves to develop the photosynthetic machinery that will make the seedling autotrophic. ABA, GA, NO and ROS are central players in this transition phase in Arabidopsis and barley. ABA establishes and maintains dormancy. NO and GA function to bring about a loss of dormancy and to promote germination. Later, GA action results in an increase in ROS production and, in barley, a weakening of cellular defences against ROS. NO also acts as an antioxidant in barley to mitigate the effects of ROS and to postpone PCD. Eventually, however, ROS contribute to the death of the nutrient-storing aleurone layer that then becomes, like the dead starchy endosperm, a source of nutrient-rich corpses. Hence, NO and ROS play important roles as signalling molecules, antioxidants and oxidants at this critical phase of the plant life cycle. Other functions in signalling and response are likely for both NO and ROS, and these will be uncovered as additional details of the molecular and biochemical events controlling dormancy, germination and early seedling growth are obtained.

References

Beligni, M.V. and Lamattina, L. (1999a) Is nitric oxide toxic or protective? *Trends Plant Sci.* **4:** 299–300.

Beligni, M.V. and Lamattina, L. (1999b) Nitric oxide counteracts cytotoxic processes mediated by reactive oxygen species in plant tissues. *Planta* **208:** 337–344.

Beligni, M.V. and Lamattina, L. (2002) Nitric oxide interferes with plant photo-oxidative stress by detoxifying reactive oxygen species. *Plant Cell Environ.* **25:** 737–748.

Beligni, M.V., Fath, A., Bethke, P.C., Lamattina, L. and Jones, R.L. (2002) Nitric oxide acts as an antioxidant and delays programmed cell death in barley aleurone layers. *Plant Physiol.* **129:** 1642–1650.

Bethke, P.C. and Jones, R.L. (2001) Cell death of barley aleurone protoplasts is mediated by reactive oxygen species. *Plant J.* **25:** 19–29.

Bethke, P.C., Swanson, S.J., Hillmer, S. and Jones, R.L. (1998) From storage compartment to lytic organelle: the metamorphosis of the aleurone protein storage vacuole. *Ann. Bot.* **82:** 399–412.

Bethke, P.C., Badger, M.R. and Jones, R.L. (2004a) Apoplastic synthesis of nitric oxide by plant tissues. *Plant Cell* **16:** 332–341.

Bethke, P.C., Gubler, F., Jacobsen, J.V. and Jones, R.L. (2004b) Dormancy of Arabidopsis seeds and barley grains can be broken by nitric oxide. *Planta* **219:** 847–855.

Bethke, P.C., Libourel, I.G.L. and Jones, R.L. (2006a) Nitric oxide reduces seed dormancy in Arabidopsis. *J. Exp. Bot.* **57:** 517–526.

Bethke, P.C., Libourel, L.G.L., Reinohl, V. and Jones, R.L. (2006b) Sodium nitroprusside, cyanide, nitrite, and nitrate break Arabidopsis seed dormancy in a nitric oxide-dependent manner. *Planta* **223:** 805–812.

Bethke, P.C., Libourel, I.G.L., Aoyama, N., Chung, Y.Y., Still, D.W. and Jones, R.L. (2007) The Arabidopsis aleurone layer responds to nitric oxide, gibberellin, and abscisic acid and is sufficient and necessary for seed dormancy. *Plant Physiol.* **143:** 1173–1188.

Bewley, J.D. (1997) Seed germination and dormancy. *Plant Cell* **9:** 1055–1066.

Bewley, J.D. and Black, M. (1994) *Seeds: Physiology of Development and Germination.* Plenum Press, New York, NY.

Bush, D.S., Cornejo, M.J., Huang, C.N. and Jones, R.L. (1986) Ca^{2+}-stimulated secretion of alpha-amylase during development in barley aleurone protoplasts. *Plant Physiol.* **82:** 566–574.

Crawford, N.M. and Guo, F.Q. (2005) New insights into nitric oxide metabolism and regulatory functions. *Trends Plant Sci.* **10:** 195–200.

Delledonne, M. (2005) NO news is good news for plants. *Curr. Opin. Plant Biol.* **8:** 390–396.

Eastmond, P.J. and Jones, R.L. (2005) Hormonal regulation of gluconeogenesis in cereal aleurone is strongly cultivar-dependent and gibberellin action involves SLENDER1 but not GAMYB. *Plant J.* **44:** 483–493.

Fath, A., Bethke, P.C. and Jones, R.L. (2001) Enzymes that scavenge reactive oxygen species are down-regulated prior to gibberellic acid-induced programmed cell death in barley aleurone. *Plant Physiol.* **126:** 156–166.

Finch-Savage, W.E. and Leubner-Metzger, G. (2006) Seed dormancy and the control of germination. *New Phytol.* **171:** 501–523.

Fincher, G.B. (1989) Molecular and cellular biology associated with endosperm mobilization in germinating cereal-grains. *Annu. Rev. Plant Physiol. Plant Mol. Biol.* **40:** 305–346.

Finkelstein, R.R., Gampala, S.S.L. and Rock, C.D. (2002) Abscisic acid signaling in seeds and seedlings. *Plant Cell* **14:** S15–S45.

Gladwin, M. T., Schechter, A. N., Kim-Shapiro, D. B., ***et al.*** (2005) The emerging biology of the nitrite anion. *Nature Chem. Biol.* **1:** 308–314.

Gubler, F., Millar, A.A. and Jacobsen, J.V. (2005) Dormancy release, ABA and pre-harvest sprouting. *Curr. Opin. Plant Biol.* **8:** 183–187.

Hendricks, S. and Taylorson, R. (1974) Promotion of seed germination by nitrate, nitrite, hydroxylamine, and ammonium salts. *Plant Physiol.* **54:** 304–309.

Higgins, T.J.V., Jacobsen, J.V. and Zwar, J.A. (1982) Gibberellic acid and abscisic acid modulate protein synthesis and mRNA levels in barley aleurone layers. *Plant Mol. Biol.* **1:** 191–215.

Jones, R.L. (1969) Effect of ultracentrifugation on fine structure and alpha-amylase production in barley aleurone cells. *Plant Physiol.* **44:** 1428–1438.

Jones, R.L. and Jacobsen, J.V. (1991) Regulation of synthesis and transport of secreted proteins in cereal aleurone. *Int. Rev. Cytol. Surv. Cell Biol.* **126:** 49–88.

Jones, R.L. and Price, J.M. (1970) Gibberellic acid and fine structure of barley aleurone cells. III. Vacuolation of aleurone cell during phase of ribonuclease release. *Planta* **94:** 191–202.

Kermode, A.R. (2005) Role of abscisic acid in seed dormancy. *J. Plant Growth Reg.* **24:** 319–344.

Koornneef, M., Alonso-Blanco, C., Bentsink, L., Blankestijn-De Vries, H., Debeaujon, I., Hanhart, C.J., Leon-Kloosterziel, K.M., Peeters, A.J.M. and Raz, V. (2000) The genetics of seed dormancy in *Arabidopsis thaliana*. In: *Dormancy in Plants* (ed. J.D.V.A.J. Crabbe). CAB International, Wallingford.

Libourel, I.G.L., Bethke, P.C., De Michele, R. and Jones, R.L. (2006) Nitric oxide gas stimulates germination of dormant Arabidopsis seeds: use of a flow-through apparatus for delivery of nitric oxide. *Planta* **223:** 813–820.

Peng, J.R. and Harberd, N.P. (2002) The role of GA-mediated signalling in the control of seed germination. *Curr. Opin. Plant Biol.* **5:** 376–381.

Shapiro, A.D. (2005) Nitric oxide signaling in plants. *Plant Hormones* **72:** 339–398.

3

Ageing and oxidants in the nematode *Caenorhabditis elegans*

David Gems

1 Testing the oxidative damage theory of ageing in *C. elegans*

The biology of ageing remains poorly understood. For example, it remains unclear what sort of biological processes are the primary determinants of ageing, or of the differences in ageing rate between different animal species. The extent to which ageing involves the same or different processes in different animal taxa also remains unclear. Many forms of pathology are associated with increased levels of molecular damage (involving, for example, oxidation), and ageing is no exception (Halliwell and Gutteridge, 2007).

Among contemporary mechanistic theories of ageing, many view accumulation of molecular damage, particularly oxidative damage, as the primary cause. It has been postulated that a major cause of such damage are reactive oxygen species (ROS), particularly the free radical O_2^- (superoxide) and its derivatives (reviewed by Balaban *et al.*, 2005; Raha and Robinson, 2000). O_2^- is generated in the cell by diverse processes, for example, as a by-product of the activity of the mitochondrial electron transport chain. Such mitochondrial ROS have been viewed as a possible mechanistic basis of another hypothetical ageing mechanism. The 'rate of living' theory postulates that the rate of energy metabolism determines the rate of ageing. It has been suggested that this could be because the rate of production of ROS by mitochondria, and therefore the rate of accrual of molecular damage, is higher when metabolic rate is higher (Sohal and Weindruch, 1996). In the following discussion, I will critically assess in turn whether metabolic rate, mitochondria, and reactive oxygen species are determinants of ageing in the nematode *Caenorhabditis elegans*. To this end, I will survey the numerous investigations of these issues that have been carried out using this model organism.

C. elegans is a free-living nematode that can be found in soil rich in organic matter, particularly compost. Experimentally, it has the advantage of being a complex animal, with a nervous system, reproductive system and alimentary canal, yet one that is so small (adults reach only ~1.2 mm in length) that it may be handled like a microorganism, with the convenience and low cost that this implies. For studying ageing it has two particular advantages: it has a very short lifespan (usually 2–3 weeks), and its lifespan is unaffected by inbreeding effects, which have complicated studies of the genetics of ageing in *Drosophila* and the mouse (Johnson and Hutchinson, 1993). There are also the many advantages associated with an established genetic model system, including well-characterized mutations in large numbers of genes, availability of a well-annotated genome sequence, and powerful molecular genetic methodologies. The latter include construction of transgenic animals, use of fluorescent proteins to visualize gene expression within the transparent body of the nematode, and RNA-mediated interference to knock down gene expression.

Several types of approach have been taken to investigate the role of oxidative stress in ageing in *C. elegans*. These include testing for correlations between ageing and various aspects of oxidative metabolism. Studies of this sort typically either examine age changes in wild-type nematodes, or differences between wild-type nematodes and mutants with altered ageing rates. Some attempts have also been made to test theories of ageing more directly by manipulating selected aspects of the relevant biology (e.g. antioxidant defence) and examining the resulting effects on ageing. Most studies have been of the first type, which is, arguably, the least informative.

One of the strengths of *C. elegans* as a model for studying ageing is the ease with which classical genetic approaches may be applied. Many genes have been identified where loss of function due to mutation or RNAi leads to altered lifespan. A problem with studies of short-lived strains is that a reduction in lifespan can result either from accelerated ageing (progeria) or from pathologies unrelated to normal ageing, and it can be difficult to distinguish the two. However, methods have been developed to identify likely instances of progeria (Garigan *et al.*, 2002; Gerstbrein *et al.*, 2005), and some short-lived mutant studies have been informative. For example, the gene *mev-1* encodes a subunit of complex II in the electron transport chain (Ishii *et al.*, 1998). Mutation of *mev-1* results in hypersensitivity to oxidative stress, elevated production of mitochondrial ROS and shortened lifespan (Ishii *et al.*, 1990; Senoo-Matsuda *et al.*, 2001; reviewed in Ishii *et al.*, 2006).

Many more studies have focused on genes where mutation increases lifespan, for example, those encoding elements of the insulin/IGF-1 signalling (IIS) pathway, which include the *daf-2* insulin/IGF-1 receptor and the *age-1* phosphatidylinositol 3-kinase (Kimura *et al.*, 1997; Morris *et al.*, 1996). Mutation of *age-1*, for example, can increase the mean and maximum lifespan of *C. elegans* adults by up to 10-fold (Ayyadevara *et al.*, 2007). Long-lived IIS mutants also show resistance to oxidative stress and increased levels of the antioxidant enzymes superoxide dismutase (SOD) and catalase (Larsen, 1993; Vanfleteren, 1993; Vanfleteren and De Vreese, 1995). Numerous studies, in *C. elegans* and other species, report correlations between oxidative damage and ageing. While this might seem to suggest that the oxidative damage theory must be true, it is not safe to conclude this. To date, there is little direct evidence demonstrating, for example, control of normal ageing in *C. elegans* by superoxide or SOD, or hydrogen peroxide (H_2O_2) or catalase. In fact, relatively few studies have been conducted that directly test oxidative damage theories of ageing in *C. elegans*. Many more studies of this sort have been conducted in other models. For example, numerous studies of the effects on ageing of over-expression of SOD and catalase have been conducted in *Drosophila* (Orr and Sohal, 2003; Parkes *et al.*, 1998; Sun and Tower, 1999). Ultimately, theories of ageing in *C. elegans* may only be verified or falsified reliably by means of such direct testing.

2 Metabolic rate and ageing in *C. elegans*

Central to many discussions of the rate of living theory is the idea that increased metabolic rate will lead to increased production of O_2^- and, consequently, an increased rate of accumulation of molecular damage and of ageing (Beckman and Ames, 1998; Finkel and Holbrook, 2000). This view has informed many studies of metabolic rate and ageing in *C. elegans*, as elsewhere. However, this view is not necessarily correct. At lower rates of metabolism, the inner mitochondrial membrane potential increases, which can increase O_2^- production. As metabolic rate increases, membrane potential and O_2^- production drop (Brand, 2000). Thus, all else being equal, one might expect lifespan to increase with increasing metabolic rate.

2.1 *Effects of temperature on lifespan*

C. elegans does show striking rate-of-living effects insofar as lifespan is shorter at higher temperatures. For example, the median lifespan of wild-type hermaphrodites is 24 days at 15°C compared to 16 days at 22.5°C (Gems *et al.*, 1998). This implies that processes whose rate determines the rate of ageing occur faster at higher temperatures; but the identity of these critical processes remain undetermined. To date, studies of rate-of-living effects have focused on energy metabolism and production of O_2^- by mitochondria. In *C. elegans*, metabolic rate does increase with increasing temperature (Van Voorhies and Ward, 1999), but a causal role of metabolic rate in determining ageing rate has not been demonstrated.

2.2 *Metabolic rate in long-lived nematodes*

In order to test the rate of living theory, metabolic rate has been measured in long-lived nematodes, including *age-1* and *daf-2* mutants, *clk-1* mutants (with defects in ubiquinone biosynthesis, see below), and nematodes subjected to various forms of dietary restriction. The instructive power of such tests is rather limited, however, for the following reasons. If metabolic rate were the sole determinant of longevity in *C. elegans*, then one should see a reduction in metabolic rate in long-lived nematodes, though such an observation would give no indication of causality. If long-lived nematodes show no change in metabolic rate, or even a small increase, this does not demonstrate that metabolic rate is not a determinant of ageing, since more than one mechanism may contribute to mutant longevity.

In the main, reductions in metabolic rate in long-lived *C. elegans* have not been detected in IIS mutants or animals subjected to dietary restriction, but they have been in some strains with mitochondrial defects. In *age-1* and *daf-2* mutants, oxygen consumption rate shows no reduction and even slight increases (Braeckman *et al.*, 2002a, 2002b; Houthoofd *et al.*, 2005a, 2005b; Vanfleteren and De Vreese, 1996). Lower levels of heat production and higher levels of ATP were also seen in *daf-2* mutants, which might imply a higher level of mitochondrial coupling in this mutant. Consistent with this, levels of O_2^- production in isolated mitochondria are higher in *daf-2(e1370)* mutants than in wild type (Brys *et al.*, 2007). One study also reported a decline in metabolic rate in *age-1* and *daf-2* mutants, and concluded that the rate of living theory is supported (Van Voorhies and Ward, 1999). This study measured CO_2 production, which could explain the discrepancy with other studies (for further discussion of methodological issues in metabolic studies of *C. elegans*, see Braeckman *et al.*, (2002a, 2002b) and Van Voorhies (2002)).

How dietary restriction (DR) affects metabolic rate in *C. elegans* depends on how DR is exerted. For example, DR by bacterial dilution has no effect on oxygen consumption rate, while DR by means of axenic culture (i.e. no microbial food source) or an *eat-2* mutation (which reduces feeding rate) increases oxygen consumption rate and heat production (Houthoofd *et al.*, 2002b, 2002c).

Strains with alterations in mitochondrial function vary in terms of metabolic rate, and *clk-1* has little effect on metabolic rate (Braeckman *et al.*, 1999, 2002c; Felkai *et al.*, 1999). However, metabolic rate was reduced in *isp-1* mutants, which affect complex III of the electron transport chain (ETC; Feng *et al.*, 2001) and animals subjected to RNAi knockdown of several ETC genes (Dillin *et al.*, 2002). Thus, it is possible that reduced metabolic rate somehow contributes to the increased longevity of long-lived forms of *C. elegans* in some cases. However, the effect of metabolic rate on ageing in *C. elegans* remains unknown.

2.3 *Differences in energy metabolism between* C. elegans *and vertebrates*

As always with studies of ageing in *C. elegans* a major concern is whether any given findings are relevant to mammalian ageing. In terms of energy metabolism, *C. elegans* is clearly different from higher animals in several respects. For example, *C. elegans* possesses the glyoxylate pathway, absent in higher animals, which makes possible the conversion of acetyl CoA to glucose. Expression of the main glyoxylate enzyme, which has both malate synthase and isocitrate lyase activity, is up-regulated in *daf-2* and *age-1* mutant adults, and in the very long-lived, diapausal dauer larva stage (McElwee *et al.*, 2006a, 2006b; Vanfleteren and De Vreese, 1995). *C. elegans* also synthesizes the disaccharide trehalose, also lacking in higher animals.

Nematodes are also capable of anaerobic respiration using an alternative electron acceptor, rhodoquinone (Takamiya *et al.*, 1999), and the malate dismutation pathway (Tielens *et al.*, 2002). When cultured under anoxic conditions, *C. elegans* excretes lactate, acetate, succinate and propionate (Foll *et al.*, 1999). It has been suggested that such anaerobic respiration might reduce O_2^- production levels, thereby increasing lifespan (Rea and Johnson, 2003). Transcript profile studies suggest that this pathway might be up-regulated in dauer larvae and *daf-2* mutants (Holt and Riddle, 2003; McElwee *et al.*, 2006a, 2006b). However, increased anaerobic respiration in *daf-2* mutants would be expected to generate heat, which would increase their calorimetric/respirometric (C/R) ratio. In fact, the C/R ratio is reduced in mutants with reduced insulin/IGF-1 signalling (Houthoofd *et al.*, 2005b).

In conclusion, while rate-of-living effects are seen in *C. elegans*, the biochemical process whose rate is so strongly determinative of ageing remains unclear. Evidence for the importance of O_2 consumption is weak, to say the least. One alternative ageing rate-determining process that is affected by temperature is protein synthesis. Several studies have recently shown that reduction of function of various genes linked to protein biosynthesis increases lifespan in *C. elegans* (Hansen *et al.*, 2007; Henderson *et al.*, 2006; Pan *et al.*, 2007; Syntichaki *et al.*, 2007).

3 Mitochondria and ageing in *C. elegans*

3.1 *Mitochondria, superoxide and ageing*

Investigations of the cause of cellular oxidative damage often focus on O_2^- produced as a by-product of the reduction of O_2 by the ETC of mitochondria. Isolated mitochondria or sub-mitochondrial particles can generate substantial amounts of O_2^-. For example, O_2^- production by isolated rat liver mitochondria respiring in State 4 accounts for around 1–2% of oxygen consumed (Boveris, 1977). However, levels of mitochondrial O_2^- production *in vivo* are much lower, in the 0.1–0.3% range (St-Pierre *et al.*, 2002; Staniek and Nohl, 2000), and the relevance of mitochondrial O_2^- to ageing remains unclear (Imlay and Fridovich, 1991; Nohl and Hegner, 1978). In addition, the relative importance of other sources of ROS as contributors to molecular damage and ageing is unknown; ROS, including O_2^- and H_2O_2, are also produced in other ways, such as O_2^- by cytochrome P450 oxidases, xanthine oxidase and membrane-associated NADPH oxidase. The hypothesis that mitochondrial O_2^- causes ageing is far from proven.

The mitochondria of *C. elegans* are largely similar to those of higher animals. For example, their mitochondrial DNA is similar in terms of gene content and overall size (Murfitt *et al.*, 1976; Okimoto *et al.*, 1992). However, there are some significant differences (see below) so, as always with *C. elegans*, one should generalize cautiously. Little is known about levels of mitochondrial O_2^- production *in vivo* in *C. elegans* and whether it contributes to ageing, though it is at least clear that isolated *C. elegans* mitochondria do produce O_2^- (Senoo-Matsuda *et al.*, 2001).

In mammalian cells, levels of mitochondrial O_2^- production increase with age, for example, there is a 25% increase with age in isolated rat heart mitochondria (Nohl and Hegner, 1978). One study has reported that there is no age increase in mitochondrial O_2^- in *C. elegans* (Yasuda *et al.*, 2006). A more recent study has even reported a decline with age in mitochondrial ROS production (measured as H_2O_2; Brys *et al.*, 2007). Consistent with this, complex I activity drops by 60% between day 4 and day 12 (Yasuda *et al.*, 2006).

As worms grow old, their oxygen consumption rate drops dramatically (De Cuyper and Vanfleteren, 1982; Suda *et al.*, 2005; Vanfleteren and De Vreese, 1996; Yasuda *et al.*, 2006). For example, a recent study measured a drop in oxygen consumption from around 200 pl min^{-1} per worm in early adulthood to around 25 pl min^{-1} per worm by 9 days old (Suda *et al.*, 2005). Taken together, these results suggest that *in vivo* levels of mitochondrial O_2^- production decrease substantially with age in *C. elegans*. It therefore seems unlikely that the age increase in oxidative damage in *C. elegans* is due to increased O_2^- production later in life.

3.2 *Effects of mitochondrial electron transport chain defects on ageing*

The oxidative damage theory could imply that mutations affecting components of the ETC would either decrease or increase lifespan. Defects in electron transport might increase production of O_2^- and reduce lifespan; alternatively, overall reduction in electron flow might lower O_2^- and increase lifespan. In *C. elegans*, both effects of disruption of ETC genes on lifespan have been seen. However, it remains unclear whether this has anything to do with O_2^- generation (reviewed by Anson and Hansford, 2004).

Genes encoding mitochondrial proteins predominated in several large-scale RNAi screens for genes with effects on lifespan (Dillin *et al.*, 2002; Hamilton *et al.*, 2005; Hansen *et al.*, 2005; Lee *et al.*, 2003a). In particular, RNAi of many mitochondrial and nuclear genes encoding proteins of ETC complexes I–V caused substantial increases in lifespan. RNAi affecting other mitochondrial proteins, such as mitochondrial carriers, also increased lifespan. The combination of mitochondrial defects and increased lifespan is sometimes referred to as the Mit phenotype (Rea, 2005). In most cases Mit animals also show delayed development, reductions in body size, fertility, activity level and feeding rate (Dillin *et al.*, 2002), as well as abnormalities in mitochondrial morphology (Lee *et al.*, 2003a). For some genes, Mit animals have normal body size and feeding rates, but increased lifespan (Hansen *et al.*, 2005), implying that life extension is not causally connected to reduced body size or feeding rate. The loss of a single protein component of the large ETC protein complexes may cause accumulation of unfolded proteins in the mitochondria and, in many cases, Mit animals accumulate the mitochondrial chaperone heat shock protein 6 (HSP6) (Hamilton *et al.*, 2005; Yoneda *et al.*, 2004). Mit mutants have also been identified with mutations that either affect ETC genes directly (Feng *et al.*, 2001; Tsang *et al.*, 2001) or, in the case of *lrs-2*, indirectly. *lrs-2* encodes a unique mitochondrial leucyl-tRNA synthetase, which is required for the expression of the 12 mitochondrially encoded polypeptides. The *lrs-2* mutation is predicted to block expression of all 12 of these polypeptides; maternally rescued mutants form small, sterile, long-lived adults (Lee *et al.*, 2003a). Severe loss of ETC function often causes larval arrest and lethality (Lee *et al.*, 2003a; Tsang *et al.*, 2001).

What mechanisms might underlie the extension of lifespan in Mit animals? One interpretation is that it is due to reduced metabolic rate and perhaps also lowered production of O_2^-. Several observations are consistent with the first view: in Mit animals there is usually a reduction in O_2 consumption rate (Lee *et al.*, 2003a) and ATP levels can be reduced to as little as 20% of wild type (Dillin *et al.*, 2002). This might suggest that ATP levels limit the rate of processes that promote ageing. However, an interesting study by Dillin *et al.* (2002) suggests that something more complex is going on. To test the timing of effects of Mit defects on ageing, expression of ETC genes was selectively knocked down during larval development or in adulthood. Knock down in larvae alone increased adult lifespan (Dillin *et al.*, 2002). This is perhaps not surprising since mitochondrial number may be programmed during development: in the transition from L4 to adulthood alone there is a sixfold increase in the number of mitochondria (Tsang and Lemire, 2002).

More surprisingly, adult-specific knock down of ETC gene expression reduced ATP levels but did not increase lifespan. The authors postulated that there exists in *C. elegans* a system that registers the rate of respiration during development, and adjusts the subsequent rate of ageing accordingly (Dillin *et al.*, 2002). The reason this is quite surprising is that the timing of action of insulin/IGF-1 signalling and dietary restriction are exactly the opposite: during adulthood and not development. This finding warrants further investigation: for example, how does life-long, larva-specific and adult-specific RNAi of ETC genes compare in terms of effects on mitochondrial O_2^- production, O_2 consumption, mitochondrial number and morphology, and HSP6 expression?

It is unlikely that life extension in Mit animals involves the insulin/IGF-1 pathway, since knock down of ETC genes increases lifespan both in *daf-16* and *daf-2* mutant animals (Dillin *et al.*, 2002; Hamilton *et al.*, 2005; Hansen *et al.*, 2005; Lee *et al.*, 2003a). Lifespan is also increased by RNAi of several genes encoding glycolytic enzymes such as phosphoglycerate mutase (F57B10.3; Lee *et al.*, 2003a) and glucose-6-phosphate isomerase (Y87G2A.8; Hansen *et al.*, 2005), suggesting that glycolysis somehow reduces lifespan. This appears to involve different mechanisms relative to Mit animals, since animals develop normally, body size is not reduced, mitochondria show normal morphology, and the extension in lifespan requires DAF-16 (Hansen *et al.*, 2005; Lee *et al.*, 2003a).

Ubiquinone (coenzyme, or CoQ) plays a major role in the ETC. Production of O_2^- by the mitochondrial ETC appears to be largely the result of transfer of electrons from ubisemiquinone to oxygen (Boveris, 1977; Raha and Robinson, 2000). Deficiency in CoQ can also increase lifespan. For example, *clk-1* encodes a mitochondrial protein necessary for the final step in CoQ biosynthesis (Ewbank *et al.*, 1997; Felkai *et al.*, 1999; Stenmark *et al.*, 2001; reviewed by Stepanyan *et al.*, 2006). Mutation of *clk-1* causes accumulation of the precursor of nematode CoQ, demethoxy-ubiquinone-9 (DMQ9; Miyadera *et al.*, 2001), and increased lifespan (Wong *et al.*, 1995). CoQ varies between species in the number of isoprene units in its side chain. *Escherichia coli* have an eight-unit side chain (CoQ_8), *C. elegans* have CoQ_9, and mammals CoQ_{10}. Likewise, if *C. elegans* are fed on *E. coli* lacking CoQ_8, this increases their lifespan too (Larsen and Clarke, 2002). The above findings might suggest that lowering CoQ levels reduces flux through the ETC, thereby lowering ROS production and increasing lifespan (but see below).

Mutations affecting ETC proteins lead to a shortening of lifespan in a minority of cases. *mev-1(kn1)* is a point mutation in the gene for succinate dehydrogenase cytochrome *b* in complex II, and causes hypersensitivity to oxidative stress and shortened lifespan (Ishii *et al.*, 1990). The mutation compromises electron transfer from succinate to ubiquinone and results in increased electron leakage to oxygen. In wild-type mitochondria, O_2^- production results from electron leak at complex I and particularly III (Raha and Robinson, 2000). *mev-1(kn1)* disrupts complex II and results in O_2^- production from complex II (Senoo-Matsuda *et al.*, 2001). *gas-1* encodes a subunit of complex I and, like *mev-1*, mutation of *gas-1* results in hypersensitivity to oxidative stress and reduced lifespan under normoxia (Hartman *et al.*, 2001). Unlike *mev-1, gas-1* does not increase nuclear mutation rate.

An early study proposed that the short lifespan and sensitivity to pro-oxidants of *mev-1* animals was due to the fact that SOD levels are half that of wild type (Ishii *et al.*, 1990). Consistent with this, deletion of *sod-1*, the major Cu/Zn SOD in *C. elegans*, shortens lifespan (J.J. McElwee and D. Gems, unpublished). However, although it was reported that *mev-1* lifespan can be extended by administration of chemical mimetics of SOD (Melov *et al.*, 2000), a further study was unable to replicate this finding (F. Matthijssens and J.R. Vanfleteren, personal communication).

Insofar as they shorten rather than increase lifespan, *mev-1* and *gas-1* are atypical among genetic interventions affecting mitochondria and lifespan. Here it is worth bearing in mind that *mev-1(kn1)* is a reduction-of-function allele, and not a null; RNAi of *mev-1* results in a high level of embryonic lethality (Ichimiya *et al.*, 2002). *mev-1(kn1)* reduces activity of the ETC by 80%, but does not affect succinate dehydrogenase (SDH) activity (Ishii *et al.*, 1998). The MEV-1 subunit of complex II contains a binding site for CoQ. Potentially, reduced affinity of CoQ to MEV-1 protein leads to increased mobility of CoQ and an electron leak to oxygen. By contrast, in most cases knock down of expression of genes encoding ETC proteins may simply reduce electron flow and O_2^- formation.

3.3 *Superoxide production in mitochondrial effects on ageing*

As mutational studies have demonstrated, mitochondria can influence ageing in *C. elegans*. One interpretation is that this reflects altered electron flux through the ETC and altered O_2^- levels. If this were true, one would expect an accompanying alteration in metabolic rate. Conversely, one would not expect an increase in somatic maintenance mechanisms (e.g. antioxidant defence).

Is retarded ageing in Mit animals attributable to reduced electron flux and reduced O_2^- production? This possibility has been extensively investigated in studies of *clk-1* and CoQ. Several findings suggest that the above view is an oversimplification. Firstly, if the longevity of *clk-1* mutants were due to an effect of lowered CoQ levels on electron flux, then this strain should have a reduced metabolic rate. In fact, neither metabolic rate nor ATP levels are lower in *clk-1* animals (Braeckman *et al.*, 1999, 2002c; Felkai *et al.*, 1999), though RNAi of other mitochondrial genes does lower ATP levels (Dillin *et al.*, 2002). Moreover, succinate-cytochrome *c* reductase activity is almost normal in *clk-1* mutants, implying that $DMCoQ_9$ (perhaps supplemented with bacterially derived CoQ_8) functions as well as CoQ_9 (Felkai *et al.*, 1999; Miyadera *et al.*, 2001). It has been suggested that $DMCoQ_9$ produces less O_2^- than CoQ_9 (Miyadera *et al.*, 2001), but this has not been tested directly.

Interpretation of the role of *clk-1* and CoQ in ageing is complicated by the fact that the reduced (quinol) form of CoQ can act as a lipid-soluble antioxidant which protects against lipid peroxidation (Kwong *et al.*, 2002; Lass and Sohal, 2000; Miyadera *et al.*, 2002), which explains why it is marketed as a human dietary supplement. Consistent with this, CoQ_{10} supplementation increases lifespan in *C. elegans* in both wild-type and *mev-1* animals (Ishii *et al.*, 2004), and also reduces O_2^- production in isolated mitochondria.

These results appear to conflict with the finding that feeding *C. elegans* with *E. coli* lacking CoQ_8 increases their lifespan (Larsen and Clarke, 2002). How may these findings be reconciled? One possibility is that different forms of CoQ have different effects on lifespan. The increases in lifespan seen by Ishii *et al.* resulted from supplementation with CoQ_{10}, which may somehow promote longevity more than CoQ_9 (Ishii *et al.*, 2004); possibly, CoQ_8 increases superoxide production more than CoQ_9. To explore this, *E. coli* strains were engineered which produce CoQ_7, CoQ_8, CoQ_9 or CoQ_{10}. *E.coli* producing CoQ_9 or CoQ_{10} partially suppressed the reduced fertility of a weak *clk-1* mutant, but effects of these *E. coli* strains on lifespan were not reported (Jonassen *et al.*, 2003).

An alternative scenario is that $DMCoQ_9$ generates less O_2^- than CoQ_9 (Miyadera *et al.*, 2002). RNAi of *sod-1* (cytosolic Cu/Zn SOD) and mutation of *sod-4* (putative extracellular Cu/Zn SOD) can partially suppress some *clk-1* mutant phenotypes. This, it has been suggested, may reflect reduced O_2^- production by $DMCoQ_9$ that interferes with signalling pathways in which O_2^- acts as a secondary messenger (Shibata *et al.*, 2003; Stepanyan *et al.*, 2006). However, it seems unlikely that the presence of $DMCoQ_9$ causes increased lifespan, since mutation of *rte-2* suppresses *clk-1* longevity without reducing levels of $DMCoQ_9$ (Branicky *et al.*, 2006).

If the increase in lifespan associated with the Mit phenotype reflects a reduction in metabolism and ROS production, one would not expect any associated increase in stress resistance. However, this prediction is not well supported. For many genes encoding mitochondrial proteins, long-lived animals subjected to RNAi proved to be resistant to H_2O_2 and heat stress, although not paraquat (Lee *et al.*, 2003b). Moreover, mutation of *isp-1* in complex III elevates *sod-3* expression and increases paraquat resistance (Feng *et al.*, 2001), and *clk-1* mutant animals show resistance to ultraviolet light (Murakami and Johnson, 1996) and increased catalase levels (but reduced SOD activity levels; Braeckman *et al.*, 2002c). This could imply that disruption of mitochondria stimulates stress resistance pathways, perhaps due to increased O_2^- production, a mechanism dubbed 'mitohormesis' (Rea, 2005). The term hormesis usually refers to situations where brief or low-level exposure to stressors induces a response that results in stress resistance.

Other findings are consistent with the occurrence of mitohormesis. For example, treatment with the drug antimycin A, which blocks complex III, increases O_2^- production from isolated *C. elegans* mitochondria (Senoo-Matsuda *et al.*, 2001) and, interestingly, appears to increase lifespan (Dillin *et al.*, 2002), although the effect is not large. Mitochondrial ROS production (measured as H_2O_2) is also elevated by mutation of *daf-2* (Brys *et al.*, 2007). There is also evidence that the SKN-1-dependent antioxidant system is activated in *clk-1* mutants (Rea, 2005). The most direct evidence that increased mitochondrial O_2^- production can contribute to longevity comes from a recent study by Schulz *et al.* (2007) who induced a state resembling dietary glucose restriction in *C. elegans* by means of a chemical inhibitor of glycolysis, 2-deoxy-D-glucose (DOG). Treatment with DOG was found to increase both ROS production, resistance to oxidative stress (paraquat and sodium azide) and lifespan. Additional treatment with the antioxidant *N*-acetyl-cysteine (NAC) blocked the DOG-induced increases in ROS levels and stress resistance and lifespan (Schulz *et al.*, 2007).

3.4 *Uncoupling proteins and ageing*

Mitochondrial O_2^- production is predicted to be highest when the ETC is fully reduced, in state 4. Uncoupling proteins (UCPs) or chemical protonophores such as dinitrophenol can uncouple electron transport from ATP synthesis, which increases heat production and lowers O_2^- production (Brand, 2000). A prediction of the oxidative damage theory is that increased uncoupling should reduce ROS production, thereby increasing lifespan. This has been investigated a little in *C. elegans*. The worm genome contains a single gene encoding a protein with sequence homology to mammalian UCPs, *ucp-4*, which is strongly expressed in muscle (Iser *et al.*, 2005). Absence of *ucp-4* function resulted in increased levels of ATP and cold sensitivity, consistent with function as an uncoupling protein. However, only a very slight increase in mitochondrial membrane potential was seen, and lifespan was not affected.

It has also been suggested that mitochondria from *daf-2* mutants have a higher level of coupling, given the lower calorimetric/respirometric ratio and the higher levels of ATP and ROS production (Brys *et al.*, 2007; Houthoofd *et al.*, 2005a, 2005b). However, it is worth noting that this lowering of the calorimetric/respirometric ratio is not suppressed by mutation of *daf-16*, which does suppress *daf-2* longevity (Houthoofd *et al.*, 2005a; Kenyon *et al.*, 1993). Thus, one can at least say that this metabolic shift is not enough in itself to increase lifespan. Further studies seem warranted to establish the effects of mitochondrial uncoupling on ageing in *C. elegans*.

In conclusion, a number of independent screens for genes with effects on ageing have all pointed to the importance of mitochondria in ageing. Disruption of mitochondrial function usually increases lifespan in *C. elegans*, but the mechanisms involved are unknown. One possibility is that O_2^- production in Mit animals is reduced, but this remains largely unexplored. In principle, reduced ATP production might seem a strong candidate mechanism, potentially linking the Mit phenotype, dietary restriction and rate-of-living effects; for example, ATP feeds growth, including protein synthesis, which promotes ageing (Hansen *et al.*, 2007; Pan *et al.*, 2007; Syntichaki *et al.*, 2007). However, the importance of ATP levels in ageing is not experimentally supported (e.g. see Dillin *et al.*, 2002). An alternative possibility that is finding increasing empirical support is that disruption of mitochondrial function activates somatic maintenance processes via the hormetic effects of increased ROS production (Rea, 2005). The support for the mitohormesis theory provided by Schulz *et al.* (2007), in particular, suggests that the standard oxidative damage theory of ageing may be very wrong, at least as far as *C. elegans* is concerned.

4 Reactive oxygen species and ageing in *C. elegans*

4.1 *Alterations of pro-oxidant levels*

According to the standard oxidative damage theory, ageing is caused by ROS. If this is correct, then manipulating ROS levels should affect ageing rate: lowering ROS should increase lifespan and *vice versa*. The effects of ambient oxygen concentration on lifespan and mortality rate have been tested in wild-type and *mev-1* mutant populations (Honda *et al.*, 1993). In the wild type, these parameters were unaltered in 2%, 8% and 40% oxygen relative to 21% oxygen. This is a striking result, as it implies either that levels of O_2^- production are unaltered over this range, or that ROS are not a determinant of ageing. Nevertheless, if large enough, changes in O_2 concentration can affect ageing in the wild type. In 1% O_2, mean lifespan was increased by 15% and the Gompertz component of mortality was decreased (Honda *et al.*, 1993). Whether this effect is mediated by changes in O_2^- production, metabolic rate or some other factor is unclear. In 60% O_2, wild-type mean lifespan was slightly reduced (by 14%), probably due to increased oxidative damage. In contrast to the wild type, in *mev-1* populations there was a direct relationship between oxygen concentration and lifespan; the mutation rate in *mev-1* mutants was also hypersensitive to effects of elevated oxygen (Hartman *et al.*, 2004).

Taken together, these findings imply that, under conditions of normoxia, O_2^- levels determine ageing in *mev-1* but not wild-type populations. The possibilities that O_2^- does not cause normal ageing, while elevated O_2^- levels can accelerate ageing, are by no means contradictory. Ageing in both cases may involve molecular damage, but resulting from different causes. Indeed, other observations suggest mechanistic differences between ageing in *mev-1* and wild type, as follows. In otherwise wild-type *C. elegans*, prevention of apoptosis (programmed cell death) by mutation of the gene *ced-3* does not extend lifespan (Garigan *et al.*, 2002). Thus, apoptosis does not contribute to normal ageing. By contrast, mutation of *ced-3* increases lifespan of *mev-1* populations, apparently by preventing O_2^--induced apoptosis (Senoo-Matsuda *et al.*, 2003). However, this extension is the result of suppression of early mortality, and late-life survival was unchanged. *mev-1* mutants also have elevated lactic acid levels, suggesting that lactic acidosis might contribute to their mortality (Senoo-Matsuda *et al.*, 2001).

A common means to test the effects of ROS on *C. elegans* is administration of redox cycling compounds such as juglone or, more commonly, paraquat (methyl viologen; Arkblad *et al.*, 2005; Ayyadevara *et al.*, 2005b; Henderson and Johnson, 2001; Keaney *et al.*, 2004; Vanfleteren, 1993), which generate O_2^- *in vivo*. *In vivo*, redox cyclers receive electrons from NADH or NADPH via the action of diaphorase enzymes, and this activity has been detected in *C. elegans* (Blum and Fridovich, 1983). O_2^- production by redox cyclers can be measured as an increase in cyanide-independent O_2 consumption. While 1 mM paraquat does not detectably increase cyanide-independent O_2 consumption by *C. elegans* (Blum and Fridovich, 1983), 2 mM paraquat does increase it, and this concentration is just sufficient to decrease adult lifespan (Keaney *et al.*, 2004). These results strongly imply that *C. elegans* lifespan can be shortened by elevated levels of O_2^-, but the extent to which this effect involves an acceleration of processes occurring during normal ageing is unknown.

4.2 *Effects of ROS on age changes in molecular damage*

If ROS causes normal ageing, experimental elevation of ROS should accelerate age changes in molecular damage seen in normal ageing. This prediction has been little explored, though one report described increased blue fluorescence under hyperoxia (Hosokawa *et al.*, 1994).

Isolated mitochondria from *mev-1* animals show elevated levels of O_2^- production (Senoo-Matsuda *et al.*, 2001). Thus, O_2^- production might be elevated *in vivo* and might account for the shortened lifespan of *mev-1* under normoxia. The increased levels of protein oxidation in *mev-1* animals supports this (Adachi *et al.*, 1998; Yasuda *et al.*, 1999). *mev-1* has also been reported to elevate levels of blue fluorescence (Hosokawa *et al.*, 1994). However, a recent study saw no such effect either in *mev-1* or *gas-1* animals (Gerstbrein *et al.*, 2005).

4.3 *Effects of antioxidant defence on ageing*

Cellular defences against oxidative damage include both chemical and enzymatic antioxidants. If oxidative damage causes ageing, then one might expect a correlation between antioxidant defence and longevity. Moreover, experimental enhancement of antioxidant defence should retard ageing. Many studies have tested both of these expectations, yet in each case, establishing a causal role of oxidative damage in ageing is difficult. For example, a correlation between level of an antioxidant agent and longevity could be coincidental. If experimentally induced elevation in levels of an antioxidant agent increases lifespan, the possibility remains that this occurs by some other mechanism than protection against molecular damage. Moreover, if increases in lifespan are not seen, it remains possible that multiple antioxidant defence mechanisms act in concert to protect against ageing, or that antioxidant mechanisms act in concert with other prolongevity mechanisms.

4.4 *Non-catalytic antioxidants*

Over the years, many studies have been conducted examining the effects on ageing of non-catalytic antioxidants (particularly vitamin E), often generating inconclusive findings. Vitamin E studies have employed its constituents, α-tocopherol and tocotrienols, and the α-tocopherol derivative α-tocopherolquinone (α-TQ). An early study found that α-tocopherol and α-TQ both increase lifespan of *C. briggsae* (a sister species of *C. elegans*) by 31% (Epstein and Gershon, 1972). Similarly, vitamin E increased lifespan in *C. elegans* (Zuckerman and Geist, 1983). However, in both studies nematodes were cultured in an axenic medium (i.e. without *E. coli*), which is nutritionally suboptimal; moreover, the effects of vitamin E on lifespan were exerted during development, not adulthood. Thus, these findings may reflect a nutritional effect on growth in axenic medium. In another study, vitamin E increased *C. elegans* lifespan by around 20%, but also reduced fecundity and delayed the timing of reproduction (Harrington and Harley, 1988). Here, the authors concluded that effects on ageing could reflect slight toxicity, which slowed development, growth and ageing. Yet another study compared the effects of α-tocopherol and tocotrienols on levels of protein oxidation, resistance to oxidative damage (exerted by ultraviolet B irradiation) and longevity. While α-tocopherol had no effect, tocotrienols had a protective effect against damage and stress, and caused a slight increase in mean but not maximum lifespan (Adachi and Ishii, 2000). A more recent report described a single trial where vitamin E increased lifespan in wild-type (+11%) but not *mev-1* animals (Ishii *et al.*, 2004). Overall, and taking into account the tendency to publish only results showing positive effects, these studies provide little persuasive evidence that vitamin E supplementation protects against ageing.

A large number of genes and processes contribute to protection against oxidative damage (Halliwell and Gutteridge, 2007; Mathers *et al.*, 2004), any one of which may limit the rate of age accumulation of molecular damage, and its impact on homeostasis and survival. In the first line of defence are enzymes which detoxify primary pro-oxidant molecules. For example, superoxide dismutases (SOD) convert O_2^- into H_2O_2 (Fridovich, 1995), and this is converted into water and O_2 by catalases and glutathione peroxidases (GPX). Numerous proteins affect ROS production levels, such as metal trafficking proteins. Free metal ions such as Fe^{3+} stimulate production of very damaging forms of ROS such as OH^-, and metallothioneins and ferritins will counteract this. The forms of molecular damage that can occur are extremely diverse, as are the enzymes that detoxify, repair or remove damaged moieties. For example, peroxidized lipids are targets for numerous glutathione lipid hydroperoxidases and glutathione S-transferases (GSTs). In proteins, oxidation of just the amino acid methionine can be repaired by methionine sulphoxide reductase. Effects of oxidative damage to protein on protein function can, to some extent, be restored by the action of molecular chaperones. Finally, oxidized proteins can be removed by cellular turnover processes such as proteasome-dependent protein degradation and autophagy. Any of these enzymes and processes could, in principle, contribute to longevity assurance by protecting against oxidative damage.

4.5 *Antioxidant enzymes: superoxide dismutase and catalase*

The superoxide dismutases and catalases of *C. elegans* are unusual in a number of ways. For one, *C. elegans* possesses more isoforms of these enzymes than higher animals. Instead of one cytosolic Cu/Zn SOD there are two, encoded by *sod-1* and *sod-5* (Giglio *et al.*, 1994; Jensen and Culotta, 2005; Larsen, 1993), and instead of one mitochondrial Mn SOD there are also two, encoded by *sod-2* and *sod-3* (Giglio *et al.*, 1994; Hunter *et al.*, 1997; Suzuki *et al.*, 1996). A combination of SOD activity assays in *sod* mutants, and studies of levels of mRNA and reporter expression imply that *sod-1* and *sod-2* are the major isoforms expressed during reproductive development, while *sod-3* and *sod-5* are dauer up-regulated isoforms (Honda and Honda, 1999; Jensen and Culotta, 2005; Wang and Kim, 2003; R. Doonan, J.J. McElwee and D. Gems, unpublished). Why there should be dauer-specific isoforms is unclear. SOD-2 and SOD-3 Mn SODs have similar specific activities (Hunter *et al.*, 1997), and either SOD-1 or SOD-5 Cu/Zn SOD can rescue the paraquat sensitivity of SOD-deficient yeast (Jensen and Culotta, 2005), suggesting that reproductive and dauer isoforms are not functionally different.

There are other ways in which the SOD-1 and SOD-5 Cu/Zn SODs are different. To mature, Cu/Zn SODs need to incorporate copper, and in all other eukaryotes, whether animals, fungi or plants, this requires the copper chaperone of SOD protein (CCS). Unusually, *C. elegans* does not possess a CCS, and Cu/Zn SOD maturation does not require it, but instead depends on an unknown glutathione-dependent pathway (Jensen and Culotta, 2005). Studies of SOD-1 and SOD-5 expressed in yeast also hint that, unlike other eukaryotes that have been studied, *C. elegans* may not have Cu/Zn SOD in the mitochondrial intermembrane space, though the evidence here is not conclusive (Jensen and Culotta, 2005).

sod-4 encodes a Cu/Zn SOD that resembles mammalian extracellular Cu/Zn SODs (Fujii *et al.*, 1998). However, SOD-4 is also distinctive in that there are two predicted isoforms, products of alternative splicing of mRNA. SOD4-1 resembles a typical secreted Cu/Zn SOD, but SOD4-2 has an additional C-terminal sequence resembling a transmembrane domain. This suggests that this unique SOD is secreted through the plasma membrane, but then remains tethered at the cell surface (Fujii *et al.*, 1998).

The catalases of *C. elegans* are also unusual. The *C. elegans* genome contains a tandem array of three genes encoding catalases *ctl-1*, *ctl-2* and *ctl-3* (Petriv and Rachubinski, 2004). By contrast, other metazoans have only a single catalase, while *S. cerevisiae* has a peroxisomal and a cytosolic catalase. CTL-2 is the *C. elegans* peroxisomal catalase, and is responsible for around 80% of total catalase activity; it also has a lower pH optimum for activity and higher peroxidase activity than mammalian peroxisomal catalases (Petriv and Rachubinski, 2004; Taub *et al.*, 1999; Togo *et al.*, 2000). Much of the *ctl-1* and *ctl-3* gene sequences are 100% identical. Studies of a CTL-1::GFP fusion protein imply that CTL-1 is a cytosolic catalase (Taub *et al.*, 1999). One possibility is that CTL-1 evolved as a cytosolic H_2O_2 scavenger because *C. elegans* lacks an H_2O_2-scavenging glutathione peroxidase (Vanfleteren, 1993; J.R. Vanfleteren, personal communication; see below). A promoter fusion test implies that *ctl-3* is expressed in pharyngeal muscle and neurones (Petriv and Rachubinski, 2004). More work is needed to confirm and define the cellular localization of CTL-1 and CTL-3. In summary, given its very short lifespan *C. elegans* has a surprisingly elaborate arsenal of SODs (six) and catalases (three) to protect itself against ROS.

Is the increased longevity of *daf-2* and *age-1* attributable to any degree to increased levels of SOD and catalase? Long-lived IIS mutants do show age increases in SOD and catalase activity levels, and in resistance to oxidative stress (e.g. paraquat and H_2O_2), increases which are not seen in the wild type (Honda and Honda, 1999; Larsen, 1993; Vanfleteren, 1993; Vanfleteren and De Vreese, 1995). Northern blot studies have shown a large increase in *sod-3* mRNA levels in *daf-2* mutants (Honda and Honda, 1999; Yanase *et al.*, 2002), and microarray studies reveal additional, smaller increases in *sod-1* and *sod-5* mRNA (McElwee *et al.*, 2003, 2004; Murphy *et al.*, 2003). *sod-3* levels are elevated throughout the life course in *daf-2* mutants, even in the developing embryo (Honda and Honda, 1999). Microarray studies also show increases in expression of at least one catalase gene in *daf-2* mutants, but because of the similarity between *ctl* gene sequences, one cannot say which. This also complicates interpretation of RNAi studies (Murphy *et al.*, 2003). Levels of SOD and catalase are also elevated in *C. elegans* subjected to dietary restriction and, in contrast to insulin/IGF-1 signalling mutants, this increase does not depend on *daf-16* (Houthoofd *et al.*, 2003). In dauer larvae, levels of SOD activity are four- to fivefold higher than in young adults, and levels of *sod-3* mRNA are elevated (Anderson, 1982; Honda and Honda, 1999; Larsen, 1993). Catalase levels also seem to be elevated in dauer larvae (Houthoofd *et al.*, 2002a), though here there is conflicting evidence (Larsen, 1993).

It seems likely that the elevated levels of antioxidant enzymes contribute to oxidative stress resistance, at least to some degree, but whether they contribute to longevity remains unclear. The effects on ageing of manipulations of SOD and catalase levels have been investigated in *C. elegans*, though not as systematically as in *Drosophila*. RNAi knock down of expression of *sod-3* has been reported to very weakly suppress *daf-2* longevity (Murphy *et al.*, 2003) but, surprisingly, RNAi of *sod-5* had the opposite effect (McElwee *et al.*, 2003). More surprisingly, deletion of *sod-2* and *sod-3*, alone or in combination, has no effect on adult lifespan (J.J. McElwee and D. Gems, unpublished).

Deletion of *ctl-1* (the cytosolic catalase) has no effect on lifespan, while deletion of *ctl-2* (the peroxisomal catalase) shortens lifespan (Petriv and Rachubinski, 2004). The authors interpreted this life shortening as progeria, though more evidence would be required to establish this with certainty. *ctl-2* mutants show abnormalities in peroxisomal morphology. Surprisingly, protein oxidation (protein carbonyl levels) increases more rapidly with age in wild-type than in *ctl-1* or *ctl-2* animals (Petriv and Rachubinski, 2004). It was at one point claimed that *ctl-1* is required for the longevity of *daf-2* mutants (Taub *et al.*, 1999), but the study concerned was subsequently retracted (Taub *et al.*, 2003).

There have been few reports of the effects of over-expression of *sod* genes. Overexpression of a *sod-3::gfp* fusion protein did not affect lifespan but, as the authors made clear, SOD activity level was not examined in this strain (Henderson *et al.*, 2006). In one study, it was observed that loss of heat shock factor 1 (HSF-1) suppressed *daf-2* mutant longevity without suppressing the elevation in *sod-3* expression (Hsu *et al.*, 2003). This suggests, at least, that elevated *sod-3* expression does not increase lifespan in HSF-1 deficient animals.

Several studies have examined the effects on *C. elegans* of administering the SOD mimetic salen manganese compounds EUK-8 and EUK-134. This results in significant increases in SOD activity levels (e.g. a fivefold increase in mitochondrial SOD activity) and resistance to paraquat (Keaney *et al.*, 2004; Sampayo *et al.*, 2003). Although one study reported that these compounds also increased lifespan in *C. elegans* (Melov *et al.*, 2000), other workers did not see this effect in *C. elegans* (Keaney and Gems, 2003; Keaney *et al.*, 2004), in *Drosophila* (Magwere *et al.*, 2006) or in houseflies (Bayne and Sohal, 2002). Levels of EUK-8 that were optimal for protection against paraquat had no effect on lifespan, suggesting again (*c.f.* ambient oxygen studies) that O_2^- does not contribute to normal ageing in *C. elegans*.

4.6 *Other antioxidant defences*

While the role of SOD and catalase in *C. elegans* ageing remains poorly understood, even less is known about the role in ageing of other antioxidant defence mechanisms.

One possibility is that metal-trafficking proteins play a role in longevity assurance. Exogenous iron shortens lifespan in *C. elegans* (Gourley *et al.*, 2003), and *daf-2* and *age-1* mutants are resistant to heavy metals (e.g. cadmium and copper) and show elevated expression of the metallothionein *mtl-1* (Barsyte *et al.*, 2001). RNAi of *mtl-1* slightly reduces *daf-2* mutant longevity (Murphy *et al.*, 2003). The *ftn-1* ferritin heavy chain is also strongly up-regulated in *daf-2* mutants (McElwee *et al.*, 2004).

In many species, GPX is a major hydrogen peroxide scavenger. In one study GPX activity was not detected in *C. elegans* using BuOOH (Vanfleteren, 1993) and H_2O_2 (J.R. Vanfleteren, personal communication). However, the *C. elegans* genome contains a number of GPX-like proteins, some or all of which could, in principle, be lipid hydroperoxidases. The possible function of these proteins, for example in stress resistance and ageing, remains unexplored.

In recent years, global changes in gene expression in *daf-2* and *age-1* mutants have been studied using DNA microarrays (Golden and Melov, 2004; McElwee *et al.*, 2003, 2004; Murphy *et al.*, 2003). One study showed that at least 2348 genes are up- or down-regulated in *daf-2* animals relative to normal-lived *daf-16;daf-2* controls; in other words, some 12% of genes in the *C. elegans* genome (McElwee *et al.*, 2004). That so many genes are regulated by IIS constitutes a difficult obstacle to understanding how IIS controls ageing. For one, it renders relatively uninformative studies that show correlation of expression of individual genes (e.g. *sod-3*) with IIS mutant longevity. In fact, among such large lists of IIS-regulated genes, it is possible to find evidence supporting most theories of ageing (Gems and McElwee, 2005).

This type of problem of bias in data interpretation can, to a degree, be avoided by identifying gene classes showing a significant level of over-representation among differentially expressed genes. One study combined this approach with a comparison of array data from *daf-2* mutants (compared to *daf-16;daf-2*) and dauer larvae (compared to recovered dauer larvae; Gems and McElwee, 2005; McElwee *et al.*, 2004; Wang and Kim, 2003). This was based on the plausible supposition that *daf-2* mutants are long-lived because they heterochronically express dauer longevity assurance mechanisms. Among the small number of longevity-associated gene classes were several specifying the phase 1/phase 2 biotransformation system (i.e. xenobiotic metabolism or drug detoxification). Glutathione S-transferases were also strongly over-represented among genes up-regulated in *daf-2* mutants, but not dauers. Biotransformation involves a complex system of enzymes and pumps involved in detoxification and clearance of a wide spectrum of endobiotic and xenobiotic toxins, and some biosynthetic processes (Gibson and Skett, 2001). Thus, comparisons of transcript profiles from *daf-2* mutants and dauers implies that the biotransformation system is activated in these long-lived milieus. This suggests that these detoxification processes might contribute to longevity, a possibility that is easy to rationalize given that biotransformation provides defence against molecular damage (Gems and McElwee, 2005). A recent comparison of transcript profiles from long-lived IIS mutant *C. elegans*, *Drosophila* and mice showed up-regulation of three classes of biotransformation enzymes (particularly GSTs) in all three species (McElwee *et al.*, 2007). Such evolutionary conservation in the correlation of increased biotransformation and increased longevity could reflect a role for this system in longevity assurance.

The glutathione S-transferases are a highly diverse, rapidly evolving enzyme class; the *C. elegans* genome includes 51 putative GST-encoding genes. Among other things,

GSTs use glutathione conjugation to detoxify endobiotic and xenobiotic toxins, including the products of oxidative damage (Hayes and McLellan, 1999). A screen for genes up-regulated upon exposure to the O_2^- generator paraquat identified *gst-4* (Tawe *et al.*, 1998). Over-expression of *gst-4* resulted in increased resistance to paraquat, but not increased lifespan (Leiers *et al.*, 2003). However, RNAi of *gst-4* slightly reduces *daf-2* mutant longevity (Murphy *et al.*, 2003), and microarray data implies that *gst-4* expression is increased in *daf-2* mutants.

One of the few *C. elegans* GSTs to have been well characterized is GST-10 which detoxifies 4-hydroxynon-2-enal (HNE), an abundant lipid peroxidation product resulting from oxidative stress (Engle *et al.*, 2001). *daf-2* mutants show up-regulation of *gst-10*, and RNAi of *gst-10* increased sensitivity to HNE toxicity and reduced lifespan in both wild-type and *daf-2* mutant populations (Ayyadevara *et al.*, 2005a). The effect of *gst-10* RNAi on *daf-2* mutant lifespan has been independently verified (D. Weinkove and D. Gems, unpublished). RNAi of *gst-5, gst-6, gst-8* or *gst-24* also increased sensitivity to HNE toxicity, but of these genes only RNAi of *gst-5* reduced lifespan (Ayyadevara *et al.*, 2007). Over-expression in *C. elegans* of either *gst-10* or murine *mGsta4* (which also detoxifies HNE) lead to increased levels of HNE-conjugating activity, increased resistance to oxidative stress (e.g. paraquat and H_2O_2) and lowered levels of HNE-protein adducts. Interestingly, over-expression of *gst-10* or *mGsta4* increased median lifespan by 22% and 13%, respectively (Ayyadevara *et al.*, 2005b). This is a rare example of evidence that robustly supports a role for oxidative damage in *C. elegans* ageing, though effects on ageing of HNE in its capacity as a secondary messenger have not been excluded.

Mitochondrial nicotinamide nucleotide transhydrogenase (NNT) is another enzyme that contributes to oxidative stress resistance. It does so by catalysing the reduction of $NADP^+$ by NADH, providing NADPH for reduction of glutathione within mitochondria. This is important in animal mitochondria, which lack catalase. H_2O_2 generated by SOD is usually detoxified instead by mitochondrial GPX. Reduced glutathione (GSH) in mitochondria is also a substrate for phospholipid hydroperoxidases. In *C. elegans*, *nnt-1* is widely expressed (e.g. in intestinal, hypodermal and neuronal cells). Deletion of *nnt-1* leads to a lowered ratio of reduced to oxidized GSH (58 *vs* 12 in wild type *vs* mutant; Arkblad *et al.*, 2005). The large magnitude of this effect implies that cytosolic as well as mitochondrial GSH pools are affected. This results in increased sensitivity to paraquat though not H_2O_2. Interestingly, there is no effect on lifespan, which implies that a shift in cellular redox state towards more oxidative conditions does accelerate ageing (Arkblad *et al.*, 2005).

In conclusion, ageing in *C. elegans* is accompanied by an accumulation of molecular damage, but why this accumulation occurs is unclear. It is also unclear to what extent this damage is caused by ROS, or O_2^- in particular, or how important is damage caused by O_2^- (as opposed to other agents of molecular damage). Perhaps the strongest evidence that O_2^- does contribute to *C. elegans* ageing is that over-expression of HNE-conjugating GSTs can increase longevity, since O_2^- contributes to HNE formation. If O_2^- can contribute to ageing, then why do SOD mimetics not increase lifespan? One possible explanation is that in *C. elegans* intramitochondrial O_2^- does not contribute to ageing: mutants lacking Mn SOD have a normal lifespan (J.J. McElwee and D. Gems, unpublished), and SOD mimetics are concentrated within mitochondria (Keaney *et al.*, 2004).

5 Conclusions

The possibilities that damage from mitochondrial O_2^- is a determinant of ageing, and that metabolic rate affects the rate of ageing, have been tested extensively in *C. elegans*. The results of these tests have, in the main, been either negative, inconclusive or mixed. In particular, there is little clear evidence that metabolic rate is a determinant of ageing, though the possibility that metabolic rate can affect ageing has not been excluded. Molecular damage clearly accumulates with age, but it remains uncertain whether this is a primary cause of ageing, and the mechanisms that determine the rate of damage accumulation remain unclear.

Mitochondria can clearly exert powerful effects on ageing in *C. elegans*, yet it is far from clear that O_2^- production plays any role in this. Besides oxidative phosphorylation and ATP production, mitochondria play many roles in the cell, including calcium homeostasis, steroid biogenesis, pyrimidine biosynthesis and fatty acid metabolism. The effects of CoQ on ageing are especially difficult to interpret, since it also affects many processes both in mitochondria and elsewhere, and can act either as a pro-oxidant or antioxidant (Miyadera *et al.*, 2002). The idea that damage from O_2^- causes ageing remains unconfirmed, and a recent study even implies that increased mitochondrial ROS can retard ageing by inducing stress defence mechanisms (so called mitohormesis; Schulz *et al.*, 2007).

One possibility is that oxidative damage plays a major role in some organisms but not others. There are reasons for being suspicious that ageing in *C. elegans* may involve at least some mechanisms that are not evolutionarily conserved, that is which are private rather than public (Martin *et al.*, 1996). For example, instead of causing rapid death, as one might expect, disruption of the electron transport chain usually extends lifespan in *C. elegans*, raising a worry that this is a nematode peculiarity; in *C. elegans* O_2 consumption and O_2^- production decreases rather than increases with age; and age accumulation of protein oxidation is largely restricted to the mitochondria. Arguably, though, these concerns should not be taken as an argument against using *C. elegans* as a model for studies of ageing. Rather, they underscore the importance of distinguishing between public and private mechanisms of ageing to maximize the utility of model organisms.

Acknowledgements

We thank J.R. Vanfleteren and F. Matthijssens for communication of unpublished information. This work was supported by the European Union and the Wellcome Trust.

References

Adachi, H. and Ishii, N. (2000) Effects of tocotrienols on life span and protein carbonylation in *Caenorhabditis elegans*. *J. Gerontol. A* **55:** B280–B285.

Adachi, H., Fujiwara, Y. and Ishii, N. (1998) Effects of oxygen on protein carbonyl and aging in *Caenorhabditis elegans* mutants with long (*age-1*) and short (*mev-1*) life spans. *J. Gerontol. A* **53A:** B240–B244.

Anderson, G.L. (1982) Superoxide dismutase activity in dauer larvae of *Caenorhabditis elegans* (Nematoda: Rhabditidae). *Can. J. Zool.* **60:** 288–291.

Anson, R.M. and Hansford, R.G. (2004) Mitochondrial influence on aging rate in *Caenorhabditis elegans*. *Aging Cell* **3:** 29–34.

Arkblad, E.L., Tuck, S., Pestov, N.B., Dmitriev, R.I., Kostina, M.B., Stenvall, J., Tranberg, M. and Rydstrom, J. (2005) A *Caenorhabditis elegans* mutant lacking functional nicotinamide nucleotide transhydrogenase displays increased sensitivity to oxidative stress. *Free Radic. Biol. Med.* **38:** 1518–1525.

Ayyadevara, S., Dandapat, A., Singh, S.P., Benes, H., Zimniak, L., Reis, R.J. and Zimniak, P. (2005a) Lifespan extension in hypomorphic *daf-2* mutants of *Caenorhabditis elegans* is partially mediated by glutathione transferase CeGSTP2-2. *Aging Cell* **4:** 299–307.

Ayyadevara, S., Engle, M.R., Singh, S.P., Dandapat, A., Lichti, C.F., Benes, H., Shmookler Reis, R.J., Liebau, E. and Zimniak, P. (2005b) Lifespan and stress resistance of *Caenorhabditis elegans* are increased by expression of glutathione transferases capable of metabolizing the lipid peroxidation product 4-hydroxynonenal. *Aging Cell* **4:** 257–271.

Ayyadevara, S., Dandapat, A., Singh, S.P., Siegel, E.R., Shmookler Reis, R.J., Zimniak, L. and Zimniak, P. (2007) Life span and stress resistance of *Caenorhabditis elegans* are differentially affected by glutathione transferases metabolizing 4-hydroxynon-2-enal. *Mech. Ageing Dev.* **128:** 196–205.

Balaban, R.S., Nemoto, S. and Finkel, T. (2005) Mitochondria, oxidants, and aging. *Cell* **120:** 483–495.

Barsyte, D., Lovejoy, D. and Lithgow, G. (2001) Longevity and heavy metal resistance in *daf-2* and *age-1* long-lived mutants of *Caenorhabditis elegans*. *FASEB J.* **15:** 627–634.

Bayne, A.C. and Sohal, R.S. (2002) Effects of superoxide dismutase/catalase mimetics on life span and oxidative stress resistance in the housefly, *Musca domestica*. *Free Radic. Biol. Med.* **32:** 1229–1234.

Beckman, K.B. and Ames, B.N. (1998) The free radical theory of aging matures. *Physiol. Rev.* **78:** 547–581.

Blum, J. and Fridovich, I. (1983) Superoxide, hydrogen peroxide, and oxygen toxicity in two free-living nematode species. *Arch. Biochem. Biophys.* **222:** 35–43.

Boveris, A. (1977) Mitochondrial production of superoxide radical and hydrogen peroxide. In: *Tissue Hypoxia and Ischemia* (eds. M. Reivich, R. Coburn, S. Lahiri, and B. Chance). Plenum Press, New York, NY.

Braeckman, B.P., Houthoofd, K., De Vreese, A. and Vanfleteren, J.R. (1999) Apparent uncoupling of energy production and consumption in long-lived Clk mutants of *Caenorhabditis elegans*. *Curr. Biol.* **9:** 493–496.

Braeckman, B., Houthoofd, K., De Vreese, A. and Vanfleteren, J. (2002a) Assaying metabolic activity in ageing *Caenorhabditis elegans*. *Mech. Ageing Dev.* **123:** 105–119.

Braeckman, B., Houthoofd, K. and Vanfleteren, J. (2002b) Assessing metabolic activity in aging *Caenorhabditis elegans*: concepts and controversies. *Aging Cell* **1:** 82–88.

Braeckman, B.P., Houthoofd, K., Brys, K., Lenaerts, I., De Vreese, A., Van Eygen, S., Raes, H. and Vanfleteren, J.R. (2002c) No reduction of energy metabolism in Clk mutants. *Mech. Ageing Dev.* **123:** 1447–1456.

Brand, M.D. (2000) Uncoupling to survive? The role of mitochondrial inefficiency in ageing. *Exp. Gerontol.* **35:** 811–820.

Branicky, R., Nguyen, P.A. and Hekimi, S. (2006) Uncoupling the pleiotropic phenotypes of clk-1 with tRNA missense suppressors in *Caenorhabditis elegans*. *Mol. Cell. Biol.* **26:** 3976–3985.

Brys, K., Vanfleteren, J.R. and Braeckman, B.P. (2007) Testing the rate-of-living/oxidative damage theory of aging in the nematode model *Caenorhabditis elegans*. *Exp. Gerontol.* **42:** 845–851.

De Cuyper, C. and Vanfleteren, J.R. (1982) Oxygen consumption during development and aging of the nematode *Caenorhabditis elegans*. *Comp. Biochem. Physiol.* **73A:** 283–289.

Dillin, A., Hsu, A., Arantes-Oliveira, N., Lehrer-Graiwer, J., Hsin, H., Fraser, A., Kamath, R., Ahringer, J. and Kenyon, C. (2002) Rates of behavior and aging specified by mitochondrial function during development. *Science* **298:** 2398–2401.

Engle, M., Singh, S., Nanduri, B., Ji, X. and Zimniak, P. (2001) Invertebrate glutathione transferases conjugating 4-hydroxynonenal: CeGST 5.4 from *Caenorhabditis elegans*. *Chemico-Biol. Int.* **133:** 244–248.

Epstein, J. and Gershon, D. (1972) Studies on ageing in nematodes. IV. The effect of anti-oxidants on cellular damage and life span. *Mech. Ageing Dev.* **1:** 257–264.

Ewbank, J.J., Barnes, T.M., Lakowski, B., Lussier, M., Bussey, H. and Hekimi, S. (1997) Structural and functional conservation of the *Caenorhabditis elegans* timing gene *clk-1*. *Science,* **275:** 980–983.

Felkai, S., Ewbank, J.J., Lemieux, J., Labbe, J.C., Brown, G.G. and Hekimi, S. (1999) CLK-1 controls respiration, behavior and aging in the nematode *Caenorhabditis elegans*. *EMBO J.* **18:** 1783–1792.

Feng, J., Bussiere, F. and Hekimi, S. (2001) Mitochondrial electron transport is a key determinant of life span in *Caenorhabditis elegans*. *Dev. Cell* **1:** 633–644.

Finkel, T. and Holbrook, N.J. (2000) Oxidants, oxidative stress and the biology of ageing. *Nature* **408:** 239–247.

Foll, R., Pleyers, A., Lewandovski, G., Wermter, C., Hegemann, V. and Paul, R. (1999) Anaerobiosis in the nematode *Caenorhabditis elegans*. *Comp. Biochem. Physiol. B Biochem. Mol. Biol.* **124:** 269–280.

Fridovich, I. (1995) Superoxide radical and superoxide dismutases. *Ann. Rev. Biochem.* **64:** 97–112.

Fujii, M., Ishii, N., Joguchi, A., Yasuda, K. and Ayusawa, D. (1998) Novel superoxide dismutase gene encoding membrane-bound and extracellular isoforms by alternative splicing in *Caenorhabditis elegans*. *DNA Res.* **5:** 25–30.

Garigan, D., Hsu, A., Fraser, A., Kamath, R., Ahringer, J. and Kenyon, C. (2002) Genetic analysis of tissue aging in *Caenorhabditis elegans*: a role for heat-shock factor and bacterial proliferation. *Genetics* **161:** 1101–1112.

Gems, D. and McElwee, J.J. (2005) Broad spectrum detoxification: the major longevity assurance process regulated by insulin/IGF-1 signaling? *Mech. Ageing Dev.* **126:** 381–387.

Gems, D., Sutton, A.J., Sundermeyer, M.L., Larson, P.L., Albert, P.S., King, K.V., Edgley, M. and Riddle, D.L. (1998) Two pleiotropic classes of *daf-2* mutation affect larval arrest, adult behavior, reproduction and longevity in *Caenorhabditis elegans*. *Genetics* **150:** 129–155.

Gerstbrein, B., Stamatas, G., Kollias, N. and Driscoll, M. (2005) *In vivo* spectrofluorimetry reveals endogenous biomarkers that report healthspan and dietary restriction in *Caenorhabditis elegans*. *Aging Cell* **4:** 127–137.

Gibson, G.G. and Skett, P. (2001) *Introduction to drug metabolism.* Nelson Thornes, Bath.

Giglio, M.P., Hunter, T., Bannister, J.V., Bannister, W.H. and Hunter, G.J. (1994) The manganese superoxide dismutase gene of *Caenorhabditis elegans*. *Biochem. Mol. Biol. Int.* **33:** 37–40.

Golden, T.R. and Melov, S. (2004) Microarray analysis of gene expression with age in individual nematodes. *Aging Cell* **3:** 111–124.

Gourley, B.L., Parker, S.B., Jones, B.J., Zumbrennen, K.B. and Leibold, E.A. (2003) Cytosolic aconitase and ferritin are regulated by iron in *Caenorhabditis elegans*. *J. Biol. Chem.* **278:** 3227–3234.

Halliwell, B. and Gutteridge, J.M.C. (2007) *Free radicals in biology and medicine.* Oxford University Press, Oxford.

Hamilton, B., Dong, Y., Shindo, M., Liu, W., Odell, I., Ruvkun, G. and Lee, S.S. (2005) A systematic RNAi screen for longevity genes in *C. elegans*. *Genes Dev.* **19:** 1544–1555.

Hansen, M., Hsu, A.L., Dillin, A. and Kenyon, C. (2005) New genes tied to endocrine, metabolic, and dietary regulation of lifespan from a *Caenorhabditis elegans* genomic RNAi screen. *PLoS Genet.* **1:** 119–128.

Hansen, M., Taubert, S., Crawford, D., Libina, L., Lee, S.J. and Kenyon, C. (2007) Lifespan extension by conditions that inhibit translation in *Caenorhabditis elegans*. *Aging Cell* **6:** 95–110.

Harrington, L.A. and Harley, C.B. (1988) Effect of vitamin E on lifespan and reproduction in *Caenorhabditis elegans*. *Mech. Ageing Dev.* **43:** 71–78.

Hartman, P.S., Ishii, N., Kayser, E.B., Morgan, P.G. and Sedensky, M.M. (2001) Mitochondrial mutations differentially affect aging, mutability and anesthetic sensitivity in *Caenorhabditis elegans*. *Mech. Ageing Dev.* **122:** 1187–1201.

Hartman, P., Ponder, R., Lo, H.H. and Ishii, N. (2004) Mitochondrial oxidative stress can lead to nuclear hypermutability. *Mech. Ageing Dev.* **125:** 417–420.

Hayes, J.D. and Mclellan, L.I. (1999) Glutathione and glutathione-dependent enzymes represent a co-ordinately regulated defence against oxidative stress. *Free Radic. Res.* **31:** 273–300.

Henderson, S.T. and Johnson, T.E. (2001) *daf-16* integrates developmental and environmental inputs to mediate aging in the nematode *Caenorhabditis elegans*. *Curr. Biol.* **11:** 1975–1980.

Henderson, S.T., Bonafe, M. and Johnson, T.E. (2006) *daf-16* protects the nematode *Caenorhabditis elegans* during food deprivation. *J. Gerontol. A* **61:** 444–460.

Holt, S. and Riddle, D. (2003) SAGE surveys *C. elegans* carbohydrate metabolism: evidence for an anaerobic shift in the long-lived dauer larva. *Mech. Ageing Dev.* **124:** 779–800.

Honda, Y. and Honda, S. (1999) The *daf-2* gene network for longevity regulates oxidative stress resistance and Mn-superoxide dismutase gene expression in *Caenorhabditis elegans*. *FASEB J.* **13:** 1385–1393.

Honda, S., Ishii, N., Suzuki, K. and Matsuo, M. (1993) Oxygen-dependent perturbation of life span and aging rate in the nematode. *J. Gerontol. A* **48:** B57–B61.

Hosokawa, H., Ishii, N., Ishida, H., Ichimori, K., Nakazawa, H. and Suzuki, K. (1994) Rapid accumulation of fluorescent material with ageing in an oxygen-sensitive mutant *mev-1* of *Caenorhabditis elegans*. *Mech. Ageing Dev.* **74:** 161–170.

Houthoofd, K., Braeckman, B., Lenaerts, I., Brys, K., De Vreese, A., Van Eygen, S. and Vanfleteren, J. (2002a) Ageing is reversed, and metabolism is reset to young levels in recovering dauer larvae of *C. elegans*. *Exp. Gerontol.* **37:** 1015–1021.

Houthoofd, K., Braeckman, B., Lenaerts, I., Brys, K., De Vreese, A., Van Eygen, S. and Vanfleteren, J. (2002b) Axenic growth up-regulates mass-specific metabolic rate, stress resistance, and extends life span in *Caenorhabditis elegans*. *Exp. Gerontol.* **37:** 1371–1378.

Houthoofd, K., Braeckman, B., Lenaerts, I., Brys, K., De Vreese, A., Van Eygen, S. and Vanfleteren, J. (2002c) No reduction of metabolic rate in food restricted *Caenorhabditis elegans*. *Exp. Gerontol.* **37:** 1359–1369.

Houthoofd, K., Braeckman, B., Johnson, T. and Vanfleteren, J. (2003) Life extension via dietary restriction is independent of the Ins/IGF-1 signalling pathway in *Caenorhabditis elegans*. *Exp. Gerontol.* **38:** 947–954.

Houthoofd, K., Braeckman, B.P., Lenaerts, I., Brys, K., Matthijssens, F., De Vreese, A., Van Eygen, S. and Vanfleteren, J.R. (2005a) DAF-2 pathway mutations and food restriction in aging *Caenorhabditis elegans* differentially affect metabolism. *Neurobiol. Aging* **26:** 689–696.

Houthoofd, K., Fidalgo, M.A., Hoogewijs, D., *et al.* (2005b) Metabolism, physiology and stress defense in three aging Ins/IGF-1 mutants of the nematode *Caenorhabditis elegans*. *Aging Cell* **4:** 87–95.

Hsu, A., Murphy, C. and Kenyon, C. (2003) Regulation of aging and age-related disease by DAF-16 and heat-shock factor. *Science* **300:** 1142–1145.

Hunter, T., Bannister, W.H. and Hunter, G.J. (1997) Cloning, expression, and characterization of two manganese superoxide dismutases from *Caenorhabditis elegans*. *J. Biol. Chem.* **272:** 28652–28659.

Ichimiya, H., Huet, R.G., Hartman, P., Amino, H., Kita, K. and Ishii, N. (2002) Complex II inactivation is lethal in the nematode *Caenorhabditis elegans*. *Mitochondrion* **2:** 191–198.

Imlay, J.A. and Fridovich, I. (1991) Assay of metabolic superoxide production in *Escherichia coli*. *J. Biol. Chem.* **266:** 6957–6965.

Iser, W.B., Kim, D., Bachman, E. and Wolkow, C. (2005) Examination of the requirement for *ucp-4*, a putative homolog of the mammalian uncoupling proteins, for stress tolerance and longevity in *C. elegans*. *Mech. Ageing Dev.* **126:** 1090–1096.

Ishii, N., Takahashi, K., Tomita, S., Keino, T., Honda, S., Yoshino, K. and Suzuki, K. (1990) A methyl viologen-sensitive mutant of the nematode *Caenorhabditis elegans*. *Mutation Res.* **237:** 165–171.

Ishii, N., Fujii, M., Hartman, P.S., Tsuda, M., Yasuda, K., Senoo-Matsuda, N., Yanase, S., Ayusawa, D. and Suzuki, K. (1998) A mutation in succinate dehydrogenase cytochrome *b* causes oxidative stress and ageing in nematodes. *Nature* **394:** 694–697.

Ishii, N., Senoo-Matsuda, N., Miyake, K., Yasuda, K., Ishii, T., Hartman, P.S. and Furukawa, S. (2004) Coenzyme Q10 can prolong *C. elegans* lifespan by lowering oxidative stress. *Mech. Ageing Dev.* **125:** 41–46.

Ishii, N., Ishii, T. and Hartman, P.S. (2006) The role of the electron transport gene SDHC on lifespan and cancer. *Exp. Gerontol*, **41:** 952–956.

Jensen, L.T. and Culotta, V.C. (2005) Activation of CuZn superoxide dismutases from *Caenorhabditis elegans* does not require the copper chaperone CCS. *J. Biol. Chem.* **280:** 41373–41379.

Johnson, T.E. and Hutchinson, E.W. (1993) Absence of strong heterosis for life span and other life history traits in *Caenorhabditis elegans*. *Genetics* **134:** 465–474.

Jonassen, T., Davis, D.E., Larsen, P.L. and Clarke, C.F. (2003) Reproductive fitness and quinone content of *Caenorhabditis elegans clk-1* mutants fed coenzyme Q isoforms of varying length. *J. Biol. Chem.* **278:** 51735–51742.

Keaney, M. and Gems, D. (2003) No increase in lifespan in *Caenorhabditis elegans* upon treatment with the superoxide dismutase mimetic EUK-8. *Free Radic. Biol. Med.* **34:** 277–282.

Keaney, M., Matthijssens, F., Sharpe, M., Vanfleteren, J.R. and Gems, D. (2004) Superoxide dismutase mimetics elevate superoxide dismutase activity *in vivo* but do not retard aging in the nematode *Caenorhabditis elegans*. *Free Radic. Biol. Med.* **37:** 239–250.

Kenyon, C., Chang, J., Gensch, E., Rudener, A. and Tabtiang, R. (1993) A *C. elegans* mutant that lives twice as long as wild type. *Nature* **366:** 461–464.

Kimura, K.D., Tissenbaum, H.A., Liu, Y. and Ruvkun, G. (1997) *daf-2*, an insulin receptor-like gene that regulates longevity and diapause in *Caenorhabditis elegans*. *Science* **277:** 942–946.

Kwong, L.K., Kamzalov, S., Rebrin, I., Bayne, A.C., Jana, C.K., Morris, P., Forster, M.J. and Sohal, R.S. (2002) Effects of coenzyme Q(10) administration on its tissue concentrations, mitochondrial oxidant generation, and oxidative stress in the rat. *Free Radic. Biol. Med.* **33:** 627–638.

Larsen, P.L. (1993) Aging and resistance to oxidative stress in *Caenorhabditis elegans*. *Proc. Natl Acad. Sci. USA* **90:** 8905–8909.

Larsen, P. and Clarke, C.F. (2002) Extension of life-span in *Caenorhabditis elegans* by a diet lacking coenzyme Q. *Science* **295:** 120–123.

Lass, A. and Sohal, R.S. (2000) Effect of coenzyme Q(10) and alpha-tocopherol content of mitochondria on the production of superoxide anion radicals. *FASEB J.* **14:** 87–94.

Lee, S., Lee, R., Fraser, A., Kamath, R., Ahringer, J. and Ruvkun, G. (2003a) A systematic RNAi screen identifies a critical role for mitochondria in *C. elegans* longevity. *Nat. Genet.* **33:** 40–48.

Lee, S.S., Kennedy, S., Tolonen, A.C. and Ruvkun, G. (2003b) DAF-16 target genes that control *C. elegans* life-span and metabolism. *Science* **300:** 644–647.

Leiers, B., Kampkotter, A., Grevelding, C., Link, C., Johnson, T. and Henkle-Duhrsen, K. (2003) A stress-response glutathione S-transferase confers resistance to oxidative stress in *Caenorhabditis elegans*. *Free Radic. Biol. Med.* **34:** 1405–1415.

Magwere, T., West, M., Riyahi, K., Murphy, M.P., Smith, R.A. and Partridge, L. (2006) The effects of exogenous antioxidants on lifespan and oxidative stress resistance in *Drosophila melanogaster*. *Mech. Ageing Dev.* **127:** 356–370.

Martin, G.M., Austad, S.N. and Johnson, T.E. (1996) Genetic analysis of ageing: role of oxidative damage and environmental stresses. *Nat. Genet.* **13:** 25–34.

Mathers, J., Fraser, J.A., Mcmahon, M., Saunders, R.D., Hayes, J.D. and Mclellan, L.I. (2004) Antioxidant and cytoprotective responses to redox stress. *Biochem. Soc. Symp.* **71:** 157–176.

McElwee, J., Bubb, K. and Thomas, J. (2003) Transcriptional outputs of the *Caenorhabditis elegans* forkhead protein DAF-16. *Aging Cell* **2:** 111–121.

McElwee, J.J., Schuster, E., Blanc, E., Thomas, J.H. and Gems, D. (2004) Shared transcriptional signature in *C. elegans* dauer larvae and long-lived *daf-2* mutants implicates detoxification system in longevity assurance. *J. Biol. Chem.* **279:** 44533–44543.

McElwee, J.J., Schuster, E., Blanc, E. and Gems, D. (2006a) Partial reiteration of dauer larva metabolism in long lived *daf-2* mutant adults in *Caenorhabditis elegans. Mech. Ageing Dev.* **127:** 458–472.

McElwee, J.J., Schuster, E., Blanc, E., Thornton, J. and Gems, D. (2006b) Erratum to 'Diapause-associated metabolic traits reiterated in long-lived *daf-2* mutants in the nematode *Caenorhabditis elegans*' [*Mech. Ageing Dev.* **127**: (2006) 458–472]. *Mech. Ageing Dev.* **127:** 922–936.

McElwee, J.J., Schuster, E., Blanc, E., *et al.* (2007) Evolutionary conservation of regulated longevity assurance mechanisms. *Genome Biol.* **8:** R132.

Melov, S., Ravenscroft, J., Malik, S., *et al.* (2000) Extension of life-span with superoxide dismutase/catalase mimetics. *Science* **289:** 1567–1569.

Miyadera, H., Amino, H., Hiraishi, A., *et al.* (2001) Altered quinone biosynthesis in the long-lived *clk-1* mutants of *Caenorhabditis elegans. J. Biol. Chem.* **276:** 7713–7716.

Miyadera, H., Kano, K., Miyoshi, H., Ishii, N., Hekimi, S. and Kita, K. (2002) Quinones in long-lived *clk-1* mutants of *Caenorhabditis elegans. FEBS Lett.* **512:** 33–37.

Morris, J.Z., Tissenbaum, H.A. and Ruvkun, G. (1996) A phosphatidylinositol-3-OH kinase family member regulating longevity and diapause in *Caenorhabditis elegans. Nature* **382:** 536–538.

Murakami, S. and Johnson, T.E. (1996) A genetic pathway conferring life extension and resistance to UV stress in *Caenorhabditis elegans. Genetics* **143:** 1207–1218.

Murfitt, R., Vogel, K. and Sanadi, D. (1976) Characterization of the mitochondria of the free-living nematode *Caenorhabditis elegans. Comp. Biochem. Physiol.* **53B:** 423–430.

Murphy, C.T., Mccarroll, S.A., Bargmann, C.I., Fraser, A., Kamath, R.S., Ahringer, J., Li, H. and Kenyon, C.J. (2003) Genes that act downstream of DAF-16 to influence the lifespan of *C. elegans. Nature* **424:** 277–284.

Nohl, H. and Hegner, D. (1978) Do mitochondria produce oxygen radicals in vivo? *Eur. J. Biochem.* **82:** 563–567.

Okimoto, R., Macfarlane, J.L., Clary, D.O. and Wolstenholme, D.R. (1992) The mitochondrial genomes of two nematodes, *Caenorhabditis elegans* and *Ascaris suum. Genetics* **130:** 471–498.

Orr, W. and Sohal, R. (2003) Does overexpression of Cu,Zn-SOD extend life span in *Drosophila melanogaster? Exp. Gerontol.* **38:** 227–230.

Pan, K.Z., Palter, J.E., Rogers, A.N., Olsen, A., Chen, D., Lithgow, G.J. and Kapahi, P. (2007) Inhibition of mRNA translation extends lifespan in *Caenorhabditis elegans. Aging Cell* **6:** 111–119.

Parkes, T.L., Elia, A.J., Dickinson, D., Hilliker, A.J., Phillips, J.P. and Boulianne, G.L. (1998) Extension of *Drosophila* lifespan by overexpression of human SOD1 in motorneurons. *Nat. Genet.* **19:** 171–174.

Petriv, O.I. and Rachubinski, R.A. (2004) Lack of peroxisomal catalase causes a progeric phenotype in *Caenorhabditis elegans*. *J. Biol. Chem.* **279:** 19996–20001.

Raha, S. and Robinson, B.H. (2000) Mitochondria, oxygen free radicals, disease and ageing. *Trends Biochem. Sci.* **25:** 502–508.

Rea, S.L. (2005) Metabolism in the *Caenorhabditis elegans* Mit mutants. *Exp. Gerontol.* **40:** 841–849.

Rea, S. and Johnson, T.E. (2003) A metabolic model for lifespan determination in *Caenorhabditis elegans*. *Dev. Cell* **5:** 197–203.

Sampayo, J.N., Olsen, A. and Lithgow, G.J. (2003) Oxidative stress in *Caenorhabditis elegans*: protective effects of superoxide dismutase/catalase mimetics. *Aging Cell* **2:** 319–326.

Schulz, T.J., Zarse, K., Voigt, A., Urban, N., Birringer, M. and Ristow, M. (2007) Glucose restriction extends *Caenorhabditis elegans* life span by inducing mitochondrial respiration and increasing oxidative stress. *Cell Metab.* **6:** 280–293.

Senoo-Matsuda, N., Hartman, P.S., Akatsuka, A., Yoshimura, S. and Ishii, N. (2003) A complex II defect affects mitochondrial structure, leading to *ced-3*- and *ced-4*-dependent apoptosis and aging. *J. Biol. Chem.* **278:** 22031–22036.

Senoo-Matsuda, N., Yasuda, K., Tsuda, M., Ohkubo, T., Yoshimura, S., Nakazawa, H., Hartman, P.S. and Ishii, N. (2001) A defect in the cytochrome *b* large subunit in complex II causes both superoxide anion overproduction and abnormal energy metabolism in *Caenorhabditis elegans*. *J. Biol. Chem.* **276:** 41553–41558.

Shibata, Y., Branicky, R., Landaverde, I.O. and Hekimi, S. (2003) Redox regulation of germline and vulval development in *Caenorhabditis elegans*. *Science* **302:** 1779–1782.

Sohal, R.S. and Weindruch, R. (1996) Oxidative stress, caloric restriction, and aging. *Science* **273:** 59–63.

St-Pierre, J., Buckingham, J.A., Roebuck, S.J. and Brand, M.D. (2002) Topology of superoxide production from different sites in the mitochondrial electron transport chain. *J. Biol. Chem.* **277:** 44784–44790.

Staniek, K. and Nohl, H. (2000) Are mitochondria a permanent source of reactive oxygen species? *Biochim. Biophys. Acta* **1460:** 268–275.

Stenmark, P., Grunler, J., Mattsson, J., Sindelar, P.J., Nordlund, P. and Berthold, D.A. (2001) A new member of the family of di-iron carboxylate proteins. Coq7 (*clk-1*), a membrane-bound hydroxylase involved in ubiquinone biosynthesis. *J. Biol. Chem.* **276:** 33297–33300.

Stepanyan, Z., Hughes, B., Cliche, D.O., Camp, D. and Hekimi, S. (2006) Genetic and molecular characterization of CLK-1/mCLK1, a conserved determinant of the rate of aging. *Exp. Gerontol.* **41:** 940–951.

Suda, H., Shouyama, T., Yasuda, K. and Ishii, N. (2005) Direct measurement of oxygen consumption rate on the nematode *Caenorhabditis elegans* by using an optical technique. *Biochem. Biophys. Res. Commun.* **330:** 839–843.

Sun, J. and Tower, J. (1999) FLP recombinase-mediated induction of Cu/Zn-superoxide dismutase transgene expression can extend the life span of adult *Drosophila melanogaster* flies. *Mol. Cell. Biol.* **19:** 216–228.

Suzuki, N., Inokuma, K., Yasuda, K. and Ishii, N. (1996) Cloning, sequencing and mapping of a manganese superoxide dismutase gene of the nematode *Caenorhabditis elegans. DNA Res.* **3:** 171–174.

Syntichaki, P., Troulinaki, K. and Tavernarakis, N. (2007) eIF4E function in somatic cells modulates ageing in *Caenorhabditis elegans. Nature* **445:** 922–926.

Takamiya, S., Matsui, T., Taka, H., Murayama, K., Matsuda, M. and Aoki, T. (1999) Free-living nematodes *Caenorhabditis elegans* possess in their mitochondria an additional rhodoquinone, an essential component of the eukaryotic fumarate reductase system. *Arch. Biochem. Biophys.* **371:** 284–289.

Taub, J., Lau, J.F., Ma, C., Hahn, J.H., Hoque, R., Rothblatt, J. and Chalfie, M. (1999) A cytosolic catalase is needed to extend lifespan in *C. elegans daf-C* and *clk-1* mutants. *Nature* **399:** 162–166.

Taub, J., Lau, J., Ma, C., Hahn, J., Hoque, R., Rothblatt, J. and Chalfie, M. (2003) A cytosolic catalase is needed to extend adult lifespan in *C. elegans daf-C* and *clk-1* mutants. *Nature* **421:** 764.

Tawe, W., Eschbach, M., Walter, R. and Henkle-Duhrsen, K. (1998) Identification of stress-responsive genes in *Caenorhabditis elegans* using RT-PCR differential display. *Nucleic Acids Res.* **26:** 1621–1627.

Tielens, A., Rotte, C., Van Hellemond, J. and Martin, W. (2002) Mitochondria as we don't know them. *Trends Biochem. Sci.* **27:** 564–572.

Togo, S.H., Maebuchi, M., Yokota, S., Bun-Ya, M., Kawahara, A. and Kamiryo, T. (2000) Immunological detection of alkaline-diaminobenzidine-negative peroxisomes of the nematode *Caenorhabditis elegans*: purification and unique pH optima of peroxisomal catalase. *Eur. J. Biochem.* **267:** 1307–1312.

Tsang, W.Y. and Lemire, B.D. (2002) Mitochondrial genome content is regulated during nematode development. *Biochem. Biophys. Res. Commun..* **291:** 8–16.

Tsang, W.Y., Sayles, L.C., Grad, L.I., Pilgrim, D.B. and Lemire, B.D. (2001) Mitochondrial respiratory chain deficiency in *Caenorhabditis elegans* results in developmental arrest and increased life span. *J. Biol. Chem.* **276:** 32240–32246.

Van Voorhies, W. (2002) The influence of metabolic rate on longevity in the nematode *Caenorhabditis elegans. Aging Cell* **1:** 91–101.

Van Voorhies, W. and Ward, S. (1999) Genetic and environmental conditions that increase longevity in *Caenorhabditis elegans* decrease metabolic rate. *Proc. Natl Acad. Sci. USA* **96:** 11399–11403.

Vanfleteren, J.R. (1993) Oxidative stress and ageing in *Caenorhabditis elegans. Biochem. J.* **292:** 605–608.

Vanfleteren, J.R. and De Vreese, A. (1995) The gerontogenes *age-1* and *daf-2* determine metabolic rate potential in aging *Caenorhabditis elegans. FASEB J.* **9:** 1355–1361.

Vanfleteren, J.R. and De Vreese, A. (1996) Rate of aerobic metabolism and superoxide production rate potential in the nematode *Caenorhabditis elegans. J. Exp. Zool.* **274:** 93–100.

Wang, J. and Kim, S. (2003) Global analysis of dauer gene expression in *Caenorhabditis elegans. Development* **130:** 1621–1634.

Wong, A.E., Boutis, P. and Hekimi, S. (1995) Mutations in the *clk-1* gene of *Caenorhabditis elegans* affect developmental and behavioral timing. *Genetics* **139:** 1247–1259.

Yanase, S., Yasuda, K. and Ishii, N. (2002) Adaptive responses to oxidative damage in three mutants of *Caenorhabditis elegans* (*age-1, mev-1* and *daf-16*) that affect life span. *Mech. Ageing Dev.* **123:** 1579–1587.

Yasuda, K., Adachi, H., Fujiwara, Y. and Ishii, N. (1999) Protein carbonyl accumulation in aging dauer formation-defective (daf) mutants of *Caenorhabditis elegans. J. Gerontol. A* **54:** B47–51; discussion B52–53.

Yasuda, K., Ishii, T., Suda, H., Akatsuka, A., Hartman, P.S., Goto, S., Miyazawa, M. and Ishii, N. (2006) Age-related changes of mitochondrial structure and function in *Caenorhabditis elegans. Mech. Ageing Dev.* **127:** 763–770.

Yoneda, T., Benedetti, C., Urano, F., Clark, S.G., Harding, H.P. and Ron, D. (2004) Compartment-specific perturbation of protein handling activates genes encoding mitochondrial chaperones. *J. Cell. Sci.* **117:** 4055–4066.

Zuckerman, B.M. and Geist, M.A. (1983) Effects of vitamin E on the nematode *Caenorhabditis elegans. Age (Omaha, Nebr.)* **6:** 1–4.

4

The perception of reactive oxygen species in plants: The road to signal transduction

John T. Hancock, Radhika Desikan, Ian Wilson and Steven J. Neill

1 Introduction

Reactive oxygen species (ROS) are produced by aerobic organisms, sometimes as a consequence of metabolic processes, but often in a dedicated fashion where their increased presence is the main aim of the synthesis. In both cases, ROS may be generated in order to regulate the functionality of various aspects of cellular biochemistry and are often produced either in response to environmental change or during a host-pathogen interaction. In such cases the presence of ROS acts as a defined signal to which the cell must respond if it is to both adapt and thrive. However, before ROS can elicit any cellular responses their presence must be perceived. Such recognition of their presence allows the message contained therein to be carried forward along a signal transduction pathway. Here, some of the mechanisms by which ROS may be perceived by cells and examples of the proteins involved are discussed.

For many years, ROS have been thought of as damaging molecules. The study of their use and roles in biological systems has primarily been centred on host defence and much of the early work was carried out using mammalian systems that involved cells such as neutrophils and macrophages. These phagocytic cells engulf pathogens producing a burst of ROS, which is thought to be instrumental to the killing process (Segal, 2005). However, it is also recognised that ROS, in particular hydrogen peroxide (H_2O_2), have a major role in the cell signalling mechanisms used by a wide range of organisms, including plants (reviewed by Dröge, 2002; Foyer and Noctor, 2005b; Neill *et al.*, 2002). ROS in plants may be generated via the leakage of electrons from redox pathways in either the mitochondria or chloroplast, but can also be synthesised by a range of more dedicated enzymes such as the NADPH oxidases or the peroxidases (Neill *et al.*, 2002). The ability of chloroplasts to generate H_2O_2 was reported first by Mehler in 1957 but its effects on photosynthesis were only considered much later (Charles and Halliwell, 1980; Kaiser, 1979). The rationale behind the production of these supposedly harmful compounds is only now coming to light.

Once produced, ROS may function as part of the signalling repertoire of plant cells. However, their exact action is very likely to be modulated by the complement of antioxidants either in or surrounding the cell. Some of the main physiological processes in plants that are controlled, at least in part, by ROS include stomatal closure (Pei *et al.*, 2000), root growth (Foreman *et al.*, 2003) and programmed cell death (Desikan *et al.*, 1998), including the hypersensitive response.

2 Do ROS and potential ROS-sensitive proteins make ideal signals?

A signal used by a cell, and in fact often the molecular components of a signalling cascade, such as proteins, usually have some general characteristics:

- a signal is usually produced only when and where needed, either by the production of a new molecule or a change within an existing one;
- a signal is often made from a readily obtainable substrate, or alteration of an existing signalling component such as a protein;
- once the action from a signal has resulted in the correct response, the signal is either removed or its action is reversed and, thus, the activity of the cell is permitted restoration to the ground state.

Thus, a key question concerns whether ROS, and, importantly, the proteins that may perceive them, fulfil the above criteria. Certainly ROS fulfil the above criteria (reviewed by Neill *et al.*, 2002), as they are often produced in a regulated way, are simply reduction products of oxygen, and they can easily be removed. However, much less is known regarding the proteins that potentially perceive ROS and whether they are able to transmit an appropriate message regarding the alteration in ROS levels. Such ROS-sensing proteins would need to be appropriately expressed, both developmentally and spatially, in the cell and would need to respond to the concentrations of ROS that are likely to be present. Furthermore, the perception process would need to lead to the propagation of a reversible signalling cascade. ROS might exist in cellular hotspots (Neill *et al.*, 2002) and the location of proteins that perceive ROS may, therefore, be diverse. Some may exist near the cell surface to sense extracellular ROS, while others may be deeper inside the cell to sense those generated intracellularly.

3 Molecular events modulated by the presence of ROS

The best way to test the potential involvement of novel components in signalling pathways is to modulate their levels in the cells of interest and then to monitor the levels of activity of the other components of the pathway. For ROS, one of the families of pathways that have been the subject of particular focus has been that of the mitogen-activated protein kinases (MAPKs). For example, when H_2O_2 is added to Arabidopsis cell-suspension cultures *in-gel* assays indicate that the activity of MAPKs increases (Desikan *et al.*, 1999). In Arabidopsis, AtMPK3 and AtMPK6 appear to be particularly involved (Desikan *et al.*, 2001a; Kovtun *et al.*, 2000). However, it does not necessarily follow that MAPKs are the direct target for ROS. Although this may be the case, it is more likely that an indirect signalling cascade leads to their activation. Thus, identifying the proteins which perceive the presence and increase in the level of

ROS and which initiate the downstream signalling pathways that involve MAPKs is a research priority.

Other kinases have also been implicated in ROS signalling. One of note has been suitably named oxidative signal-inducible 1 (OXI1) kinase and although not thought to be the protein that directly perceives ROS, interestingly has been shown to be upstream of the MAPKs AtMPK3 and AtMPK6 (Rentel *et al.*, 2004). Other kinases involved include the MAPK kinase kinases (MAPKKKs) ANP1 and OMTK1, which are activated by the addition of H_2O_2 (Kovtun *et al.*, 2000; Nakagami *et al.*, 2004). Thus, kinase cascades that may transduce the ROS signal within the cell have been identified, but the exact mechanisms by which this signalling is initiated remain elusive.

The activation of kinases leads to an increase in phosphorylation. However, a similar effect can be obtained by decreasing the dephosphorylation step in the signalling cascade and it is no major surprise, therefore, to find that ROS have a negative effect on the activity of phosphatases. Mammalian tyrosine phosphatases have been shown to be inactivated by the addition of H_2O_2 (Cho *et al.*, 2004; Salmeen *et al.*, 2003; van Montfort *et al.*, 2003). Similarly, phosphatases also appear to be involved in H_2O_2 signalling in plants, and in Arabidopsis the protein tyrosine phosphatase AtPTP1 and the protein phosphatase 2C enzymes ABI1 and ABI2 appear to be involved (Gupta and Luan, 2003; Meinhard and Grill, 2001).

4 ROS interactions with nitric oxide

In plants, nitric oxide (NO) is very often produced in the same location and at the same time as ROS. NO is similarly involved in a plethora of physiological and biochemical processes (reviewed by Neill *et al.*, 2003) and many of these overlap with those regulated by ROS. Of particular importance is the fact that ROS and NO can react together, in particular with the formation of peroxynitrite. Thus, one might modulate the bioavailability of the other (Vanin *et al.*, 2004). Furthermore, as is discussed below, ROS and NO may partake in very similar chemistry. Therefore, any discussion concerning the role of ROS in the modulation of cellular activity needs to consider the impact of NO.

5 Thiol (-SH) group modification is likely to be key

As discussed above, several proteins have their functional activity altered by the presence of altered levels of ROS. This alteration may occur either directly or as the result of downstream signalling by other proteins, which perceive the ROS and initiate the signal transduction cascade.

For a protein to detect the presence of a ROS such as H_2O_2, there needs to be a mechanism that allows for specific substrate recognition. The physical recognition of H_2O_2 via the model ligand and receptor mechanism may occur, but because of its oxidising nature it is thought more likely that this occurs in the form of a chemical reaction. Certainly the reactive nature of H_2O_2 would allow it to directly modify -SH groups in proteins. This has been suggested by a number of researchers (Cooper *et al.*, 2002; Foyer and Noctor, 2005a; Vranová *et al.*, 2002) and although the -SH groups of cysteine residues are considered the most likely targets, other amino acids, including tyrosine, tryptophan and histidine, may be similarly oxidised (Dröge, 2002). However,

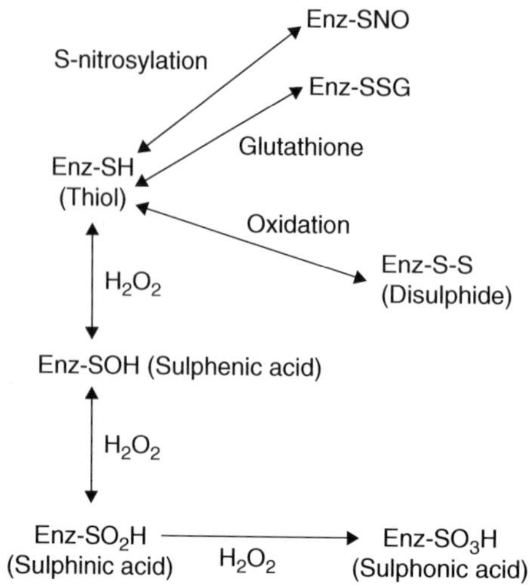

Figure 1. *ROS perception is achieved partly by the modification of -SH groups. With the presence of two proximal cysteine -SH groups disulphide bridges may be formed either within or between proteins. Alternatively, isolated -SH groups may be modified by varying degrees of oxidation. Lower levels of oxidation (-SOH) allow reversible protein modification to occur, but the highest levels (-SO_3H) probably result in irreversible modification and perhaps permanent alterations in protein function.*

as they are all likely to contain these amino acids, all proteins, therefore, might be expected to be potential targets for ROS. Obviously, this cannot be the case and those proteins that perceive ROS must have some specific characteristic that enables them to do so and to propagate the resulting signal. This in some way must be related not only to the amino acid sequence of the protein but also its topology in the cell.

An increase in ROS and resultant protein oxidation may yield a variety of products. The oxidation of methionine (Met) residues produces methionine sulphoxide (MetSO), which can be reduced back to Met by protein MetSO reductase (PMSR; Sadanandom *et al.*, 2000). This enzyme has two types, A and B, with the former using thioredoxin (TRX) as the reductant (Romero *et al.*, 2004).

The oxidation of cysteine residues by H_2O_2 yields several potential modified amino acid products (*Figure 1*). If two cysteine residues are spatially able to come together for oxidation the result may be the formation of a disulphide bridge (S-S). If, on the other hand, either only a single cysteine is present or others are spatially too distant, the individual -SH group may be oxidised to varying degrees to either sulphenic acid (-SOH) or further to sulphinic acid (-SO_2H) and sulphonic acid (-SO_3H; *Figure 1*). As the levels of oxidising ROS increase, the thiol will be modified to the most oxidised state that is the -SO_3H form. However, although the -SOH group can easily be reversibly reduced back to the thiol, the higher oxidation states are more difficult to re-reduce, and it is thought that the formation of the -SO_3H group is an end point from which the modified thiol does not return. Of course, any oxidation of a -SH group would be dependent on its mid-point redox potential and its availability to the oxidant. Thus, in reality it is likely that only a small proportion of any -SH groups within a protein would be expected to be modified. Additionally, different -SH groups will sit in the varied environments occupied by the proteins in which they reside. Thus, there is likely to be a redox hierarchy among proteins with the exposed -SH groups of individual proteins having different susceptibilities to oxidation dependent on variations in their mid-point potentials. It is possible that other signalling pathways also impact on the accessibility of thiols and their mid-point potentials in proteins. A conformational change in a protein brought about by the impact of an upstream signalling pathway, perhaps for example through phosphorylation, may alter the protein topology sufficiently to allow subsequent oxidation and signalling from that protein. There is, therefore, scope for the differential regulation of the activity of some cell proteins by alterations in the intracellular redox status and this may be one way that oxidants such as H_2O_2 can exert a subtle control over cell physiology.

The formation of oxidised forms of cysteine has been shown in some proteins. For example, in mammalian tyrosine phosphatase, an enzyme known to be inhibited directly by H_2O_2, the -SOH group is formed. However, this product is relatively unstable and subsequent cyclisation occurs within the peptide structure leading to the formation of a sulphenyl-amide intermediate (Salmeen *et al.*, 2003; van Montfort *et al.*, 2003). This cyclic form can then be decyclised to regenerate the original -SH group. Thus, there exists a rapid and reversible protein modification that could potentially impact on phosphorylation-mediated signalling cascades. Such a modification would seem to lend itself well in terms of a mechanism for ROS signalling.

As mentioned previously, any ROS-mediated protein modification involved in the initiation of a cell-signalling cascade must be reversible. In the example of the mammalian phosphatase, cyclisation and rearrangement ultimately lead to the restoration of the original -SH group. However, the ability to reverse other -SH group modifications must also exist if this potential signalling mechanism is to receive proper consideration. Fortunately, such processes of reversal are entirely feasible. For example, in addition to their ability to reduce groups such as -SOH (Collin *et al.*, 2004; Rouhier *et al.*, 2001), TRXs and glutaredoxins can also act as protein S-S reductases (Lemaire, 2004; Schürmann and Jacquot, 2000). Any $-SO_2H$ groups can be reduced to -SOH groups by sulphiredoxin, an ATP-dependent enzyme first described in yeast (Biteau *et al.*, 2003). The resulting -SOH groups can then be reduced further by TRXs or glutaredoxins to regenerate the original -SH groups. Thioedoxins are important players in the control of redox function in cells and, as such, have been found to interact with a variety of proteins (Yano *et al.*, 2001). The roles and influences of thioredoxins in plant cells have been reviewed recently by Vieira Dos Santos and Rey (2006), while for a review on glutaredoxins see Rouhier *et al.* (2004).

It is also possible that some of the proteins involved in the reversal of -SH group modification may themselves form part of the H_2O_2 perception mechanisms. For instance, it has been suggested that 2-cys peroxiredoxins may act as redox sensors in chloroplasts (Dietz *et al.*, 2006; Konig *et al.*, 2002, 2003) and, giving reversibility to the system, that thioredoxins are able to reduce oxidised 2-cys peroxiredoxins.

The modification of -SH groups is in fact much more complex than this. In addition to their modification by ROS, in the process of S-nitrosylation NO may also react with protein -SH groups to yield S-nitrosothiol (-S-NO) groups. In plants, many proteins have been found that are modified in this way and many of the proteins identified were also those which could potentially be modified by H_2O_2 (Lindermayr *et al.*,

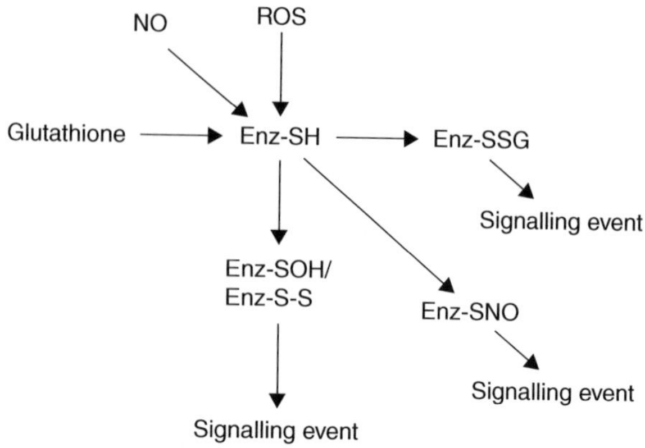

Figure 2. *Modification of thiol groups in proteins may be a competition. Thiol groups in proteins, depending on their availability and mid-point potentials, may be modified through different chemical routes, and the signalling that ensues will depend on the final outcome.*

2005). Thus, it is likely that NO and H_2O_2 may potentially regulate the activity of similar complements of cell proteins. Therefore, there may well be competition between H_2O_2 and NO for the same -SH groups (*Figure 2*) and the exact modification which occurs to these may well determine how the topology and resulting function of the protein is altered and, thus, the exact signalling process which predominates.

As well as reacting with ROS and NO, -SH groups can also undergo cellular reactions with other prevalent compounds such as glutathione (GSH/GSSG; *Figure 2*). In plant cells GSH can reach millimolar concentrations (Ball *et al.*, 2004), with a small proportion being present in the oxidised form, glutathione disulphide (GSSG). In a process called S-thiolation, the reaction of -SH groups with low molecular weight thiols such as glutathione may lead to the formation of what are termed mixed disulphides. Many proteins have been shown to be able to undergo this process. For example, by using a combination of *in vitro* and *in vivo* labelling techniques, 2-D gel electrophoresis and the analysis of isolated protein spots by mass spectroscopy, 79 Arabidopsis polypeptides that could be modified by glutathione were identified (Dixon *et al.*, 2005). Included among these were proteins such as alcohol dehydrogenase that had previously been shown to be susceptible to oxidative stress (Echave *et al.*, 2003). Another protein shown to be capable of S-thiolation and also modification by ROS and NO was glyceraldehyde 3-phosphate dehydrogenase (GAPDH). In *Saccharomyces cerevisiae* this process appears to be isoform specific: the Tdh3 isoform of GAPDH can be S-thiolated but the Tdh2 isoform cannot (Grant *et al.*, 1999). Perhaps such an observation provides initial evidence for the existence of redox-dependent, differential regulation mechanisms. Regardless, it can clearly be seen that there are redox-susceptible amino-acid side chains within some proteins that can toggle between different redox states as influenced by the presence or absence of ROS/NO. The modification of such side chains could potentially alter the structures of these proteins, which may thus be ideally suited to partake in H_2O_2-mediated signalling mechanisms (Buchanan and Balmer, 2005; Foyer and Noctor, 2005b).

6 Control of gene expression by H_2O_2

One of the major changes observed following the application of H_2O_2 to cells is an alteration in the patterns of accumulation of different mRNAs. Studies often observed individual transcripts using techniques such as Northern analysis to estimate whether the presence of H_2O_2 either increased or decreased their levels, and it could be seen that the accumulation of several mRNAs was altered following such exposure. However, recent technological advances in global transcript profiling, such as microarray analysis, has allowed more holistic studies to be undertaken. Using such methods it has been shown that the level of accumulation of approximately 3% of an organism's transcripts may be altered by ROS, with the levels of some being increased and some reduced (Gasch *et al.*, 2000). The use of early Arabidopsis microarrays on which fewer than all the potential transcripts were represented still revealed some 170 mRNAs that showed increased levels following exposure to ROS and a further 65 that showed a decrease (Desikan *et al.*, 2001b). Similar studies in tobacco (Vandenabeele *et al.*, 2003) revealed similar results to those observed in Arabidopsis, and a microarray study using catalase-deficient Arabidopsis grown under high light exposure revealed that the accumulation of 349 transcripts was increased and that of 88 was reduced in an H_2O_2-mediated manner (Vanderauwera *et al.*, 2005).

Although one must not ignore the possibility of the involvement of post-transcriptional gene silencing mechanisms mediated by small RNAs such as micro RNAs, these altered levels of transcript accumulation presumably reflect a general alteration in gene expression that occurs following the exposure of plants to ROS. This being the case, it can be concluded that the expression of some genes is enhanced by ROS and the expression of others repressed. Such changes in gene expression usually reflect alterations in the activity of associated transcription factors and one of the early studies in this area was an investigation into the affects of ROS on the activity of nuclear factor NF-κB in mammalian cells. In the inactive state, NF-κB resides as a complex with I-κB. Dissociation of the complex and, hence, activation of NF-κB, is enhanced by the presence of H_2O_2 (Schreck *et al.*, 1991). However, considering the antioxidant capacity of the cytoplasm, it is difficult to envisage physiologically relevant levels of externally applied H_2O_2 ever reaching NF-κB in order to achieve this effect directly. Thus, it may be that the oxidative activation of such transcription factors either results from the intracellular generation of ROS or from the action of intermediate signalling proteins. Having said this, NF-κB is not the only mediator of gene expression to be activated in response to ROS and several other transcription factors have been identified that are redox controlled. In *Escherichia coli*, the transcription factor OxyR is activated by cellular H_2O_2, which causes the formation of a disulphide bond between Cys199 and Cys208 in the protein (Lee *et al.*, 2004; reviewed by D'Autréaux and Toledano, 2007).

On exposure to oxidative stress, the yeast transcription factor Yap1 activates the expression of genes encoding antioxidant proteins. Oxidation of Yap1 is mediated by the glutathione peroxidase- (GPX-) like enzyme Gpx3 (also known as Orp1, which stands for oxidant receptor peroxidase-1), and its reduction results from the action of thioredoxins. Oxidation by H_2O_2 causes Cys36 of Gpx3 to bridge to Cys598 on Yap1. The inter-polypeptide disulphide is then rearranged to form a disulphide bridge within Yap1 itself and this results in the activation of the transcription factor. Thus, Gpx3 effectively acts as a H_2O_2 perception protein that transduces the perceived signal into an alteration in the activity of a transcription factor (Delaunay *et al.*, 2000, 2002), which thus results in altered gene expression. Homologues to this protein also exist in plants. In Arabidopsis, insertion mutants of ATGPX3 showed an increased generation of H_2O_2 in guard cells, had greater water loss during drought stress, and were more sensitive to H_2O_2 during seed germination and plant development (Miao *et al.*, 2006). Furthermore, mutants showed aberrations in ABA signalling such as ABA activation of calcium channels and expression of ABA-responsive genes. Plants in which ATGPX3 was over-expressed had higher leaf temperatures and greater drought tolerance. ATGPX3 was found to interact with phosphatases, in particular with ABI1 and more so with ABI2. Interestingly, in *in vitro* assays, if ATGPX3 was oxidised the activity of ABI2 was lowered fivefold and, moreover, ABI2 was oxidised in the presence of oxidised ATGPX3 (Miao *et al.*, 2006). For a very thorough review of the role of such peroxidases, especially in non-plant species see D'Autréaux and Toledano (2007).

In plants, the modulation of gene expression during the induction of systemic acquired resistance often results from alterations in the activity of the protein Non-Expressor of Pathogenesis Related Genes 1 (NPR1) and also the TGA transcription factors, a class of basic leucine zippers (bZIP). In the inactive state, NPR1 has an oligomeric form held together by intermolecular disulphide bonds. On activation, NPR1 becomes reduced and releases a monomeric form that accumulates in the nucleus and alters gene expression (Mou *et al.*, 2003). The importance of cysteine modification in this process has been highlighted. If either Cys82 or Cys216 in NPR1 are removed by mutation, the protein attains constitutive activity. Cysteine residues in the TGA transcription factors are also important. The binding of TGA1 and TGA4 to DNA is stimulated by their interaction with NPR1, which occurs following the reduction of cysteine S-S bridges in these bZIP proteins (Fobert and Despres, 2005).

Other proteins that control gene expression have also been suggested as ROS-sensors, for example, heat-shock transcription factors (Miller and Mittler, 2006). These proteins bind to a DNA consensus sequence seen in the promoters of many defence genes and are important for a plant's response to both biotic and abiotic stress.

7 Glyceraldehyde-3-phosphate dehydrogenase as a H_2O_2-sensing protein

As mentioned previously, GAPDH can undergo S-thiolation (Grant *et al.*, 1999) and also is potentially susceptible to S-nitrosylation by NO (Lindermayr *et al.*, 2005). It has long been known that mammalian GAPDH can be inhibited by ROS and that oxidation of a Cys was involved in this process (Brodie and Reed, 1987). Thus, the question arose as to whether or not such a vital metabolic protein could also be involved in ROS signalling? A number of attempts to identify cell proteins susceptible to ROS have adopted a competitive tagging approach. Any protein -SH groups that routinely undergo ROS-mediated oxidation may also be open to competitive covalent modification by applied reactive thiol tags. This approach has been used to identify H_2O_2-susceptible proteins in mammalian cells (Wu *et al.*, 1998). In this case, a fluorescein-labelled iodoacetamide tag was employed and anti-fluorescein antibodies used to estimate the molecular weights of tagged proteins via a western blot approach. In mammalian cells the primary target for H_2O_2 appears to be a tyrosine phosphatase

(Wu *et al.*, 1998), a protein for which the exact modification has subsequently been established (Salmeen *et al.*, 2003; van Montfort *et al.*, 2003).

Using a similar approach, 5′-iodoacetamide fluorescein (IAF) was used to competitively tag a number of proteins in cytosolic extracts from Arabidopsis cells (Hancock *et al.*, 2005). H_2O_2 competed effectively with the tag and reduced the level of protein fluorescein labelling when used as a pretreatment. ESI-Q-TOF mass spectroscopy and MALDI-TOF mass spectroscopy identified a number of the tagged ROS-susceptible proteins and included among these was GAPDH (Hancock *et al.*, 2005). Thus, an important glycolytic enzyme appears to contain -SH groups which are susceptible to multiple forms of modification. However, GAPDH is not a protein that would necessarily be expected to be involved in signalling. The GAPDH protein has been localised in parts of the cell where it would not normally be expected to perform its glycolytic function. For example, it has been reported to be present in Arabidopsis cell walls (Chivasa *et al.*, 2002) and in the peribacteroid membrane of *Lotus japonicus* (Wienkoop and Saalbach, 2003). In other systems it has also been found to have unexpected associations and activities. In mammalian cells, GAPDH has been found to interact with the enzyme phospholipase D, which it appears to activate as a result. This process has been shown to involve the oxidation of a cysteine in the GAPDH (Kim *et al.*, 2003). GAPDH also seems to have some involvement in age-induced apoptosis in neuronal cells (Ishitani *et al.*, 1996). More importantly, it has been demonstrated that it undergoes ROS-mediated translocation to the nucleus (Dastoor and Dreyer, 2001) and that here it mediates a number of NO-initiated nuclear events (Benhar and Stamler, 2005).

In the light of these observations, a H_2O_2 signalling role for GAPDH in plants may not be as unlikely as it would first appear; GADPH may be able to perceive the redox state of either the cytoplasm or other cell locations, such as the cell wall (Chivasa *et al.*, 2002), and initiate a transduction event that ultimately leads to an alteration of the cell metabolism. Certainly the inhibition of GAPDH activity by ROS appears to be reversible, which, mechanistically at least, makes it an ideal candidate to be involved in signalling processes (Hancock *et al.*, 2005, 2006).

8 Histidine kinases and H_2O_2 perception

As discussed above, modulation of gene expression is one of the recognised cellular responses to ROS (Desikan *et al.*, 2001b; Gasch *et al.*, 2000; Vandenabeele *et al.*, 2003; Vanderauwera *et al.*, 2005). Importantly, such studies have also hinted at which proteins are involved in the signal transduction process itself. Close scrutiny of Arabidopsis transcriptomics data (Desikan *et al.*, 2001b) reveals that the expression of genes encoding a number of the proteins involved in ethylene signalling are induced by H_2O_2 and this includes those encoding some of the histidine kinases. Interestingly, the involvement of histidine kinases in oxidative stress responses has also been reported in yeast (Buck *et al.*, 2001; Singh, 2000).

In general, histidine kinases are modular proteins that on activation become autophosphorylated on a histidine residue. The phosphate group is then subsequently transferred to a receptor domain, which usually involves phosphotransfer to an aspartate group. Often such proteins are involved in what are termed two-component signalling mechanisms and, in plants, one of the most well-studied stimulators of such a system is the gaseous hormone ethylene.

Although ethylene is known to stimulate plant histidine kinases, recent work has shown that this is not their only mode of activation and ROS may also signal through these pathways. The Arabidopsis ethylene-receptor histidine kinase, ETR1, has been shown to be essential for H_2O_2 perception in guard cells during stomatal closure (Desikan *et al.*, 2005). The stomata of the *etr1-1* mutant, in which Cys65 is substituted by Tyr, show reduced closure in response to H_2O_2. Thus, as might be expected if ROS do indeed signal via the modification of these groups, the -SH group of Cys65 in ETR1 appears to be important for effective H_2O_2 signalling to occur. The use of other ETR1 mutants indicated that although required for ethylene signalling, the C terminal histidine kinase domain itself was not required for H_2O_2 signalling (Desikan *et al.*, 2005). Thus, the exact mechanism by which ETR1 facilitates H_2O_2 signalling in plants requires further investigation.

The *Saccharomyces cerevisiae* mutant *TM219* lacks a functional SLN1-SSK1 two-component system and shows greater inhibition of growth in response to H_2O_2 than does the wild type (Singh, 2000). However, transformation of this mutant with Arabidopsis ETR1 restores its tolerance to ROS. As mentioned before, only the N-terminal domain of ETR1 appears to be required for ROS perception in plants and, here too, the transformation of *TM219* with just the first 128 amino acids of ETR1 restored its ROS tolerance. However, this restoration of tolerance failed to materialise if the 128-amino-acid ETR1 peptide lacked the critical Cys65 amino acid. Once again, this indicates that -SH group modification is key to the histidine-kinase-mediated ROS signalling that occurs in these yeast cells (Desikan *et al.*, 2005).

9 Are other proteins involved in H_2O_2 sensing?

A plethora of proteins could potentially be modified by ROS. Some, such as ETR1 and the transcription factors discussed above, have come to light and the confirmed list is growing. Included is the enzyme ADP-glucose pyrophosphorylase (AGPase), an important regulatory enzyme in plastid starch synthesis. In potato tubers, AGPase has been found to undergo post-translational redox modification and is also activated by reduced thioredoxin in the chloroplasts of pea leaves (Geigenberger *et al.*, 2005). Iodoacetamide tagging has also revealed several candidate proteins from the cytoplasm of Arabidopsis (Hancock *et al.*, 2005) and some of these, including alcohol dehydrogenase (Echave *et al.*, 2003) and glutamine synthetase (Butterfield *et al.*, 1997; Levine, 1983; Ortega *et al.*, 1999), have already been reported as being modified in a redox-mediated way. However, these studies were far from comprehensive and there is no doubt that many more such modified proteins are yet to be discovered.

10 Conclusion

There is no doubt that the production and presence of ROS such as H_2O_2 are important in cell signalling. Many of their physiological effects stem from the changes in gene expression that they induce and the adaptive responses that result from the subsequent alterations in the protein complement of the cell. The generation and removal of ROS are often instrumental in a variety of stress responses and the exposure of plants to one type of stress often leads to increased tolerance to further periods of stress, regardless of their origins. Thus, for example, a period of heat stress ameliorates

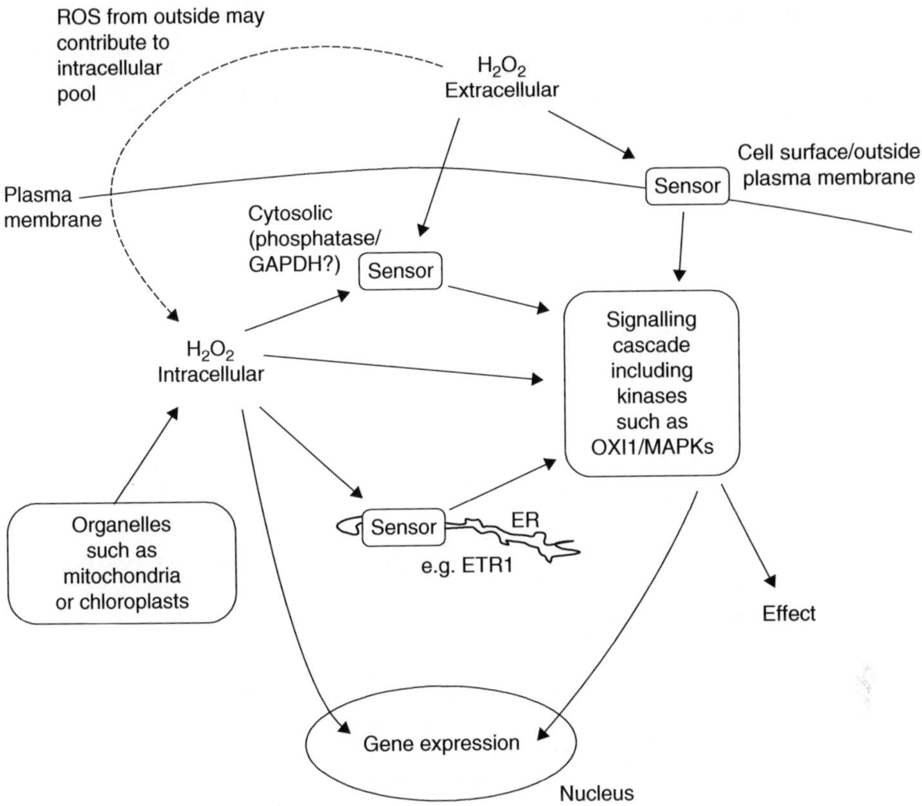

Figure 3. *Perception of hydrogen peroxide. H_2O_2 may be produced externally and arrive at the outside of the cell. Alternatively ROS can be generated intracellularly. Therefore, proteins which are involved in sensing the presence of H_2O_2 may have varied locations in the cell. Such proteins need to transduce their signal via pathways that may contain kinases such as OXI1 and MAPKs.*

the effects of chilling during the post-harvest storage of tomato fruit (Vlachonasios *et al.*, 2001). However, for ROS to act as cell signals the first vital step down the road to signal transduction must be their perception (*Figure 3*). Often this may involve protein oxidation and commonly a -SH group may be involved. Cysteine-associated -SH groups are prevalent in proteins. Thus, alone, their mere presence cannot be the key to this process. The surrounding redox environment and the mid-point potential and accessibility of the -SH group must all play a role in whether or not it is capable of ROS sensing and signalling. Such signalling must also be reversible, and in many cases the re-reduction of the modified -SH group may allow cysteine side groups to either toggle between redox states or to be re-modified by molecules such as NO. Many proteins are now being identified that may be involved in these processes and the challenge will be to determine how their individual functions align with our current understanding of how ROS signal and the effects that they have on plant biology.

Acknowledgements

We would like to thank the Biotechnology and Biological Sciences Research Council (BBSRC) UK, the Leverhulme Trust and The Wellcome Trust for funding this research.

References

Ball, L., Accotto, G.P., Bechtold, U., *et al.* (2004) Evidence for a direct link between glutathione biosynthesis and stress defence gene expression in *Arabidopsis*. *Plant Cell* **16:** 2448–2462.

Benhar, M. and Stamler, J.S. (2005) A central role for S-nitrosylation in apoptosis. *Nat. Cell Biol.* **7:** 645–646.

Biteau, B., Labarre, J. and Toledano, M.B. (2003) ATP-dependent reduction of cysteine-sulphinic acid by *S. cerevisiae* sulphiredoxin. *Nature* **425:** 980–984.

Brodie, A.E. and Reed, D.J. (1987) Reversible oxidation of glyceraldehyde 3-phosphate dehydrogenase thiols in human lung carcinoma cells by hydrogen peroxide. *Biochem. Biophys. Res. Commun.* **148:** 120–125.

Buchanan, B.B. and Balmer Y. (2005) Redox Regulation: a broadening horizon. *Annu. Rev. Plant Biol.* **56:** 187–220.

Buck, V., Quinn, J., Soto Pino, T., Martin, H., Saldanha, J., Makino, K., Morgan, B.A. and Millar, J.B.A. (2001) Peroxide sensors for the fission yeast stress-activated mitogen-activated protein kinase pathway. *Mol. Biol. Cell* **12:** 407–419.

Butterfield, A., Hensley, K., Cole, P., Subramaniam, R., Aksenov, M., Aksenova, M., Bummer, P.M., Haley, B.E. and Carney, J.M. (1997) Oxidatively induced structural alteration of glutamine synthetase assessed by analysis of spin label incorporation kinetics: relevance to Alzheimer's disease. *J. Neurochem.* **68:** 2451–2457.

Charles, S.A. and Halliwell, B. (1980) Effect of hydrogen peroxide on spinach (*Spinacia oleracea*) chloroplast fructose bisphosphatase. *Biochem. J.* **189:** 373–376.

Chivasa, S., Ndimba, B.K., Simon, W.J., Robertson, D., Yu, X.L., Knox, J.P., Bolwell, P. and Slabas, A.R. (2002) Proteomic analysis of the *Arabidopsis thaliana* cell wall. *Electrophoresis* **23:** 1754–1765.

Cho, S.-H., Lee, C.-C., Ahn, Y., Kim, H., Yang, K.-S. and Lee, S.-R. (2004) Redox regulation of PTEN and protein tyrosine phosphatase in H_2O_2-mediated cell signalling, *FEBS Letts.* **560:** 7–13.

Collin, V., Lankemeyer, P., Miginiac-Maslow, M., Hirasawa, M., Knaff, D.B., Dietz, K.J. and Issakidis-Bourguet, E. (2004) Characterization of plastidial thioredoxins belonging to the new y-type. *Plant Physiol.* **136:** 4088–4095.

Cooper, C., Patel, R.P., Brookes, P.S. and Darley-Usmar, V.M. (2002) Nanotransducers in cellular redox signalling: modification of thiols by reactive oxygen and nitrogen species. *Trends Biochem. Sci.* **27:** 489–492.

D'Autréaux, B. and Toledano, M.B. (2007) ROS as signalling molecules: mechanisms that generate specificity in ROS homeostasis. *Nature Mol. Cell Biol.* **8:** 813–824.

Dastoor, Z. and Dreyer, J.-C. (2001) Potential role of nuclear translocation of glyceraldehyde 3-phosphate dehydrogenase in apoptosis and oxidative stress. *J. Cell Sci.* **114:** 1643–1653.

Delaunay, A., Isnard, A.D. and Toledano, M.B. (2000) H_2O_2 sensing through oxidation of the Yap1 transcription factor. *EMBO J.* **19:** 5157–5166.

Delaunay, A., Pflieger, D., Barrault, M.B., Vinh, J. and Toledano, M.B. (2002) A thiol peroxidase is an H_2O_2 receptor and redox-transducer in gene activation. *Cell* **111:** 471–481.

Desikan, R., Reynolds, A., Hancock, J.T. and Neill, S.J. (1998) Harpin and hydrogen peroxide both initiate programmed cell death but have differential effects on defence gene expression in *Arabidopsis* suspension cultures. *Biochem. J.* **330:** 115–120.

Desikan, R., Clarke, A., Hancock, J.T. and Neill, S.J. (1999) H_2O_2 activates a MAP kinase-like enzyme in *Arabidopsis thaliana* suspension cultures. *J. Exp. Bot.* **50:** 1863–1866.

Desikan, R., Hancock, J.T., Ichimura, K., Shinozaki, K. and Neill, S.J. (2001a) Harpin induces activation of the Arabidopsis MAP kinases ATMPK4 and ATMPK6. *Plant Physiol.* **126:** 1579–1587.

Desikan, R., A.-H.-Mackerness, S., Hancock, J.T. and Neill, S.J. (2001b) Regulation of the Arabidopsis transcriptome by oxidative stress. *Plant Physiol.* **127:** 159–172.

Desikan, R., Hancock, J.T., Bright, J., Harrison, J., Weir, I., Hooley, R. and Neill, S.J. (2005) A role for ETR1 in hydrogen peroxide signaling in stomatal guard cells. *Plant Physiol.* **137:** 831–834.

Dietz, K.J., Jacob, S., Oelze, M.L., Laxa, M., Tognetti, V., de Miranda, S.M.N., Baier, M. and Finkemeier, I. (2006) The function of peroxiredoxins in plant organelle redox metabolism. *J. Exp. Bot.* **57:** 1697–1709.

Dixon, D.P., Skipsey, M., Grundy, N.M. and Edwards, R. (2005) Stress-induced protein S-glutathionylation in Arabidopsis. *Plant Physiol.* **138:** 2233–2244.

Dröge, W. (2002) Free radicals in the physiological control of cell function. *Physiol. Rev.* **82:** 47–95.

Echave, P., Tamarit, J., Cabiscol, E. and Ros, J. (2003) Novel antioxidant role of alcohol dehydrogenase E from *Escherichia coli*. *J. Biol. Chem.* **278:** 30193–30198.

Fobert, P.R. and Despres, C. (2005) Redox control of systemic acquired resistance. *Curr. Opin. Plant. Biol.* **8:** 378–382.

Foreman, J., Denidchik, V., Bothwell, J.H.F., *et al.* (2003) Reactive oxygen species produced by NADPH oxidase regulate plant cell growth. *Nature* **422:** 442–446.

Foyer, C.H. and Noctor, G. (2005a) Oxidant and antioxidant signalling in plants: a re-evaluation of the concept of oxidative stress in a physiological context. *Plant Cell Environ.* **28:** 1056–1071.

Foyer, C.H. and Noctor, G. (2005b) Redox homeostasis and antioxidant signaling: a metabolic interface between stress perception and physiological responses. *Plant Cell* **17:** 1866–1875.

Gasch, A.P., Spellman, P.T., Kao, C.M., Carmel-Harel, O., Eisen, M.B., Storz, G., Botstein D. and Brown, P.O. (2000) Genomic expression programs in the response of yeast cells to environmental changes. *Mol. Biol. Cell* **11:** 4241–4257.

Geigenberger, P., Kolbe, A. and Tiessen, A. (2005) Redox regulation of carbon storage and partitioning in response to light and sugars. *J. Exp. Bot.* **56:** 1469–1479.

Grant, C.M., Quinn, K.A. and Dawes, I.W. (1999) Differential protein S-thiolation of glyceraldehyde-3-phosphate dehydrogenase isoenzymes influences sensitivity to oxidative stress. *Mol. Cell Biol.* **19:** 2650–2656.

Gupta, R. and Luan, S. (2003) Redox control of protein tyrosine phosphatases and mitogen activated protein kinases in plants. *Plant Physiol.* **132:** 1149–1152.

Hancock, J.T., Henson, D., Nyirenda, M., Desikan, R., Harrison, J., Lewis, L., Hughes, J. and Neill, S.J. (2005) Proteomic identification of glyceraldehyde 3-phosphate dehydrogenase as an inhibitory target of hydrogen peroxide in *Arabidopsis*. *Plant Physiol. Biochem.* **43:** 828–835.

Hancock, J., Desikan, R., Harrison, J., Bright, J., Hooley, R. and Neill, S. (2006) Doing the unexpected: proteins involved in hydrogen peroxide perception. *J. Exp. Bot.* **57:** 1711–1718.

Ishitani, R., Sunaga, K., Hirano, A., Saunders, P., Katsube, N. and Chuang, D.M. (1996) Evidence that glyceraldehyde-3-phosphate dehydrogenase is involved in age-induced apoptosis in mature cerebellar neurons in culture. *J. Neurochem.* **66:** 928–935.

Kaiser, W.M. (1979) Reversible inhibition of the Calvin cycle and activation of oxidative pentose phosphate cycle in isolated chloroplasts by hydrogen peroxide. *Planta* **145:** 377–382.

Kim, J.H., Lee, S., Park, J.B., *et al.* (2003) Hydrogen peroxide induces association between glyceraldehyde 3-phosphate dehydrogenase and phospholipase D2 to facilitate phospholipase D2 activation in PC12 cells. *J. Neurochem.* **85:** 1228–1236.

Konig, J., Baier, M., Horling, F., Kahmann, U., Harris, G., Schurmann, P. and Dietz, K.J. (2002) The plant-specific function of 2-Cys peroxiredoxin-mediated detoxification of peroxides in the redox-hierarchy of photosynthetic electron flux. *Proc. Natl Acad. Sci. USA* **99:** 5738–5743.

Konig, J., Lotte, K., Plessow, R., Brockhinke, A., Baier, M. and Dietz, K.J. (2003) Reaction mechanism of plant 2-Cys peroxiredoxin. Role of the C terminus and the quaternary structure. *J. Biol. Chem.* **278:** 24409–24420.

Kovtun, Y., Chiu, W.L., Tena, G. and Sheen, J. (2000) Functional analysis of oxidative stress-activated mitogen-activated protein kinase cascade in plants. *Proc. Natl Acad. Sci. USA* **97:** 2940–2945.

Lee, C., Lee, S.M., Mukhopadhyay, P., Kim, S.J., Lee, S.C., Ahn, W.S., Yu, M.H., Stroz, G. and Ryu, S.E. (2004) Redox regulation of OxyR requires specific disulfide bond formation involving a rapid kinetic reaction path. *Nat. Struct. Mol. Biol.* **11:** 1179–1185.

Lemaire, S.D. (2004) The glutaredoxin family in oxygenic photosynthetic organisms. *Photosynth. Res.* **79:** 305–318.

Levine, R.L. (1983) Oxidative modification of glutamine synthetase. *J. Biol. Chem.* **258:** 11828–11833.

Lindermayr, C., Saalbach, G. and Durner, J. (2005) Proteomic identification of S-nitrosylated proteins in *Arabidopsis*. *Plant Physiol.* **137:** 921–930.

Meinhard, M. and Grill, E. (2001) Hydrogen peroxide is a regulator of ABI1, a protein phosphatase 2C from *Arabidopsis*. *FEBS Letts.* **508:** 443–444.

Miao, Y., Lv, D., Wang, P., Wang, X.C., Chen, J., Miao, C. and Song, C.P. (2006) An Arabidopsis glutathione peroxidase functions as both a redox transducer and a scavenger in abscisic acid and drought stress responses. *Plant Cell* **18:** 2749–2766.

Miller, G. and Mittler, R. (2006) Could heat shock transcription factors function as hydrogen peroxide sensors in plants? *Ann. Bot.* **98:** 279–288.

Mou, Z., Fan, W. and Dong, X. (2003) Inducers of plant systemic acquired resistance regulate NPR1 function through redox changes. *Cell* **113:** 935–944.

Nakagami, H., Kiegerl, S. and Hirt, H. (2004) OMTK1, a novel MAPKKK, channels oxidative stress signalling through direct MAPK interaction. *J. Biol. Chem.* **279:** 26959–26966.

Neill, S.J., Desikan, R. and Hancock, J.T. (2002) Hydrogen peroxide signalling. *Curr. Opin. Plant Biol.* **5:** 388–395.

Neill, S.J., Desikan, R. and Hancock, J.T. (2003) Nitric oxide signalling in plants. *New Phytol.* **159:** 11–35.

Ortega, J.L., Roche, D. and Sengupta-Gopalan, C. (1999) Oxidative turnover of soybean root glutamine synthetase. *In vitro* and *in vivo* studies. *Plant Physiol.* **119:** 1483–1495.

Pei, Z.M., Murata, Y., Benning, G., Thomine, S., Klusener, B., Allen, G.J., Grill, E. and Schroeder, J.I. (2000) Calcium channels activated by hydrogen peroxide mediate abscisic acid signalling in guard cells. *Nature* **406:** 731–734.

Rentel, M. C., Lecourieux, D., Quaked, F., *et al.* (2004) *OXI1* kinase is necessary for oxidative burst-mediated signalling in *Arabidopsis*. *Nature* **427:** 858–861.

Romero, H.M., Berlett, B.S., Jenson, P.J., Pell, E.V. and Tien, M. (2004) Investigations into the role of plastidial peptide methionine sulfoxide reductase in response to oxidative stress in *Arabidopsis*. *Plant Physiol.* **136:** 3784–3994.

Rouhier, N., Gelhaye, E., Sautiere, P.E., *et al.* (2001) Isolation and characterization of a new peroxiredoxin from poplar sieve tubes that uses either glutaredoxin or thioredoxin as a proton donor. *Plant Physiol.* **127:** 1299–1309.

Rouhier, N., Gelhaye, E. and Jacquot, J.P. (2004) Plant glutaredoxins: still mysterious reducing systems. *Cell Mol. Life Sci.* **61:** 1266–1277.

Sadanandom, A., Poghosyan, Z., Fairbairn, D.J. and Murphy, D.J. (2000) Differential regulation of plastidial and cytosolic isoforms of peptide methionine sulfoxide reductase in Arabidopsis. *Plant Physiol.* 123: 255–264.

Salmeen, A., Anderson, J.N., Myers, M.P., Meng, T.C., Hinks, J.A., Tonks, N.K. and Barford, D. (2003) Redox regulation of protein tyrosine phosphatase 1B involves a novel sulfenyl-amide intermediate. *Nature* **423:** 769–773.

Schreck, R., Rieber, P. and Baeuerle, P.A. (1991) Reactive oxygen intermediates as apparently widely used messengers in the activation of the NF-κB transcription factor and HIV-1. *EMBO J.* **10:** 2247–2258.

Schürmann, P. and Jacquot, J.P. (2000) Plant thioredoxin systems revisited. *Ann. Rev. Plant Physiol. Mol. Biol.* **51:** 371–400.

Segal, A.W. (2005) How neutrophils kill microbes. *Ann. Rev. Immunol.* **23:** 197–223.

Singh, K.K. (2000) The *Saccharomyces cerevisiae* SLN1P-SSK1P two-component system mediates response to oxidative stress and in an oxidant-specific fashion. *Free Radic. Biol. Med.* **29:** 1043–1050.

van Montfort, R.L., Congreve, M., Tisi, D., Carr, R. and Jhoti, H. (2003) Oxidation state of the active-site cysteine in protein tyrosine phosphatase 1B. *Nature* **423:** 773–777.

Vandenabeele, S., Der Kelen, K.V. and Dat, J. (2003) A comprehensive analysis of hydrogen peroxide-induced gene expression in tobacco. *Proc. Natl Acad. Sci. USA* **100:** 16113–16118.

Vanderauwera, S., Zimmermann, P., Rombauts, S., Vandenbeele, S., Langebartels, C., Gruissem, W., Inzé, D. and Van Breusegem, F. (2005) Genome-wide analysis of hydrogen peroxide-regulated gene expression in Arabidopsis reveals a high light-induced transcriptional cluster involved in anthocyanin biosynthesis. *Plant Physiol.* **139:** 806–821.

Vanin, A.F., Svistunenko, D.A., Mikoyan, V.D., Serezhenkov, V.A., Fryer, M.J., Baker, N.R. and Cooper, C.E. (2004) Endogenous superoxide production and the nitrite/nitrate ratio control the concentration of bioavailable free nitric oxide in leaves. *J. Biol. Chem.* **279:** 24100–24107.

Vieira Dos Santos, C. and Rey, P. (2006) Plant thioredoxins are key actors in the oxidative stress response. *Trends Plant Sci.* **11:** 329–334.

Vlachonasios, K.E., Kadyrzhanova, D.K. and Dilley, D.R. (2001) Heat treatment prevents chilling injury of tomato (*Lycopersicon escultentum*) fruits: heat shock genes and heat shock proteins in the resistance of tomato fruit to low temperatures. *Acta Hort. (ISHS)* **553:** 543–548.

Vranová, E., Inzé, D. and Van Breusegem, F. (2002) Signal transduction during oxidative stress. *J. Exp. Bot.* **53:** 1227–1236.

Wienkoop, S. and Saalbach, G. (2003) Proteome analysis. Novel protein identified at the peribacteroid membrane from *Lotus japonicus* root nodules. *Plant Physiol.* **131:** 1080–1090.

Wu, Y., Kwon, K.S. and Rhee, S.G. (1998) Probing cellular targets of H_2O_2 with fluorescein-conjugated iodoacetamide and antibodies to fluorescein. *FEBS Letts.* **440:** 111–115.

Yano, H., Wong, J.H., Lee, Y.M., Cho, M.J. and Buchanan, B.B. (2001) A strategy for the identification of proteins targeted by thioredoxin. *Proc. Natl Acad. Sci. USA* **98:** 4794–4799.

5

Mechanisms and genes controlling programmed cell death and Darwinian fitness in plants

Stanisław Karpiński, Christine Chang, Per Mühlenbock and Barbara Karpinska

1 Introduction

The sessile and photosynthetic nature of plants has necessitated the evolution of different stress response mechanisms in comparison to those observed in animals. Since they cannot move and so avoid stressful conditions, plants must be able to adjust metabolic processes to constantly fluctuating and changing environmental factors. In full sunlight and optimal growth conditions, only a small fraction of the total absorbed light energy is used for CO_2 fixation (Asada, 1999; Foyer and Noctor 1999, 2000, 2003; Karpiński *et al.*, 1999, 2003; Mullineaux and Karpiński 2002). The amount of absorbed light energy that is in excess of the amount needed to drive photosynthetic metabolism is termed the 'excess excitation energy' (EEE; Fryer *et al.*, 2003; Karpiński *et al.*, 1997, 1999; Niyogi, 1999, 2000). In full sunlight, EEE represents an ever-present danger for plant cell function and stability. Thus, plants have a complex system of light acclimation processes that fulfil several major compensatory functions. Firstly, acclimatory processes are required to optimize the energy status of the cell (NADPH/ATP ratio) in relation to metabolic requirements. Secondly, acclimation optimizes the use of absorbed light for photosynthesis and minimizes the probability of reactive oxygen species (ROS) formation. Thirdly, acclimatory responses monitor changes in the light quality (spectrum) and quantity (energy) as a source of information concerning diurnal or seasonal shifts (circadian or seasonal checkpoints). The extent of EEE usually varies with light intensity, increasing as light levels increase, but it can also occur at relatively low light intensities when plants are exposed to different stresses which affect photosynthetic CO_2 assimilation, such as drought, chilling or freezing stress and avirulent pathogen infection (Mullineaux and Karpiński 2002). Environmental factors are pivotal to the Darwinian fitness of plants and other photosynthetic organisms, which cannot escape stressful conditions (Kulheim *et al.*, 2002).

The plant acclimation processes necessitate accurate sensing and simultaneous processing of the incoming environmental signals, such as changes in light intensity and quality, fluctuations in temperature and relative humidity, photoperiod length, water availability, nutritional status, wounding, pathogen attack, herbivores, and so on. Information concerning changes in multiple environmental parameters must be simultaneously conveyed throughout the plant cell in order to generate the 'only correct' and optimal acclimatory and defence responses, which will ultimately allow for successful propagation (Darwinian fitness).

The reaction of a plant cell to a stimulus is assumed to be composed of at least five elementary processes: signal perception, signal transduction, signal amplification, and primary and secondary responses. Signal perception includes physical and chemical interactions between the stimulus and a receptor. Subsequently, signal amplification and transduction of changes in the receptor generates specific effectors and downstream signalling cascades, which are amplified. The ultimate responses are specific changes in primary-response gene expression, enzyme activities and subsequent changes in secondary-response gene expression and in cellular metabolism. These interactions may be quite complex and are likely to involve plant hormones such as auxin (IAA), jasmonic acid (JA), ethylene, abscisic acid (ABA) and ROS (Chang *et al.*, 2004; Fryer *et al.*, 2003; Mateo *et al.*, 2004, 2006; Mühlenbock *et al.*, 2007a, 2007b).

1.1 *Light perception in plants*

The plant cell is equipped with sophisticated light-sensing mechanisms that are localized in the plasma membrane, the cytosol, within chloroplasts and in the nucleus (e.g. Karpiński *et al.*, 2003). There are three major types of photoreceptors in higher plants: phytochromes (PHY), cryptochromes (CRY), and phototropins (PHOT; Briggs *et al.*, 2001; Clack *et al.*, 1994; Karpiński *et al.*, 2003; Lin, 2002; Schäfer and Bowler, 2002). Only PHY are active in the long wavelength region of the light spectrum at greater than 660 nm; they absorb predominantly red light and far-red light, whereas CRY and PHOT monitor the blue-light spectra and the ultraviolet-A region respectively (Ahmad *et al.*, 1998). Presumably CRY function by mediating a light-dependent redox reaction (Yang *et al.*, 2001). Recently, it was also demonstrated that CRY regulate part of the EEE responses (Kleine *et al.*, 2007). PHY mediate the light control of all major phases of a plant development, from seed germination, through seedling establishment (de-etiolation, leaf expansion and chloroplast maturation) and determination of the architecture of the mature plant, to the induction of flowering and dormancy (Schäfer and Bowler, 2002). PHOT control three distinct types of light-induced movement responses that are thought to optimize photosynthesis: phototropism, chloroplast relocation and stomatal opening. In *Arabidopsis thaliana*, there are two CRY genes (Lin, 2002), two PHOT genes (Lin, 2002) and five PHY photoreceptor genes (Schäfer and Bowler, 2002). The photosystems (PSI and PSII) in the chloroplasts are also important sensors of light quantity and quality, through redox changes in specific photosynthetic electron-transport carriers (Havaux and Niyogi, 1999; Karpinska *et al.*, 2000; Karpiński *et al.*, 1999, 2003; Mateo *et al.*, 2004, 2006; Muller *et al.*, 2001; Mullineaux *et al.*, 2000; Niyogi 1999, 2000; Pfannschmidt *et al.*, 1999).

1.2 *Photo-oxidative stress and programmed cell death*

Plants almost always have the capacity to absorb more light energy than they need to drive photosynthesis (Asada 1999; Karpiński *et al.*, 1999, 2003; Mullineaux and Karpiński, 2002; Noctor and Foyer, 1998). Many different light harvesting and electron-transport components, from the water splitting complex to ferredoxin, have the capacity to generate ROS including singlet oxygen ($^1O_2$), superoxide anions (O_2^-), hydrogen peroxide (H_2O_2) and the hydroxyl radical (OH). Accumulation of ROS leads to photo-inhibition, photo-oxidative damage and can eventually trigger programmed cell death (PCD; Allan and Fluhr 1997; Karpiński *et al.*, 1999; Karpinska *et al.*, 2000, 2001; Kasahara *et al.*, 2002; Kliebenstein et al., 1998; Mateo *et al.*, 2004; Mittler *et al.*, 2004; Mühlenbock *et al.*, 2007a, 2007b; 2008). Superoxide dismutases play an important role in this process, since they regulate intracellular and extracellular homeostasis between O_2^- and H_2O_2 (e.g. Karpinska et al., 2001; Kliebenstein et al., 1998).

Dissipation of EEE is achieved by a combination of non-photochemical and photochemical quenching processes. Non-photochemical quenching (NPQ) processes transfer excitation energy from antenna chlorophyll to carotenoids and zeaxanthin instead of the reaction centres in the photosystems in a process that converts EEE into heat. This process is regulated by conformational changes in the PSII antenna and by the reversible conversion of violaxanthin to zeaxanthin in the xanthophyll cycle. Both processes are dependent on the pH gradient across the thylakoid membrane. Additionally, carotenoids can quench $^1O_2$ and/or excited chlorophyll. Excited chlorophyll in the triplet state can interact with the triplet ground state of $^3O_2$ and induces a spin conversion to produce $^1O_2$ (Asada 1999; Havaux and Niyogi, 1999; Muller *et al.*, 2001; Niyogi, 1999, 2000).

Photochemical quenching of EEE is the collective term for processes that increase the consumption of photosynthetic electrons by the deployment of additional metabolic sinks, like the ascorbate-glutathione cycle, water-water cycle (Asada, 1999; Foyer and Halliwell 1976; Foyer and Noctor 2000, 2003; Mullineaux *et al.*, 2000) or photorespiration (Kozaki and Takeba, 1996; Mateo *et al.*, 2004; Willekens *et al.*, 1997; Wingler *et al.*, 2000). These include the reduction of O_2 in PSII or PSI with subsequent return of the electrons to water by antioxidant enzymes (water-water cycle) and an increased photosynthetic and photorespiratory metabolism. Photorespiration starts from the oxygenase reaction of RuBisCO and results in the energy consuming release of CO_2 from fixed carbon. Both the water-water cycle and photorespiration have an added value because they recycle electron acceptors such as $NADP^+$ during the dissipation process.

1.3 *Light acclimatory responses*

Light acclimation processes serve to optimize light harvesting and CO_2 fixation for growth and development. Plants dissipate EEE or minimize its formation by diminishing the capacity of the leaf to capture light energy (Bailey *et al.*, 2001; Niyogi, 2000). Among the acclimatory pathways, a short-term chromatic adaptation known as the state-1/state-2 transition is required when the flux of electron excitation in PSII and PSI is unequal (Allen, 1995, 2003; Haldrup *et al.*, 2001). In oxygenic photosynthesis, the two separate reaction centres in PSI and PSII, which have distinct light absorption properties, are coupled in sequence by a chain of electron carriers. PSI is enriched in chlorophyll *a*, absorbing in the far-red region (absorption maximum $\lambda \approx 700$ nm) whereas PSII is enriched in chlorophyll *b*, with an absorption maximum at shorter wavelengths in the orange/red region ($\lambda \approx 650$–680 nm). Thus, spectral imbalances of the exciting light may result in an uneven excitation of the two photosystems leading to lower or higher ROS production (Fryer *et al.*, 2003; Karpiński *et al.*, 2003; Mateo *et al.*, 2004). During state transitions, a part of the PSII antenna is moved towards PSI or back to PSII to level out imbalanced excitation of the two photosystems.

Short-term processes are subsequently complemented by long-term light acclimation processes if unbalanced excitation pressure persists (Karpiński *et al.*, 2003). Common long-term EEE acclimation responses include optimization of the capacity for photosynthetic electron transport and CO_2 fixation pathways, as well as decreasing the size of the light-harvesting antennae that are associated with the photosystems (Klenell *et al.*, 2005; Mateo *et al.*, 2004; Noctor and Foyer, 1998; Pfannschmidt *et al.*, 1999). The long-term acclimation responses involve the regulation of nuclear and chloroplastic gene expression, controlled by specific redox potential changes and by proteolytic degradation of existing light-harvesting complexes and other proteins (Allen, 1993; Bailey *et al.*, 2001; Durnford and Falkowski, 1997; Pfannschmidt *et al.*, 2001a, 2001b, 2003). Although potentially damaging, ROS are also needed to trigger other protective responses, such as the down-regulation of PSII activity. ROS are cellular and systemic signals that can influence other signalling systems and so induce protective mechanisms in remote chloroplasts, including redox changes in the proximity of PSII (Karpiński *et al.*, 1999; Mullineaux *et al.*, 2000).

1.4 *Signalling in response to EEE*

Recent work from our lab and that of others shows that redox changes in the glutathione (GSH) pool and in components of the salicylic acid (SA) signalling pathway are recruited to control gene expression in response to EEE and light acclimatory processes (Alvarez *et al.*, 1998; Ball *et al.*, 2004; Hammond-Kosack and Parker 2003; Jabs *et al.*, 1996; Karpinska *et al.*, 2000; Karpiński *et al.*, 1997, 1999, 2003; Mateo *et al.*, 2006). Responses to increases in light intensity, and hence increasing EEE, not only involve changes in photosynthesis but also leaf water status and temperature, resulting in increases in heat-shock transcription factors (HSFs), ABA and changes in glutathione redox state that can activate multiple signalling pathways (Fryer *et al.*, 2003; Mateo *et al.*, 2006; Mou *et al.*, 2003). Runaway- (R-) PCD in the *lesion simulating disease 1* (*lsd1*) mutant was linked to the activity of PSII (R-PCD was induced by light-2 (680 nm) but not by light-1 (700 nm)) and to reduced stomatal conductance, and ultimately to a photorespiratory H_2O_2 burst and redox changes in the plastoquinone (PQ) pool (Mateo *et al.*, 2004, 2006). Insights into the ROS-dependent regulation of cell death have come from analysis of the *lsd1* mutant, which was initially characterized for its O_2^--dependent spreading of chlorotic/necrotic lesions (a phenotype termed R-PCD) that develop under long or continuous photoperiods or after infection with an avirulent pathogen (Dietrich *et al.*, 1994; Jabs *et al.*, 1996). R-PCD is an extreme form of PCD observed in *lsd1* null mutant plants. Once R-PCD is induced in a single leaf, it spreads to other leaves. The *lsd1* null mutant was crossed with the *chaos* (*cao*) mutant that has a reduced PSII antenna (Klenell *et al.*, 2005). The *lsd1/cao* double mutants showed reduced lesion formation indicating the importance of photosynthesis to the R-PCD phenotype (Mateo *et al.*, 2004). The *lsd1* mutant had reduced stomatal conductance and catalase activity in short-day permissive conditions and increased H_2O_2 accumulation followed by R-PCD when stomatal gas exchange was impeded. Importantly, these traits depend on the defence regulators *Enhanced Disease Susceptibility 1* (*EDS1*), *Phytoalexin Deficient 4* (*PAD4*) and *Ethylene Insensitive 2* (*EIN2*; Mateo *et al.*, 2004; Muhlenbock *et al.*, 2007a, 2007b, 2008). Mutations in *PAD4* and *EDS1* block the *lsd1*-conditioned R-PCD triggered by long photoperiods, pathogen inoculation, ROS provision, or treatment with the phenolic signalling molecule, SA (Jabs *et al.*, 1996; Rusterucci *et al.*, 2001). Furthermore, non-photorespiratory conditions retarded the propagation of lesions in *lsd1*. The results point to a multiple role of *LSD1* in reducing cellular ROS content: (i) by controlling *PAD4*- and *EDS1*-dependent stomatal closure and consequent foliar (photoresiratory) H_2O_2 production during EEE, (ii) by regulating H_2O_2 scavenging capacity and (iii) by controlling cellular ethylene/ROS homeostasis in a *PAD4*- and *EDS1*-dependent manner so that ethylene production is proportional to stomatal conductance (Mateo *et al.*, 2004; Mühlenbock *et al.*, 2007a, 2007b, 2008). Therefore, R-PCD in *lsd1* results from a limitation in CO_2 availability and concomitant activation of the photorespiratory pathway together with over-reduction of the PQ pool and, in consequence, higher ethylene and ROS accumulation (Mühlenbock *et al.*, 2007a, 2007b, 2008). We hypothesize that *LSD1*, *PAD4* and *EDS1* are important integrative regulators of responses to pathogens, EEE, wounding and hypoxia. Our recent work shows that *LSD1* is indeed a general controller of PCD in response to variety of different environmental stresses. Taken together, these data emphasize the importance of examining multiple environmental conditions and plant responses when investigating the function of genes.

1.5 *Induction of plant immunity to infection by biotrophic pathogens*

It is widely accepted that resistance to a given biotrophic pathogen can be induced by prior inoculation of the plant host with avirulent races of the same or a different biotrophic pathogen. Such races are avirulent because they produce specific molecules that activate one or more resistance (*R*) gene-driven signalling pathways leading to the successful induction of plant defences at the site of infection. These defences are commonly associated with localized PCD, called the hypersensitive response (HR), that also leads to the establishment of whole plant immunity (often called systemic acquired resistance, SAR; e.g. Asai *et al.*, 2002; Jabs *et al.*, 1996; Lamb and Dixon 1997; Levine *et al.*, 1994; Petersen *et al.*, 2000) to subsequent infections by virulent forms of the same or different biotrophic pathogens. After initial recognition of the avirulent pathogen, ROS and nitric oxide (NO) are rapidly produced (Allan and Fluhr 1997; Delledonne *et al.*, 1998). There are several sources of ROS in plant cells during the elicitation of *R* gene-mediated resistance, the best studied being from the plasma membrane, although cytoplasmic sources are also likely to be important (Allan and Fluhr 1997). In the plasma membrane, NADPH oxidases and possibly cell-wall-bound peroxidases and amine oxidases, catalyse the production of ROS (Dangl and Jones, 2001; Delledonne *et al.*, 1998; Jabs *et al.*, 1996; Keller *et al.*, 1998; Lamb and Dixon, 1997; Levine *et al.*, 1994). The burst of ROS and NO then leads to activation of a complex network of signal transduction pathways involving small molecules, such as SA, JA and ethylene. These signalling events ultimately lead, via cascades of protein phosphorylation and activated transcription factors, to changes in the expression of many defence-associated genes, some of which are also expressed in response to abiotic stresses.

The SA pathway that triggers systemic immunity is negatively controlled by MAP kinase 4 (Brodersen *et al.*, 2002, 2006; Petersen *et al.*, 2000; Rusterucci *et al.*, 2001). The SA pathway also regulates light acclimatory processes (Karpiński *et al.*, 2003; Mateo *et al.*, 2004, 2006; Mühlenbock *et al.*, 2007a, 2007b, 2008) and is genetically and functionally interconnected to redox changes in the glutathione pool and to hydrogen peroxide levels (Karpinska *et al.*, 2000; Karpiński *et al.*, 2003; Mateo *et al.*, 2006). EDS1 and PAD4 have been implicated in the amplification of ethylene and SA signals by processing ROS that are essential for the regulation of expression of cellular immunity against biotrophic pathogens (Brodersen *et al.*, 2006; Rusterucci *et al.*, 2001). This signalling pathway involves dynamic interactions between cell compartments in an effective stress-signal relay system in which EDS1 signalling complexes fulfil nucleo-cytoplasmic roles (Feys *et al.*, 2005; Mateo *et al.*, 2004). Two new components of Arabidopsis immunity have been discovered using gene expression microarrays combined with reverse genetics (Brodersen *et al.*, 2006). A flavin-dependent monooxygenase (FMO1) positively regulates the EDS1 pathway and one member (NUDT7) of a family of cytosolic Nudix hydrolases exerts negative control on EDS1 signalling. A common theme underlying the functions of counterparts of these proteins in animals and yeast is in redox-stress responses.

LSD1, *PAD4*, *EDS1* and *MPK4* regulate many aspects of plant defence and acclimation, controlling SA, JA/ethylene and other, as yet unidentified, signal pathways leading to PCD (Wiermer *et al.*, 2005), as well as systemic immunity and systemic acquired acclimation to EEE (Mateo *et al.*, 2004, 2006). It was recently discovered that lysigenous aerenchyma formation in response to root hypoxia also involves this pathway (Mühlenbock *et al.*, 2007a, 2007b, 2008). All of these features point to an intrinsic activity of *LSD1*, *EDS1* and *PAD4* in processing and transduction of redox-derived signals from different subcellular compartments and different plant organs and tissues during a variety of environmental stresses.

1.6 *Common underlying principles of whole plant acclimation to EEE and immunity to infection*

Several common features characterize the response of plants to both EEE and infection by incompatible biotrophic pathogens. These responses are characterized by a rapid increase in the foliar concentrations of ROS, the depletion of antioxidant pools, chlorosis and/or necrosis, local and systemic defence responses such as systemic acquired resistance (SAR) and induction of defence gene expression. However, in cells infected by an avirulent pathogen, the subcellular origin of the ROS burst is thought to be NADPH oxidases at the plasma membrane, which is different from EEE, which stimulates ROS production mainly in the chloroplasts and peroxisomes. It has been proposed that ROS produced during the oxidative burst act as the central regulator of the HR (Levine *et al.*, 1994). The manipulation of endogenous ROS alters the outcome of cell death signals and the induction of cell death requires the controlled attenuation of ROS detoxification (Lamb and Dixon, 1997). In situations where ROS are actively involved in the induction of cell death, there are constitutive and induced ROS detoxification systems that serve to prevent cell death. *LSD1* is considered to be a negative regulator of cell death by acting as a ROS rheostat. Short-day photoperiods and low photosynthetically active photo flux densities are permissive conditions for the growth of lsd1. We have demonstrated that *LSD1*, *EDS1*, *PAD4* and other genes considered to be involved in the regulation of immune defences are also involved in the regulation of light acclimatory responses (Karpiński *et al.*, 2003; Mateo *et al.*, 2004, 2006; Mühlenbock *et al.*, 2007a, 2007b, 2008; Mullineaux *et al.*, 2000). These observations strongly imply that the initiation of R-PCD in lsd1 is linked to the prevailing light conditions. Recently, Bart Rossel and colleagues (2007) proposed that the process of systemic acquired acclimation (SAA) to increasing light is different from the SAR to pathogens. However these authors based their conclusions solely on a comparison of mRNA profiles from wild-type and mutant plants that had de-regulated light acclimatory responses, and they did not compare the physiology of the SAR and SAA responses in their genetic system. Hence, their conclusions are not consistent with the physiological evidence (Bart Rossel *et al.*, 2007). The above discussion shows that, overall, the results from a number of studies suggest the presence of common pathways in the SAR and SAA responses.

1.7 *Plant responses to anoxia*

Every year flooding contributes to major losses in crop harvests around the globe (Dennis *et al.*, 2000; Setter and Waters, 2003). Flooding leads to the depletion of soil oxygen and hence a number of strategies for the improvement of flooding and/or water-logging tolerance in land crops have been directed towards the amelioration of tolerance to hypoxia (<4% O_2) and anoxia (close to 0% O_2). The pores in aerated soils contain the same level of oxygen as the atmosphere (21% O_2). During flooding, however, the air in these pores is replaced by water. Water has much slower diffusion rates of oxygen than air and hence the aerobic activities of plant roots and microorganisms in the soil quickly fall in response to the lower oxygen levels (Pezeshki, 1994).

Plant roots respond to hypoxia by inducing morphological and physiological changes and by inducing specific sets of genes that enhance tolerance to decreasing oxygen levels. Plants adapt to low oxygen levels in three stages. Initially, the plant rapidly induces a set of signal transduction components (Dennis *et al.*, 2000). This is followed by a slower period of metabolic adaptation involving activation of fermentation pathways. Finally, species-dependent morphological changes, such as aerenchyma and adventitious root formation, are initiated (Dennis *et al.*, 2000). Aerenchyma tissues form gas-conducting tubes that provide roots with oxygen under hypoxic conditions. Physiological responses to hypoxia include wilting and stomatal closure (Pezeshki *et al.*, 1996). These responses to hypoxia involve signalling pathways controlled by ethylene, ROS and ABA in *Arabidopsis thaliana* (L.) Heyhn. (Baxter-Burrell *et al.*, 2002; de Bruxelles *et al.*, 1996; Klok *et al.*, 2002; Peng *et al.*, 2001) and other plant species such as *Zea mays* L. (Fukao and Bailey-Serres, 2004; He *et al.*, 1996) and *Iris pseudacorus* L. (Monk *et al.*, 1987). They also involve Ca^{2+} signalling (Subbaiah and Sachs, 2003) and increased levels of haemoglobins (Dordas *et al.*, 2003). The levels of haemoglobins may be affected by auxins (Watts *et al.*, 2001) and may directly affect ethylene signalling (Manac'h-Little *et al.*, 2005).

Hypoxia-induced aerenchyma occurs in some non-wetland plants to improve the aeration of the rhizosphere (Jackson and Armstrong, 1999). Plants form aerenchyma using either schizogenous or lysigenous processes, or by a combination of the two pathways (Drew *et al.*, 2000). Schizogenous aerenchyma formation involves cell-wall reorganization and cell separation whereas lysigenous aerenchyma is formed as a consequence of PCD and cell-wall autolysis (Campbell and Drew, 1983; Evans, 2003; Gunawardena *et al.*, 2001). In *Zea mays*, ethylene has been associated with lysigenous aerenchyma formation (He *et al.*, 1996; Jackson *et al.*, 1985).

Aerenchyma is not always found in Arabidopsis, where mainly young seedlings have been studied in relation to hypoxia (Colmer *et al.*, 2004; Evans, 2003; Fukao and Bailey-Serres, 2004). Waterlogging gives rise to stratified oxygenation in soils, with top layers often maintaining higher oxygen levels (Yu and Patrick, 2003). The formation of aerenchyma in very small Arabidopsis seedlings may therefore have little advantage under natural conditions. However, in mature Arabidopsis plants, which have deeper root systems and larger rhizospheres, the ability to form aerenchyma will be of greater value.

1.8 *Wounding responses*

Plants are able to limit the damage caused by chewing insects or larger herbivores by the presence of pre-existing physical barriers. Ubiquitous mechanical stimuli such as rain or wind are known to result in morphological changes in plants. A localized injury activates defence mechanisms throughout the plant both in the directly damaged tissues (local response) and in the non-wounded areas (systemic response). In response to mechanical injury, a series of wounding-responsive genes are activated by components such as oligosaccharides, the peptide systemin, JA, ABA, ethylene and electrical signals, which together form a complex signalling network (Guan and Scandalios, 2000; Pei *et al.*, 2000). Wounding-responsive genes can be separated into three major groups according to their expression patterns. These are the JA-dependent group of genes, such as vegetative storage protein 1 (*VSP1*), and the JA-independent group of genes, including basic chitinase (*CHIB*). The last category involves both JA-dependent and independent signalling molecules, such as acyl coA oxidase (ACO). Although the wound signal-transduction pathways appear to be conserved in higher plants, there are some discrepancies in the data obtained from Arabidopsis and tomato (Leon *et al.*, 2001). It therefore appears that JA may not be the only general primary-signal transducer of wound responses in all plant species (Ellis and Turner, 2002; Leon *et al.*, 1998; Nishiuchi *et al.*, 1997; Titarenko *et al.*, 1997). R-PCD is triggered by wounding in the *lsd1* mutants suggesting that *LSD1* is a negative regulator of wounding-induced PCD in wild-type plants (Mateo *et al.*, 2004).

2 Cellular redox hub pathways govern Darwinian fitness

A holistic, multidisciplinary and integrated approach is required in order to gain new knowledge concerning the genetic and molecular mechanisms that control Darwinian fitness in plants under natural multivariable environmental conditions. The following experiments were designed to test the hypothesis that common underlying redox-sensing processes operate in the plant cell to integrate abiotic (acclimation to EEE, wounding and root hypoxia responses) and biotic stress responses (basal resistance against biotrophic pathogen infections) and that these common underlying redox processes therefore govern the plant's capability to survive in the field.

2.1 *EEE causes massive induction of acclimatory and defence-related gene expression*

To investigate the genetic mechanisms that control plant responses to EEE, a subtractive suppression cDNA library was constructed and analysed (Mühlenbock *et al.*, 2008). Messenger RNA was isolated from excess light-exposed leaves and from leaves undergoing SAA. In this study, 730 expression sequence tags (ESTs) were identified that represent 278 different genes induced in response to 40 min of excess light (Mühlenbock *et al.*, 2008). A second set of ESTs was also analysed from a second subtractive library constructed from leaves undergoing SAA (Karpiński *et al.*, 1999). From the second library, 462 EST clones were obtained that represented 148 different genes (Mühlenbock *et al.*, 2008). This included groups of genes that were only detected in excess light or SAA libraries (Mühlenbock *et al.*, 2008). The EST clones identified in both libraries were assigned to 15 different functional groups including categories and subcategories (*Figure 1*, see colour plate section). When a number of genes in each of the groups and their percentages of the total number of identified genes are compared between two

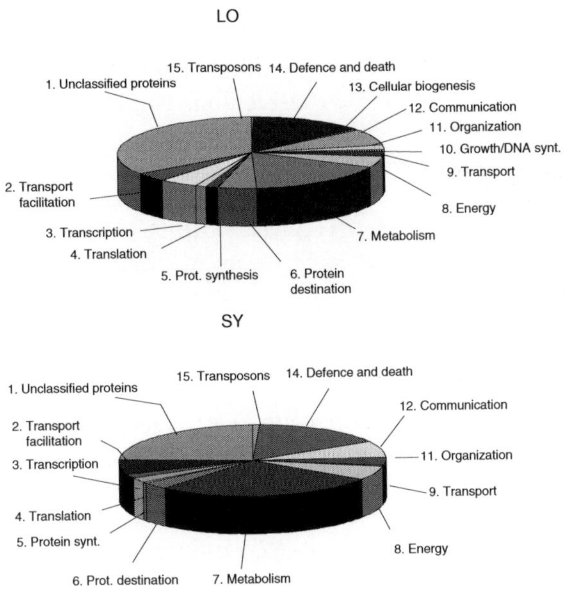

Categories	% in LO	% in SY
1. Unclassified proteins	34.5	23.8
2. Transport facilitation	4.0	6.8
3. Transcription	4.7	2.0
4. Translation	1.4	2.0
5. Protein synthesis	1.4	0.7
6. Protein destination	4.7	3.4
7. Metabolism	16.9	26.5
8. Energy	5.0	6.1
9. Cellular transport/ Transport mechanisms	2.2	0.7
10. Cell growth, cell division and DNA synthesis	0.7	0.0
11. Cellular organization	2.2	3.4
12. Cellular communication/ Signal transduction	7.6	7.5
13. Cellular biogenesis	0.7	0.0
14. Cell rescue, defence, cell death and ageing	13.7	16.3
15. Transposons	0.4	0.7

Figure 1. *Functional categorization of suppressed substructive EST libraries. mRNA for these libraries was isolated from leaves directly exposed to 40 min of excess light (2200 ± 200 μmol m^{-2} s^{-1}), 'local' (LO), and leaves undergoing systemic acquired acclimation, 'systemic' (SY), after 40 min. A list of all sequenced ESTs from these libraries is presented in Mühlenbock* et al., *(2008). Comparative distribution in percentage of different gene categories is presented in* Table 1.

different libraries, interesting differences emerge. There is a higher representation of genes involved in cell rescue, defence and programmed cell death in leaves directly exposed (local, LO) than in leaves undergoing SAA (systemic, SY). This could be due to higher requirements for more defence proteins for repair and programmed cell death regulation in leaves directly challenged with EEE. The percentages of genes involved in cell growth, cell division and DNA synthesis were also higher in such leaves.

The induction of a high percentage of transcripts involved in cell division and growth could be required in order to replace or repair damaged cells in tissues where EEE was high. Nevertheless, the biggest group of up-regulated genes falls into the category of metabolic proteins. These are represented much more in leaves undergoing SAA than in leaves directly exposed to EEE (Mühlenbock *et al.*, 2008). The differences suggest that leaves directly exposed to EEE may have a different metabolic status compared to leaves undergoing SAA. The above results confirm that both types of leaves undergo transcriptional reprogramming, which includes changes in different metabolic pathways linked to defence and acclimation (Mühlenbock *et al.*, 2008).

Knowing that the redox status of the photosynthetic electron-transport carriers plays an important regulatory role in EEE responses (Karpinska *et al.*, 2000; Karpiński *et al.*, 1997, 1999; Kruk and Karpiński, 2006), we looked at the effects of two photosynthetic electron-transport inhibitors, DCMU and DBMIB, that cause oxidation and reduction of the PQ pool respectively, on EEE-induced ESTs (Mühlenbock *et al.*, 2008). If the redox status of the photosynthetic electron-transport carriers plays an important role then DCMU and DBMIB treatments will have opposite effects on EEE-regulated ESTs. In all, we identified 25 ESTs from both libraries (*Table 1*) that were induced by excess light or SAA and that were antagonistically regulated by these inhibitors. We have thus called this new regulon, the PQ-regulon. The PQ-regulon contains genes that are robust markers for light acclimatory, pathogen defence, ozone, drought and low temperature responses and are regulated by ethylene, ROS, glutathione, SA, ABA, sugars and IAA signalling. In the PQ-regulon we detected a putative regulator of ribulose bisphosphate carboxylase/oxygenase (RuBisCO) and other nuclear-encoded chloroplast genes (*Table 1*) controlled by sugars and light signalling. Only the *LOX2* gene detected in the excess light (EL) 'local' library (*Table 1*) was actually induced by DCMU treatment. All other genes presented in *Table 1* were suppressed by DCMU and induced by DBMIB.

The *regulator of ascorbate peroxidase 1* (*rax1*) null mutant has more than or equal to 50% lower foliar glutathione than the wild type (Ball *et al.*, 2004). A mapping study revealed that *rax1-1* is an allele of *GSH1*, which encodes chloroplastic gamma-glutamylcysteine synthetase (gamma-ECS; EC 6.3.2.2), the controlling step of glutathione biosynthesis. A comparison of *rax1-1* with the GSH1 mutant, cadmium hypersensitive 2 (*cad2-1*), revealed that a suite of 25 stress-responsive genes were coordinately influenced by glutathione synthesis in *rax1-1*. It was suggested that this control may reflect the physiological state of the plant at the time of the onset of an environmental challenge and suggests that changes in glutathione metabolism may be one means of integrating the function of several signalling pathways (Ball *et al.*, 2004). We identified one gene, AT5G24770, from the *RAX1*-regulon in the PQ-regulon (*Table 1*) that is differentially regulated by DCMU and DBMIB treatments. These studies suggest that PQ-regulon members are also co-regulated by the *RAX1*-regulon of acclimation to EEE (Ball *et al.*, 2004).

Table 1. *Pool of genes regulated by changes in redox status of the plastoquinone pool (PQ-regulon) and induced by excess light (EL) and by the systemic acquired acclimation (SAA) mechanism.*

Genes detected in EL or SAA EST libraries	DB number	EL	SAA	DC	DB
Chloroplastic lipoxygenase, LOX2	AT3G45140				
Heat shock protein 3 (HSC70-3)	AT3G09440				
Beta Carotene Hydroxylase	AT5G52570				
Plastid-lipid associated protein PAP	AT4G04020				
Pathogenesis receptor kinase PR5K	AT4G36010				
Plant defensin-fusion protein, PDF1.2b	AT2G26020				
DNAJ heat shock protein, putative (J3)	AT3G44110				
Encodes an auxin-inducible GST	AT2G29450				
Glycosyl hydrolase family 17 protein	AT4G16260				
Pyruvate decarboxylase	AT5G54960				
Dehydrin Xero2	AT3G50970				
Disease resistance (TIR-NBS-LRR)	AT4G16860				
Cytochrome P450 81D1 (CYP81D1)	AT5G36220				
Phenylalanine ammonia-layse, PAL1	AT3G53260				
Glutathione S-Transferase 6, GST6	AT2G47730				
Superoxide dismutase (Cu-Zn), CSD1	AT1G08830				
Monodehydroascorbate reductase, MDR	AT3G09940				
Phospholipase D alpha 1, PLD1	AT3G15730				
Phospholipase C, PLC2	AT3G08510				
RuBisCO subunit binding-protein beta subunit	AT3G13470				
Aminotransferase, putative	AT2G24850				
Xyloglucan endo-1,4-beta-D-glucanase, SEN4	AT4G30270				
Pfam domain, PF00069: Protein kinase	AT4G23150				
Vegetative storage protein, VSP2	AT5G24770				
S-Linalool synthase	AT1G61120				

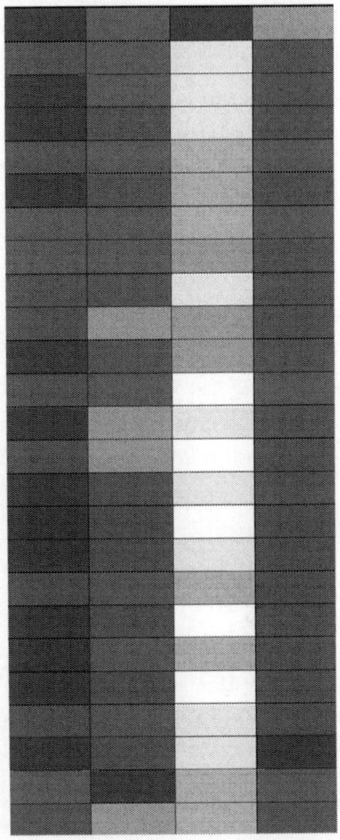

Custom gene arrays (Mühlenbock *et al.*, 2008) were hybridized to P^{33}-labelled cDNAs from *Arabidopsis thaliana* leaves exposed to EL (2200 ± 200 mmol m^{-2} s^{-1}, 40 min), leaves undergoing SAA, leaves treated with 8 mM DCMU (DC) for 3 h and leaves treated with 14 mM DBMIB (DB) in ambient low light (LL) as described by Karpiński *et al.*, (1999). Relative transcript level values represent ratios of mean mRNA abundance to mean of mRNA abundance in LL after normalization to mean actin mRNA levels. Data are representative for pooled samples of 10 leaves from four independent experiments ($N = 4$). Chi-squared analysis based on the values observed in induced and suppressed by DBMIB and DCMU, respectively, and values compared to that neutral for DCMU and DBMIB. All presented differences in relative transcript abundance are significant for 2 degrees of freedom $P<0.005$.

Treatment/LL ratios

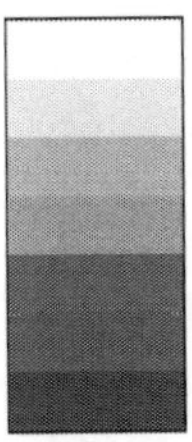

<0.1
<0.25
<0.5
1–2.5
2.5–4.5
4.5-6.5
>6.5

The above results indicate strong similarities between EEE acclimatory and pathogen defence responses. Therefore, we tested whether leaves undergoing local and systemic acclimation to EEE had altered pathogen defence responses. The restricted growth of virulent biotrophic *Pseudomonas syringae* DC3000 after infection compared to non-exposed control plants (Mühlenbock *et al.*, 2008). Leaves treated with DCMU for 3 h and then exposed to EEE for 1 h, or leaves exposed to 4 h of light-1 before infection, permitted similar growth of *P. syringae* DC3000 as control plants (Mühlenbock *et al.*, 2008). In contrast, treatment of leaves with DBMIB in ambient light conditions or 4 h exposure to light-2 caused significantly reduced growth of *P. syringae* DC3000 compared to wild-type plants (Mühlenbock *et al.*, 2008). Taken together, these data show that the redox status of the PQ-pool, at least in part, regulates acclimation to EEE and also basal defence responses to the virulent biotrophic bacteria.

3 Salicylic acid, photo-oxidative stress and stomatal conductance

Due to the agronomical relevance, pathogen defence mechanisms are among the most intensely studied environmental responses in plants. However, before this knowledge can be used in agronomy, the relationships and interdependency between pathogen defences and acclimation to abiotic stress needs to be fully characterized. Most characterized mutants displaying constitutive pathogen resistance accumulate SA. *Constitutive expresser of pathogenesis-related proteins* (*cpr*) mutants were identified in a screen for constitutive activation of the pathogen-responsive BGL2/PR2 promoter (*cpr1-1*, *cpr5-1*, *cpr6-1*; Bowling *et al.*, 1994, Clarke *et al.*, 1998). *cpr1* was mapped to an R-gene cluster on chromosome 4, thus representing the initial step of pathogen recognition (Stokes and Richards, 2002). Cpr5 is a membrane protein of unknown function. Mutations in *cpr5* have pleiotropic effects on the regulation of cell death, cell elongation and trichome development (Yoshida *et al.*, 2002). Similarly, *cpr6* has not yet been identified at the molecular level but it has been genetically defined as a negative regulator of several defence pathways (Clarke *et al.*, 2000). The *dnd1-1* (*defence no death*) mutation affects a cyclic nucleotide gated cation channel (AtCNGC2) and displays constitutive SA-dependent pathogen resistance in the absence of hypersensitive cell death (Clough *et al.*, 2000; Yu *et al.*, 1998). However, a mutant defective in SA synthesis (*sid2-2*) and transgenic NahG plants expressing a bacterial salicylate hydroxylase both display impaired pathogen defences (Bowling *et al.*, 1994; Nawrath and Metraux, 1999; Wildermuth *et al.*, 2001).

The effects of SA deficiency or constitutive SA accumulation on plant-pathogen responses and on temperature tolerance, flowering and reproductive fitness have been reported recently (Clarke *et al.*, 2004; Heidel *et al.*, 2004; Martinez *et al.*, 2004; Scott *et al.* 2004). Little is known about the effects of SA on light acclimation and photo-oxidative stress tolerance. Moreover, all mutants with constitutive SA accumulation show retarded growth, but so far it remains unclear whether the growth effect is a consequence of resource-consuming defence mechanisms or deteriorated energy capture and fixation. We assessed the influence of SA on plant growth, acclimation to high light (HL) and on the redox homeostasis of *Arabidopsis thaliana* leaves (Karpiński *et al.*, 2003; Mateo *et al.*, 2006). The dwarf phenotype displayed by mutants with high SA content (*cpr1-1*, *cpr5-1*, *cpr6-1* and *dnd1-1*) was almost fully reverted when these plants were grown in HL, suggesting that the inhibitory effect of SA on growth was partly overcome at higher light intensities. Moreover, higher SA content affected energy conversion processes in low light but did not impair acclimation to HL (Karpiński *et al.*, 2003; Mateo *et al.*, 2006). On the other hand, mutants with low foliar SA content (*NahG* and *sid2-2*) were impaired in acclimation to transient exposure to HL or EEE and thus predisposed to photo-oxidative stress. Low and high SA levels were strictly correlated to lower and higher foliar H_2O_2 content, respectively. Furthermore, high leaf SA was associated with higher leaf GSH contents, suggesting a tight correlation between SA, H_2O_2 and GSH contents in plants. These observations imply an essential role of SA in the light acclimation processes and in regulating the redox homeostasis of the cell. It is probable, therefore, that SA, GSH and ROS are genetically and functionally linked in the acclimation of plants to EEE, defence against biotrophic pathogen infections and in cellular redox and antioxidant homeostasis (Mateo *et al.*, 2006).

3.1 *SA induces PCD during EEE*

SA treatment causes rapid stomatal closure and an increase in H_2O_2 in leaves (Chen *et al.*, 1993; Manthe *et al.*, 1992; Mori *et al.*, 2001; Rao *et al.*, 1997; Shirasu *et al.*, 1997). This may be associated with SA-dependent inhibition of catalase (CAT) activity (Chen *et al.*, 1997; Conrath *et al.*, 1995; Durner and Klessig, 1996; Sanchez-Casas and Klessig, 1994). If peroxisomal CAT levels are crucial to successful acclimation or recovery processes, then wild-type leaves treated with SA would be predicted to behave in a manner similar to *lsd1* (Mateo *et al.*, 2004). Treatment of wild-type Arabidopsis leaves with SA under short-day conditions induced rapid stomatal closure within 2 h (figure 6 in Mateo *et al.*, 2004). Furthermore, such leaves showed enhanced inhibition of photosynthetic electron transport under low-light conditions (figure 6B in Mateo *et al.*, 2004) and enhanced photo-bleaching (figure 6C in Mateo *et al.*, 2004). Photo-bleaching (PCD) in wild-type leaves exposed to EEE was characterized by the appearance of a delimited area of cell death revealed by lactophenoltrypan-blue staining (figure 6D in Mateo *et al.*, 2004). These effects were strongly accelerated by a combined effect of SA treatment and excess light (EL) exposure (figure 6B, C and D in Mateo *et al.*, 2004). It is important to note that stomatal gas exchange values in SA-treated wild-type leaves before EL exposure were reduced by a factor of approximately 4. In the combined treatment, gas exchange was decreased by a factor of 12. Treatment of *lsd1* leaves with SA in low light caused reduction of stomatal gas exchange by a factor of 14, in comparison to wild-type control values. These data suggest that SA induces EEE under low light intensities so that SA-treated leaves are no longer able to tolerate EL. This notion is reinforced by the observation that high-light-acclimated leaves, that are already in the process of dissipating EL or limiting EEE, were tolerant to SA treatment (figure 6D in Mateo *et al.*, 2004) and have higher foliar SA levels (Karpiński *et al.*, 2003). We conclude that the effectiveness of photorespiration in dissipating EEE may be limited by the efficiency of photorespiratory H_2O_2 scavenging. The *lsd1* mutant failed to acclimate to light conditions that promote photo-oxidative stress. Hence, LSD1 function is required for the expression of *CAT1*, encoding a subunit of peroxisomal catalase (Mateo *et al.*, 2004). Through this regulation, LSD1 can control the effectiveness of photorespiration in dissipating EEE absorbed by PSII. The *lsd1* mutant also has reduced stomatal conductance in permissive conditions, and induced H_2O_2 accumulation together with runaway cell death when stomatal gas exchange is impeded. While the involvement of H_2O_2 had been eliminated in a previous study (Jabs *et al.*, 1996), the

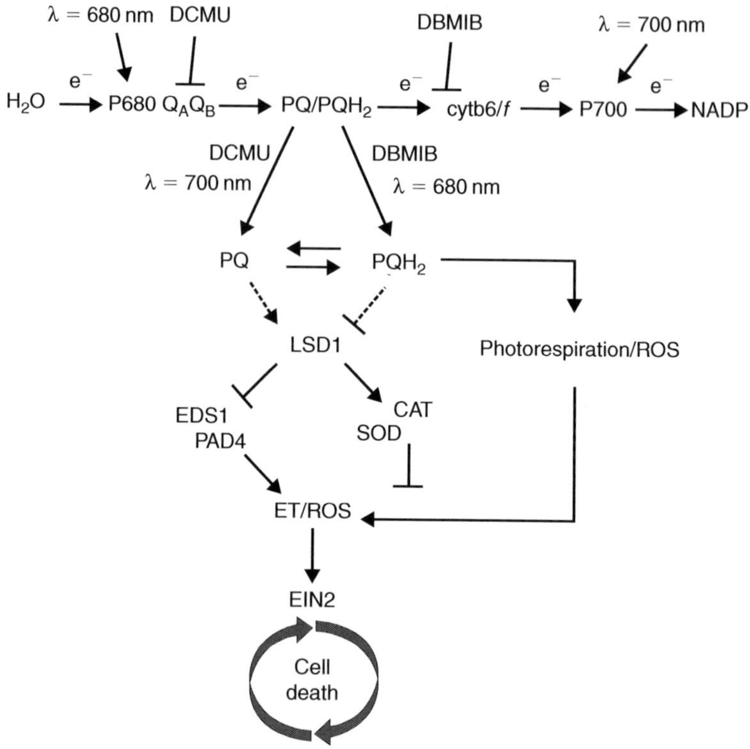

Figure 2. *Model for EEE-induced programmed cell death controlled by the chloroplast redox signalling, photorespiration and LSD1. Pro-programmed cell death redox signalling originating from redox changes in the plastoquinone pool (PQ) is negatively regulated by LSD1 that acts to limit the spread of cell death. LSD1 negatively regulates ROS from photorespiration (Mateo* et al.*, 2004, Fig. 5); PAD4- and EDS1-dependent cellular ethylene production, together with EIN2 (Mühlenbock* et al.*, 2007b, 2008), modulate ethylene- (ET) induced pro-programmed cell death signalling, acclimatory and defence responses. LSD1 positively regulates, directly or indirectly, superoxide dismutase (SOD) and catalase (CAT) gene expression and activities and thus controls cellular reactive oxygen species (ROS) production (Jabs* et al.*, 1996; Kliebenstein* et al.*, 1998; Mateo* et al.*, 2004). Additionally we suggest the putative role of LSD1 in restricting MPK4, a negative regulator of SA signalling pathway (Petersen* et al.*, 2000). We propose that LSD1, EDS1 and PAD4 constitute a ROS/ethylene homeostatic switch, controlling acclimation and defence holistic responses (after Mühlenbock* et al.*, 2008).*

data presented here show that H_2O_2 is a molecular trigger for the *lsd1* runaway cell death phenotype. The light sensitive *lsd1* phenotype depends on the defence regulators *EDS1* and *PAD4* and on PSII light-harvesting antenna size (Mateo *et al.*, 2004) and ROS/ethylene signalling (Mühlenbock *et al.*, 2008). Salicylic acid, which induces stomatal closure, inhibited catalase activity and triggered the runaway cell death phenotype of *lsd1*, also impaired acclimation of wild-type plants to conditions that promote EEE. It can be proposed that the roles of LSD1 in light acclimation and regulation of stomatal conductance, as well as in restricting pathogen-induced cell death, are functionally linked (Mateo *et al.*, 2004; *Figure 2*).

4 The role of chloroplast-derived redox signalling and ROS in regulation of wounding responses

When the wounded tissues of detached transgenic Arabidopsis leaves expressing the firefly luciferase gene (*LUC*) under the control of the *APX2* promoter were partially exposed to EL (Karpiński *et al.*, 1999), the *LUC* signal was enhanced in the veins closest to the wound (Mullineaux *et al.*, 2000). *LUC* activity was observed in the central vein and apical region of both the wounded leaf and adjacent unwounded leaves (Chang *et al.*, 2004).

These data raise the possibility that the signalling pathway controlling the expression of *APX2* may be integrated with one or more wound-signalling pathways or that these pathways could share common components, or be responsive to the same effectors molecules. Further investigations, however, led to a rather different and unexpected conclusion (Chang *et al.*, 2004). It appears that photosynthetic electron flux to O_2 is increased in the vascular regions of wounded leaves upon exposure to HL, with an increased production of H_2O_2 that, in turn, triggers induction of *APX2:LUC* expression (Chang *et al.*, 2004).

4.1 *Synergistic effect of wounding and excess light on APX2 expression*

In the *APX2:LUC* expression system, *APX2* transcript was detected rapidly (within 45 min) after wounding. The wounded plants expressed the JA-dependent wound-inducible *VSP1* and JA-independent wound-inducible *CHIB* genes (Ellis and Turner, 2001; Rojo *et al.*, 1999). Similarly, *APX2* expression was induced in leaves subjected to EL treatments at times as little as 7 min after the transition to high light (Karpiński *et al.*, 1997). The effect of combined EL and wounding stresses on gene expression was compared with that observed in undamaged leaves subjected to a fivefold excess light stress. *APX2* transcripts were detected 15 min earlier in the combined stress experiment and they accumulated to a greater than 10-fold higher level in wounded, EL-stressed leaves than in their undamaged counterparts (Chang *et al.*, 2004). We conclude that wounding and EL have synergistic effects on induced *APX2* expression. In contrast, while *VSP1* and *CHIB* expression were induced by EL treatment, no synergistic effects were observed (Chang *et al.*, 2004). Neither of the JA-dependent nor JA-independent wound-signalling pathways known in Arabidopsis (Leon *et al.*, 2001; McConn *et al.*, 1997; Nishiuchi *et al.*, 1997; Rojo *et al.*, 1998, 1999; Titarenko *et al.*, 1997) are involved in *APX2* expression in wounded leaves (Chang *et al.*, 2004). Moreover, HL-mediated induction of *VSP1* expression was not JA-dependent (*Table 1*). Thus, the JA signalling pathways appear to play no direct role in the regulation of gene expression in response to HL stress.

4.2 *Photosynthetic electron transport is required for wound-induced APX2 expression*

While *APX2* expression was always highest in leaves subjected to both wounding and EL stress, *APX2* expression was blocked in the presence of DCMU (Chang *et al.*,

2004). Thus, the mechanism of *APX2* induction in wounded Arabidopsis leaves was similar to that observed in EL-stressed leaves, being dependent on redox changes in the photosynthetic electron-transport system that generate H_2O_2 and are influenced by ABA (Fryer *et al.*, 2003; Karpiński *et al.*, 1999). The induction of *APX2* expression thus differs between wounded and EL-stressed leaves only in the degree of response.

4.3 *Wound- and excess-light-induced APX2 expression is not controlled by known wound signalling pathways in Arabidopsis*

Two wound signalling pathways that are JA-dependent and JA-independent have been described in Arabidopsis (Ellis and Turner, 2002; Rojo *et al.*, 1999; Titarenko *et al.*, 1997). These pathways are coordinately regulated by ethylene generated in damaged tissues (Rojo *et al.*, 1999). The mutant *coronatine-insensitive 1-1* (*coi1-1*; Feys *et al.*, 1994) is blocked in the perception of JA and *COI 1-1* may play a role in cross-talk between the JA pathway and those directed by ethylene and ABA (Ellis and Turner, 2002; Feys *et al.*, 1994). The *coi1-1* mutant showed no difference from wild-type plants in the abundance of *APX2* transcript when wounded or subjected to excess light (Chang *et al.*, 2004). *VSP1* transcript induction was blocked in wounded, but not in high-light-stressed *coi1-1* plants (Chang *et al.*, 2004). Similarly, wounded or high-light-treated mutants impaired in their perception of ethylene, *ethylene insensitive 2-1* and *ethylene resistant 1-3*, or JA (Staswick *et al.*, 1992) showed no inhibition of *APX2* expression (Chang *et al.*, 2004). *APX2:LUC* Arabidopsis rosettes sprayed with either JA or chitosan, an elicitor of the JA-independent pathway (Rojo *et al.*, 1999), did not induce luciferase activity under ambient light conditions (Chang *et al.*, 2004). The effects of JA and chitosan in the leaves were confirmed by *CHIB* and *VSP1* gene induction (Chang *et al.*, 2004). Taken together these data indicate that the wound induction of *APX2* expression was not regulated by either of the known JA-dependent or JA–independent pathways in Arabidopsis and, in contrast to wounding, *VSP1* induction of expression in excess-light-stressed leaves also was not JA-dependent.

5 Programmed cell death and Darwinian fitness

The analysis of defects in mutants, and the characterization of the altered responses by which such mutants are defined, is normally undertaken under highly controlled laboratory conditions. Laboratory conditions do not generally offer the multiple signals that plants would normally perceive within their natural environments and that ultimately drive Darwinian fitness. There is mounting evidence that plants are able to integrate multiple stress inputs and prioritize their responses to the prevailing environment (Ferrari *et al.*, 2003; Genoud and Métraux 1999; Karpiński *et al.*, 1999, 2003; Mateo *et al.*, 2004, 2006; Mühlenbock *et al.*, 2007a, 2007b, 2008; Mullineaux *et al.*, 2000). This enables them to survive adverse conditions, conserve energy for growth and reproduction and compete with other plants for soil space. There is therefore a need to address how particular plant lines that have defined stress-signalling mutations or alterations in the levels of stress-response regulators behave in the field.

The data illustrated in *Table 2* reveal striking differences between the phenotypes of Arabidopsis defence and abiotic stress mutants challenged with single stresses in the laboratory and those same genetic backgrounds grown in nature. These data strongly suggest that *LSD1*, *PAD4* and *EDS1* regulate plants' Darwinian fitness.

Table 2. *Darwinian fitness tested as healthy seeds production from 100 seeds planted.*

Genotype	Total seed production from 100 seeds cultivated under natural conditions	Healthy seeds production from 100 seeds cultivated under natural conditions
WT	18 870	17 662
lsd1	26 645	25 739
eds1	20 196	15 026
pad4	20 705	17 503
eds1/lsd1	8928	8876
pad4/lsd1	33 600	23 250

At the end of April 2004, 100 healthy seeds of each genotype were planted in the experimental field and plants with seeds were harvested in July 2004.

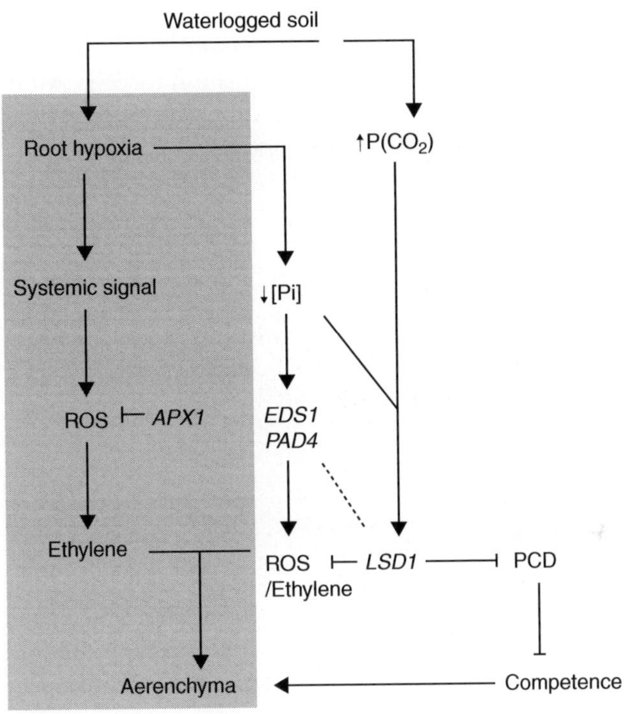

Figure 3. *Proposed working model for signalling pathways that regulate aerenchyma in Arabidopsis. The grey area denotes root-specific events. Root hypoxia produces a systemic signal that promotes ROS and ethylene formation, leading to the induction of aerenchyma (grey area). Waterlogging and root hypoxia also lead to increased* CO_2 *levels (P(*CO_2*)) and decreased phosphorous levels (Pi), that induce* LSD1, EDS1 *and* PAD4 *expression (Mühlenbock* et al., *2007a,b).* LSD1 *negatively regulates, while* EDS1 *and* PAD4 *positively regulate aerenchyma putatively through ROS (*H_2O_2*) and ethylene signalling. Additionally LSD1 is a negative regulator (by blocking PAD4 and EDS1 pro PCD regulatory mode) of foliar PCD that produces a yet unidentified signal that blocks the competence of parenchyma cells to form aerenchyma.*

The model shown in *Figure 3* illustrates the involvement of the plant defence regulators *LSD1*, *EDS1* and *PAD4*, that operate upstream of ethylene and ROS, in the production in Arabidopsis of lysigenous aerenchyma in response to root hypoxia caused by flooding (Mühlenbock *et al.*, 2007a, 2007b). *LSD1*, *EDS1* and *PAD4* control the extent and the timing of aerenchyma formation (Mühlenbock *et al.*, 2007b). These genes also have important functions in other responses (Mateo *et al.*, 2004; Mühlenbock *et al.*, 2007b). An interesting aspect of the flooding response in Arabidopsis is the predictable and complete autolysis of cell walls. Moreover, we have shown that the lysigenous aerenchyma of Arabidopsis is formed at a late stage of development, when secondary growth is advanced. It is possible that most species with profuse secondary growth, especially trees, form a similar type of aerenchyma and this is an area that desperately needs to be investigated. Since Arabidopsis is already a well-explored model system for wood formation our findings open up new possibilities of using the Arabidopsis model to specifically address issues relating to tree responses to flooding and hypoxia. The above results show that *LSD1* is indeed a general controller of environmental stress responses to a variety of different environmental stress, since it controls PCD, acclimation to EEE, wounding, pathogen defences and lysigenous aerenchyma formation. While these data emphasize the importance of examining multiple physiological conditions when investigating the function of genes, the above evidence allows us to conclude that *LSD1* is a negative regulator of Darwinian fitness in plants.

Acknowledgements

This work was supported by the Polish Ministry of Science and Higher Education with research and supervisor grants N301 075 31/2414 and N301 2550 33, respectively given to S.K.

References

Ahmad, M., Jarillo, J.A. and Cashmore, A.R. (1998) Chimeric proteins between cry1 and cry2 *Arabidopsis* blue light photoreceptors indicate overlapping functions and varying protein stability. *Plant Cell* **10:** 197–207.

Allan, A.C. and Fluhr, R. (1997) Two distinct sources of elicited reactive oxygen species in tobacco epidermal cells. *Plant Cell* **9:** 1559–1572.

Allen, J.F. (1993) Control of gene expression by redox potential and the requirement for chloroplast and mitochondrial genomes. *J. Theor. Biol.* **165:** 609–631.

Allen, J.F. (1995) Thylakoid protein phosphorylation, state 1-state 2 transitions, and photosystem stoichiometry adjustment: redox control at multiple levels of gene expression. *Physiol. Plant* **93:** 196–205.

Allen, J.F. (2003) State transitions – a question of balance. *Science* **299:** 1530–1532.

Alvarez, M.E., Pennell, R.I., Meijer, P.J., Ishikawa, A., Dixon, R.A. and Lamb, C. (1998) Reactive oxygen intermediates mediate a systemic signal network in the establishment of plant immunity. *Cell* **92:** 773–784.

Asada, K. (1999) The water-water cycle in chloroplasts: scavenging of active oxygens and dissipation of excess photons. *Annu. Rev. Plant Physiol. Plant Mol. Biol.* **50:** 601–639.

Asai, T., Tena, G., Plotnikova, J., Willmann, M.R., Chiu, W. L., Gomez-Gomez, L., Boller, T., Ausubel, F.M. and Sheen, J. (2002) MAP kinase signalling cascade in *Arabidopsis* innate immunity. *Nature* **415:** 977–983.

Bailey, S., Walters, R., Jansson, S. and Horton, P. (2001) Acclimation of *Arabidopsis thaliana* to the light environment: the existence of separate low light and high light responses. *Planta* **213:** 794–801.

Ball, L., Accotto, G.P., Bechtold, U., *et al.* (2004) An Arabidopsis mutant with raised *ASCORBATE PEROXIDASE 2* expression reveals glutathione as a direct modulator of stress responsive gene expression. *Plant Cell* **16:** 2448–2462.

Bart Rossel, J., Wilson, P.B., Hussain, D., *et al.* (2007) Systemic and intracellular responses to photooxidative stress in Arabidopsis. *Plant Cell* www.plantcell.org/cgi/doi/10.1105/tpc.106.045898.

Baxter-Burrell, A., Yang, Z., Springer, P.S. and Bailey-Serres, J. (2002) RopGAP4-dependent Rop GTPase rheostat control of *Arabidopsis* oxygen deprivation tolerance. *Science* **296:** 2026–2028.

Bowling, S.A., Guo, A., Cao, H., Gordon, A.S., Klessig, D.F. and Dong, X. (1994) A mutation in *Arabidopsis* that leads to constitutive expression of systemic acquired resistance. *Plant Cell* **6:** 1845–1857.

Briggs, W.R., Beck, C.F., Cashmore, A.R., *et al.* (2001) The phototropin family of photoreceptors. *Plant Cell* **13:** 993–997.

Brodersen, P., Petersen, M., Pike, H.M., Olszak, B., Skov, S., Odum, N., Jorgensen, L.B., Brown, R.E. and Mundy, J. (2002) Knockout of *Arabidopsis ACCELERATED-CELL-DEATH11* encoding a sphingosine transfer protein causes activation of programmed cell death and defense. *Genes Dev.* **16:** 490–502.

Brodersen, P., Petersen, M., Nielsen, H.B., Zhu, S., Newman, M.A., Shokat, K.M., Rietz, S., Parker, J. and Mundy, J. (2006) *Arabidopsis* MAP kinase 4 regulates salicylic acid- and jasmonic acid/ethylene-dependent responses via EDS1 and PAD4. *Plant J.* **47:** 532–546.

Campbell, R. and Drew, M.C. (1983) Electron microscopy of gas space (aerenchyma) formation in adventitious roots of *Zea mays* L. subjected to oxygen shortage. *Planta* **157:** 350.

Chang, C.C.-C., Ball, L., Fryer, M.J., Baker, N.R., Karpiński, S. and Mullineaux, P.M. (2004) Induction of *ASCORBATE PEROXIDASE 2* expression in wounded *Arabidopsis* leaves does not involve known wound-signalling pathways but is associated with changes in photosynthesis. *Plant J.* **38:** 499–511.

Chen, Z., Silva, H. and Klessig, D. (1993) Active oxygen species in the induction of plant systemic acquired resistance by salicylic acid. *Science* **262:** 1883–1886.

Chen, Z., Iyer, S., Caplan, A., Klessig, D.F. and Fan, B. (1997) Differential accumulation of salicylic acid and salicylic acid-sensitive catalase in different rice tissues. *Plant Physiol.* **114:** 193–201.

Clack, T., Mathews, S. and Sharrock, R. (1994) The phytochrome apoprotein family in *Arabidopsis* is encoded by five genes: the sequences and expression of *PHYD* and *PHYE*. *Plant Mol. Biol.* **25:** 413–427.

Clarke, J.D., Liu, Y., Klessig, D.F. and Dong, X. (1998) Uncoupling PR gene expression from NPR1 and bacterial resistance: characterization of the dominant *Arabidopsis cpr6-1* mutant. *Plant Cell* **10:** 557–569.

Clarke, J.D., Volko, S.M., Ledford, H., Ausubel, F.M. and Dong, X. (2000) Roles of salicylic acid, jasmonic acid, and ethylene in *cpr*-induced resistance in *Arabidopsis. Plant Cell* **12:** 2175–2190.

Clarke, S.M., Mur, L.A., Wood, J.E. and Scott, I.M. (2004) Salicylic acid dependent signaling promotes basal thermotolerance but is not essential for acquired thermotolerance in *Arabidopsis thaliana. Plant J.* **38:** 432–447.

Clough, S.J., Fengler, K.A., Yu, I.C., Lippok, B., Smith, R.K. Jr and Bent A.F. (2000) The *Arabidopsis dnd1* 'defense, no death' gene encodes a mutated cyclic nucleotide-gated ion channel. *Proc. Natl Acad. Sci. USA* **97:** 9323–9328.

Colmer, T.D., Peeters, A.J.M., Wagemaker, C.A.M., Vriezen, W.H., Ammerlaan, A. and Voesenek, L.A.C.J. (2004) Expression of alpha-expansin genes during root acclimations to O_2 deficiency in *Rumex palustris. Plant Mol. Biol.* **56:** 423.

Conrath, U., Chen, Z., Ricigliano, J. and Klessig, D. (1995) Two inducers of plant defense responses, 2,6-dichloroisonicotinec acid and salicylic acid, inhibit catalase activity in tobacco. *Proc. Natl Acad. Sci. USA* **92:** 7143–7147.

Dangl, J.L. and Jones, J.D.G. (2001) Plant pathogens and integrated defence responses to infection. *Nature* **411:** 826–833.

de Bruxelles, G.L., Peacock, W.J., Dennis, E.S. and Dolferus, R. (1996) Abscisic acid induces the alcohol dehydrogenase gene in *Arabidopsis. Plant Physiol.* **111:** 381–391.

Delledonne, M., Xia, Y., Dixon, R.A. and Lamb, C. (1998) Nitric oxide functions as a signal in plant disease resistance. *Nature* **394:** 585–588.

Dennis, E.S., Dolferus, R., Ellis, M., *et al.* (2000) Molecular strategies for improving waterlogging tolerance in plants. *J. Exp. Bot.* **51:** 89–97.

Dietrich, R., Delaney, T., Uknes, S., Ward, E., Ryals, J. and Dangl, J. (1994) *Arabidopsis* mutants simulating disease resistance response. *Cell* **77:** 565–577.

Dordas, C., Rivoal, J., and Hill, R.D. (2003). Plant haemoglobins, nitric oxide and hypoxic stress. *Ann Bot* **91**, 173-178.

Drew, M.C., He, C.J. and Morgan, P.W. (2000) Programmed cell death and aerenchyma formation in roots. *Trends Plant Sci.* **5:** 123–127.

Durner, J. and Klessig, D.F. (1996) Salicylic acid is a modulator of tobacco and mammalian catalases. *J. Biol. Chem.* **271:** 28492–28501.

Durnford, D. and Falkowski, P. (1997) Chloroplast redox regulation of nuclear gene transcription during photoacclimation. *Photosynth. Res.* **53:** 229–241.

Ellis, C. and Turner, J.G. (2001) The *Arabidopsis* mutant *cev1* has constitutively active jasmonate and ethylene signal pathways and enhanced resistance to pathogens. *Plant Cell* **13:** 1025–1033.

Ellis, C. and Turner, J.G. (2002) A conditionally fertile *coi1* allele indicates cross-talk between plant hormone signalling pathways in *Arabidopsis thaliana* seeds and young seedlings. *Planta* **215:** 549–556.

Evans, D.E. (2003) Aerenchyma formation. *New Phytol.* **161:** 35–49.

Ferrari, S., Vairo, D., Ausubel, F.M., Cervone, F. and Lorenzo, G.D. (2003) Tandemly duplicated *Arabidopsis* genes that encode polygalacturonase-inhibiting proteins are regulated coordinately by different signal transduction pathways in response to fungal infection. *Plant Cell* **15:** 93–106.

Feys, B., Benedetti, C.E., Penfold, C.N., and Turner, J.G. (1994). Arabidopsis mutants selected for resistance to the phytotoxin coronatine are male sterile, insensitive to methyl jasmonate, and resistant to a bacterial pathogen. *Plant Cell* **6**, 751–759.

Feys, B.J., Wiermer, M., Bhat, R.A., Moisan, L.J., Medina-Escobar, N., Neu, C., Cabral A. and Parker J.E. (2005) *Arabidopsis senescence associated gene 101* stabilizes and signals within an *enhanced disease susceptibility 1* complex in plant innate immunity. *Plant Cell* **17:** 2601–2613.

Foyer, C. and Halliwell, B. (1976) The presence of glutathione and glutathione reductase in chloroplasts: a proposed role in ascorbic acid metabolism. *Planta* **133:** 21–25.

Foyer, C. and Noctor, G. (1999) Leaves in the dark see the light. *Science* **284:** 599–601.

Foyer, C.H. and Noctor, G. (2000) Tansley Review No. 112. Oxygen processing in photosynthesis: regulation and signalling. *New Phytol.* **146:** 359–388.

Foyer, C.H. and Noctor, G. (2003) Redox sensing and signalling associated with reactive oxygen in chloroplasts, peroxisomes and mitochondria. *Physiol. Plant* **119:** 355–364.

Fryer, M.J., Ball, L., Oxborough, K., Karpiński, S., Mullineaux, P.M. and Baker, N.R. (2003) Control of *ascorbate peroxidase 2* expression by hydrogen peroxide and leaf water status during excess light stress reveals a functional organisation of *Arabidopsis* leaves. *Plant J.* **33:** 691–705.

Fukao, T. and Bailey-Serres, J. (2004) Plant responses to hypoxia – is survival a balancing act? *Trends Plant Sci.* **9:** 449–456.

Genoud, T. and Métraux, J.P. (1999) Crosstalk in plant cell signaling: structure and function of the genetic network. *Trends Plant Sci.* **4:** 503–507.

Guan, L.M. and Scandalios, J.G. (2000) Hydrogen peroxide-mediated catalase gene expression in response to wounding. *Free Radic. Biol. Med.* **28:** 1182–1190.

Gunawardena, A.H.L.A.N., Pearce, D.M., Jackson, M.B., Hawes, C.R. and Evans, D.E. (2001) Characterisation of programmed cell death during aerenchyma formation induced by ethylene or hypoxia in roots of maize (*Zea mays* L.). *Planta* **212:** 205.

Haldrup, A., Jensen, P.E., Lunde, C. and Scheller, H.V. (2001) Balance of power: a view of the mechanism of photosynthetic state transitions. *Trends Plant Sci.* **6:** 301–305.

Hammond-Kosack, K.E. and Parker, J.E. (2003) Deciphering plant-pathogen communication: fresh perspectives for molecular resistance breeding. *Curr. Opin. Biotechnol.* **14:** 177–193.

Havaux, M. and Niyogi, K. (1999) The violaxanthin cycle protects plants from photooxidative damage by more than one mechanism. *Proc. Natl Acad. Sci. USA* **96:** 8762–8767.

He, C.J., Morgan, P.W. and Drew, M.C. (1996) Transduction of an ethylene signal is required for cell death and lysis in the root cortex of maize during aerenchyma formation induced by hypoxia. *Plant Physiol.* **112:** 463–472.

Heidel, A.J., Clarke, J.D., Antonovics, J. and Dong, X. (2004) Fitness costs of mutations affecting the systemic acquired resistance pathway in *Arabidopsis thaliana*. *Genetics* **168:** 2197–2206.

Jabs, T., Dietrich, R. and Dangl, J. (1996) Initiation of runaway cell death in an *Arabidopsis* mutant by extracellular superoxide. *Science* **273:** 1853–1856.

Jackson, M.B. and Armstrong, W. (1999) Formation of aerenchyma and the processes of plant ventilation in relation to soil flooding and submergence. *Plant Biol.* **1:** 274–287.

Jackson, M.B., Fenning, T.M., Drew, M.C., and Saker, L.R. (1985) Stimulation of ethylene production and gas-space (aerenchyma) formation in adventitious roots of *Zea mays* L. by small partial pressures of oxygen. *Planta* **165:** 486.

Karpinska, B., Wingsle, G. and Karpiński, S. (2000) Antagonistic effects of hydrogen peroxide and glutathione on acclimation to excess excitation energy in *Arabidopsis. IUBMB Life* **50:** 21–26.

Karpinska, B., Karlsson, M., Schinkel, H., Streller, S., Suss, K.H., Melzer, M. and Wingsle, G. (2001) A novel superoxide dismutase with a high isoelectric point in higher plants. Expression, regulation, and protein localization. *Plant Physiol.* **126:** 1668–1677.

Karpiński, S., Escobar, C., Karpinska, B., Creissen, G. and Mullineaux, P. (1997) Photosynthetic electron transport regulates the expression of cytosolic ascorbate peroxidase genes in *Arabidopsis* during excess light stress. *Plant Cell* **9:** 627–640.

Karpiński, S., Reynolds, H., Karpinska, B., Wingsle, G., Creissen, G. and Mullineaux, P. (1999) Systemic signaling and acclimation in response to excess excitation energy in *Arabidopsis. Science* **284:** 654–657.

Karpiński, S., Gabrys, H., Mateo, A., Karpinska, B. and Mullineaux, P.M. (2003) Light perception in plant disease defence signalling. *Curr. Opin. Plant Biol.* **6:** 390–396.

Kasahara, M., Kagawa, T., Oikawa, K., Suetsugu, N., Miyao, M. and Wada, M. (2002) Chloroplast avoidance movement reduces photodamage in plants. *Nature* **420:** 829–832.

Keller, T., Damude, H.G., Werner, D., Doerner, P., Dixon, R.A. and Lamb, C. (1998) A plant homolog of the neutrophil NADPH oxidase gp91*phox* subunit gene encodes a plasma membrane protein with Ca^{2+} binding motifs. *Plant Cell* **10:** 255–266.

Kleine, T., Kindgren, P., Benedict, C., Hendricksen, L. and Strand, Å. (2007) Genome wide gene expression analysis reveals a critical role for *CRYPTOCHROME1* in the response of *Arabidopsis* to high irradiance. *Plant Physiol.* **144:** 1391–1406.

Klenell, M., Morita, S., Tiemblo-Olmo, M., Mühlenbock, P., Karpiński, S. and Karpinska, B. (2005) Photosystem II light harvesting antenna organization controlled by cpSRP43 plays an essential role in regulation of *Arabidopsis thaliana* photooxidative stress tolerance. *Plant Cell Physiol.* **46:** 118–126.

Kliebenstein, D.J., Monde, R.A. and Last, R.L. (1998) Superoxide dismutase in *Arabidopsis*: an eclectic enzyme family with disparate regulation and protein localization. *Plant Physiol.* **118:** 637–650.

Klok, E.J., Wilson, I.W., Wilson, D., Chapman, S.C., Ewing, R.M., Somerville, S.C., Peacock, W.J., Dolferus, R. and Dennis, E.S. (2002) Expression profile analysis of the low-oxygen response in *Arabidopsis* root cultures. *Plant Cell* **14:** 2481–2494.

Kozaki, A. and Takeba, G. (1996) Photorespiration protects C3 plants from photooxidation. *Nature* **384:** 557–560.

Kruk, J. and Karpiński, S. (2006) An HPLC-based method of estimation of the total redox state of plastoquinone in chloroplasts, the size of the photochemically active plastoquinone-pool and its redox state in thylakoids of *Arabidopsis. Biochim. Biophys. Acta* **1757:** 1669–1675.

Kulheim, C., Agren, J. and Jansson, S. (2002) Rapid regulation of light harvesting and plant fitness in the field. *Science* **297:** 91–93.

Lamb, C. and Dixon, R.A. (1997) The oxidative burst in plant disease resistance. *Annu. Rev. Plant Physiol. Plant Mol. Biol.* **48:** 251–275.

Leon, J., Rojo, E., Titarenko, E. and Sanchez-Serrano, J. (1998) Jasmonic acid-dependent and -independent wound signal transduction pathways are differentially regulated by Ca^{2+}/calmodulin in *Arabidopsis thaliana. Mol. Gen. Genet.* **258:** 412–419.

Leon, J., Rojo, E. and Sanchez-Serrano, J.J. (2001) Wound signalling in plants. *J. Exp. Bot.* **52:** 1–9.

Levine, A., Tenhaken, R., Dixon, R. and Lamb, C. (1994) H_2O_2 from the oxidative burst orchestrates the plant hypersensitive disease resistance response. *Cell* **79:** 583–593.

Lin, C. (2002) Blue light receptors and signal transduction. *Plant Cell* **14:** S207–S225.

Manac'h-Little, N., Igamberdiev, A.U. and Hill, R.D. (2005) Hemoglobin expression affects ethylene production in maize cell cultures. *Plant Physiol. Biochem.* **43:** 485.

Manthe, B., Schulz, M. and Schnabl, H. (1992) Effects of salicylic acid on growth and stomatal movements of *Vicia faba* L. – evidence for salicylic acid metabolization. *J. Chem. Ecol.* **18:** 1525–1539.

Martinez, C., Pons, E., Prats, G. and Leon, J. (2004) Salicylic acid regulates flowering time and links defence responses and reproductive development. *Plant J.* **37:** 209–217.

Mateo, A., Mühlenbock, P., Rusterucci, C., Chang, C.C.-C., Miszalski, Z., Karpinska, B., Parker, J.E., Mullineaux, P.M. and Karpiński, S. (2004) *LESION SIMULATING DISEASE 1* is required for acclimation to conditions that promote excess excitation energy. *Plant Physiol.* **136:** 2818–2830.

Mateo, A., Funck, D., Mühlenbock, P., Kular, B., Mullineaux, P.M. and Karpiński, S. (2006) Controlled levels of salicylic acid are required for optimal photosynthesis and redox homeostasis. *J. Exp. Bot.* **57:** 1795–1807.

McConn, M., Creelman, R.A., Bell, E., Mullet, J.E. and Browse, J. (1997) Jasmonate is essential for insect defense in *Arabidopsis. Proc. Natl Acad. Sci. USA* **94:** 5473–5477.

Mittler, R., Vanderauwera, S., Gollery, M. and Van Breusegem, F. (2004) Reactive oxygen gene network of plants. *Trends Plant Sci.* **9:** 490–498.

Monk, L.S., Fagerstedt, K.V. and Crawford, R.M.M. (1987) Superoxide dismutase as an anaerobic polypeptide – a key factor in recovery from oxygen deprivation in *Iris pseudacorus. Plant Physiol.* **85:** 1016–1020.

Mori, I.C., Pinontoan, R., Kawano, T. and Muto, S. (2001) Involvement of superoxide generation in salicylic acid-induced stomatal closure in *Vicia faba. Plant Cell Physiol.* **42:** 1383–1388.

Mou, Z., Fan, W. and Dong, X. (2003) Inducers of plant systemic acquired resistance regulate NPR1 function through redox changes. *Cell* **113:** 935–944.

Mühlenbock, P., Karpinska, B. and Karpiński, S. (2007a) Oxidative stress and redox signalling in plants. In: *Encyclopedia of Life Sciences*. John Wiley & Sons, Chichester. http://www.els.net/ (DOI: 10.1002/9780470015902.a0020135).

Mühlenbock, P., Plaszczyca, M., Mellerowicz, E. and Karpiński, S. (2007b) Lysigenous aerenchyma formation in *Arabidopsis thaliana* is controlled by *LESION SIMULATING DISEASE 1. Plant Cell* **19:** 3819–3830 .

Mühlenbock, P., Szehyńska-Hebda, M., Plaszczyca, M., Baudo, M., Mullineaux P.M., Parker, J.E., Karpinska, B. and Karpiński, S. (2008) Chloroplast Dependent Cell Death Signaling and *LESION SIMULATING DISEASE 1* Control Cross Talk Between Light Acclimation and Plant Immunity. *Plant Cell* (in press).

Muller, P., Li, X.-P. and Niyogi, K.K. (2001) Non-photochemical quenching: a response to excess light energy. *Plant Physiol.* **125:** 1558–1566.

Mullineaux, P. and Karpiński, S. (2002) Signal transduction in response to excess light: getting out of the chloroplast. *Curr. Opin. Plant Biol.* **5:** 43–48.

Mullineaux, P., Ball, L., Escobar, C., Karpinska, B., Creissen, G. and Karpiński, S. (2000) Are diverse signalling pathways integrated in the regulation of *Arabidopsis* antioxidant defence gene expression in response to excess excitation energy? *Philos. Trans. R. Soc. Lond. B. Biol. Sci.* **355:** 1531–1540.

Nawrath, C. and Metraux, J.P. (1999) Salicylic acid induction-deficient mutants of *Arabidopsis* express PR-2 and PR-5 and accumulate high levels of camalexin after pathogen inoculation. *Plant Cell* **11:** 1393–1404.

Nishiuchi, T., Hamada, T., Kodama, H. and Iba, K. (1997) Wounding changes the spatial expression pattern of the *Arabidopsis* plastid (omega)-3 fatty acid desaturase gene (*FAD7*) through different signal transduction pathways. *Plant Cell* **9:** 1701–1712.

Niyogi, K.K. (1999) Photoprotection revisited: genetic and molecular approaches. *Annu. Rev. Plant Physiol. Plant Mol. Biol.* **50:** 333–359.

Niyogi, K. (2000) Safety valves for photosynthesis. *Curr. Opin. Plant Biol.* **3:** 455–460.

Noctor, G. and Foyer, C. (1998) Ascorbate and glutathione: keeping active oxygen under control. *Annu. Rev. Plant Physiol. Plant Mol. Biol.* **49:** 249–279.

Pei, Z., Murata, Y., Benning, G., Thomine, S., Klusener, B., Allen, G., Grill, E. and Schroeder, J. (2000) Calcium channels activated by hydrogen peroxide mediate abscisic acid signalling in guard cells. *Nature* **406:** 731–734.

Peng, H.P., Chan, C.S., Shih, M.C. and Yang, S.F. (2001) Signaling events in the hypoxic induction of alcohol dehydrogenase gene in Arabidopsis. *Plant Physiol.* **126:** 742–749.

Petersen, M., Brodersen, P., Naested, H., *et al.* (2000) *Arabidopsis* MAP kinase 4 negatively regulates systemic acquired resistance. *Cell* **103:** 1111–1120.

Pezeshki, S.R. (1994) Responses of bald cypress (*Taxodium distichum*) seedlings to hypoxia – leaf protein-content, ribulose-1,5-bisphosphate carboxylase oxygenase activity and photosynthesis. *Photosynthetica* **30:** 59–68.

Pezeshki, S.R., Pardue, J.H. and DeLaune, R.D. (1996) Leaf gas exchange and growth of flood-tolerant and flood-sensitive tree species under low soil redox conditions. *Tree Physiol.* **16:** 453–458.

Pfannschmidt, T., Nilsson, A. and Allen, J.F. (1999) Photosynthetic control of chloroplast gene expression. *Nature* **397:** 625–628.

Pfannschmidt, T., Allen, J.F. and Oelmuller, R. (2001a) Principles of redox control in photosynthesis gene expression. *Physiol. Plant* **112:** 1–9.

Pfannschmidt, T., Schutze, K., Brost, M. and Oelmuller, R. (2001b) A novel mechanism of nuclear photosynthesis gene regulation by redox signals from the chloroplast during photosystem stoichiometry adjustment. *J. Biol. Chem.* **276:** 36125–36130.

Pfannschmidt, T., Schütze, K., Fey, V., Sherameti, I. and Oelmüller, R. (2003) Chloroplast redox control of nuclear gene expression – a new class of plastid signals in interorganellar communication. *Antioxid. Redox Signal.* **5:** 95–101.

Rao, M.V., Paliyath, G., Ormrod, D.P., Murr, D.P. and Watkins, C.B. (1997) Influence of salicylic acid on H_2O_2 production, oxidative stress, and H_2O_2-metabolizing enzymes (salicylic acid-mediated oxidative damage requires H_2O_2). *Plant Physiol.* **115:** 137–149.

Rojo, E., Titarenko, E., Leon, J., Berger, S., Vancanneyt, G. and Sanchez-Serrano, J.J. (1998) Reversible protein phosphorylation regulates jasmonic acid-dependent and -independent wound signal transduction pathways in *Arabidopsis thaliana*. *Plant J.* **13:** 153–165.

Rojo, E., Leon, J. and Sanchez-Serrano, J.J. (1999) Cross-talk between wound signaling pathways determines local versus systemic gene expression in *Arabidopsis thaliana*. *Plant J.* **20:** 135–142.

Rusterucci, C., Aviv, D.H., Holt, B.F. III, Dangl, J.L. and Parker, J.E. (2001) The disease resistance signaling components *EDS1* and *PAD4* are essential regulators of the cell death pathway controlled by *LSD1* in *Arabidopsis*. *Plant Cell* **13:** 2211–2224.

Sanchez-Casas, P. and Klessig, D.F. (1994) A salicylic acid-binding activity and a salicylic acid-inhibitable catalase activity are present in a variety of plant species. *Plant Physiol.* **106:** 1675–1679.

Schäfer, E. and Bowler, C. (2002) Phytochrome-mediated photoperception and signal transduction in higher plants. *EMBO Rep.* **3:** 1042–1048.

Scott, I.M., Clarke, S.M., Wood, J.E. and Mur, L.A. (2004) Salicylate accumulation inhibits growth at chilling temperature in Arabidopsis. *Plant Physiol.* **135:** 1040–1049.

Setter, T.L., and Waters, I. (2003). Review of prospects for germplasm improvement for waterlogging tolerance in wheat, barley and oats. *Plant and Soil* **253,** 1.

Shirasu, K., Nakajima, H., Rajasekhar, V.K., Dixon, R.A. and Lamb, C. (1997) Salicylic acid potentiates an agonist-dependent gain control that amplifies pathogen signals in the activation of defense mechanisms. *Plant Cell* **9:** 261–270.

Staswick, P., Su, W. and Howell, S. (1992) Methyl jasmonate inhibition of root growth and induction of a leaf protein are decreased in an *Arabidopsis thaliana* mutant. *Proc. Natl Acad. Sci. USA* **89:** 6837–6840.

Stokes, T.L. and Richards, E.J. (2002) Induced instability of two Arabidopsis constitutive pathogen-response alleles. *Proc. Natl Acad. Sci. USA* **99:** 7792–7796.

Subbaiah, C.C. and Sachs, M.M. (2003) Molecular and cellular adaptations of maize to flooding stress. *Ann. Bot.* **91:** 119–127.

Titarenko, E., Rojo, E., Leon, J. and Sanchez-Serrano, J. (1997) Jasmonic acid-dependent and -independent signaling pathways control wound-induced gene activation in *Arabidopsis thaliana*. *Plant Physiol.* **115:** 817–826.

Watts, R.A., Hunt, P.W., Hvitved, A.N., Hargrove, M.S., Peacock, W.J. and Dennis, E.S. (2001) A hemoglobin from plants homologous to truncated hemoglobins of microorganisms. *Proc. Natl Acad. Sci. USA* **98:** 10119–10124.

Wiermer, M., Feys, B.J. and Parker, J.E. (2005) Plant immunity: the EDS1 regulatory node. *Current Opinion in Plant Biology*, **8**, 383-388.

Wildermuth, M.C., Dewdney, J., Wu, G. and Ausubel, F.M. (2001) Isochorismate synthase is required to synthesize salicylic acid for plant defence. *Nature* **414:** 562–565.

Willekens, H., Chamnongpol, S., Davey, M., Schraudner, M., Langebartels, C., Van Montagu, M., Inze, D. and Van Camp, W. (1997) Catalase is a sink for H_2O_2 and is indispensable for stress defence in C3 plants. *EMBO J.* **16:** 4806–4816.

Wingler, A., Lea, P., Quick, W. and Leegood, R. (2000) Photorespiration: metabolic pathways and their role in stress protection. *Philos. Trans. R. Soc. Lond. B Biol. Sci.* **355:** 1517–1529.

Yang, H.-Q., Tang, R.H. and Cashmore, A.R. (2001) The signaling mechanism of *Arabidopsis* CRY1 involves direct interaction with COP1. *Plant Cell* **13:** 2573–2587.

Yu, I.C., Parker, J. and Bent, A.F. (1998) Gene-for-gene disease resistance without the hypersensitive response in *Arabidopsis dnd1* mutant. *Proc. Natl Acad. Sci. USA* **95:** 7819–7824.

Yu, K., and Patrick, W.H., Jr (2003) Redox range with minimum nitrous oxide and methane production in a rice soil under different pH. *Soil Sci. Soc. Am. J.* **67:** 1952–1958.

Yoshida, S., Ito, M., Nishida, I. and Watanabe, A. (2002) Identification of a novel gene HYS1/CPR5 that has a repressive role in the induction of leaf senescence and pathogen-defence responses in *Arabidopsis thaliana. Plant J.* **29:** 427–437.

6

Ageing research in the post-genome era: New technologies for an old problem

João Pedro de Magalhães

1 Introduction

'New horizons are never discovered by following old roads.' – John O'Reilly

The sequencing of a growing number of genomes, and of the human genome in particular, has paved the way for remarkable technological innovations that are revolutionizing research across the life sciences. Large-scale sequencing and emerging techniques and approaches, like microarrays and systems biology, are changing the way researchers study life and its intrinsic complexity. Cells and the molecules that compose them, in particular proteins programmed from the genome to give rise to complex interactions and functions, can now be studied in a high-throughput fashion. Genomics and its derivative technologies promise to transform biomedical research and be the preeminent area of scientific discovery during the 21st century.

New technologies are particularly useful in scientific fields that have already been the subject of considerable research, and thus for which substantial observations exist, but that remain largely unsolved (de Magalhães and Toussaint, 2004a). Biogerontology is one of such fields and the post-genome era promises to be a time of unprecedented breakthroughs in our understanding of ageing. The goal of this chapter is to review some of the key technological and methodological advances made possible by the recent sequencing of genomes and summarize their relevance to the study of complex traits like longevity, ageing and age-related diseases.

2 The genomic revolution

Clearly, genomic technologies are changing the way research on the science of ageing is conducted (Kaeberlein, 2004; Lim and Ng, 2007; Raghothama *et al.*, 2005). One fitting example of a breakthrough made possible by having the sequence of the human genome was the association of mutations in the *LMNA* gene with Hutchinson-Gilford progeria syndrome. Succinctly, using a combination of genetic techniques, Eriksson *et al.* (2003) identified a candidate interval of 4.82 Mb on chromosome 1. Though there are roughly 80 known genes in this region, the researchers immediately focused on *LMNA* because mutations in this gene had been previously associated with various disorders. Sequencing of *LMNA* revealed a heterozygous base substitution in 18 of 23 patients (Eriksson *et al.*, 2003).

The work on *LMNA* is an excellent example of how useful it is for genetic research to have the sequence of the human genome, yet it is only the tip of the iceberg. As will be further detailed ahead, having the genome sequence for a given organism allows genome-wide studies and genetic screens. One example is the powerful technology of RNA interference (RNAi), which can now be used to manipulate gene expression on a whole-genome level (Hannon, 2002). In fact, with the availability of genome sequences for several model organisms, RNAi technology can be used to screen thousands of genes for their effects on a given phenotype. Not surprisingly, large-scale RNAi longevity studies have already been conducted in *C. elegans* and have allowed the identification of new genes and pathways that influence lifespan (Hamilton *et al.*, 2005; Lee *et al.*, 2003).

In addition to facilitating genetic research, the sequencing of genomes has allowed the development of several large-scale techniques for the quantification of the basic molecular components of living organisms. Having the genome sequence of a given organism opens the door to the development of microarrays used in gene expression studies (Brown and Botstein, 1999) and modern proteomics like mass spectrometry (Aebersold and Mann, 2003; Dierick *et al.*, 2002). These technologies allow researchers to generate massive amounts of data and help understand how the molecular components of cells, and the genes that give rise to them, produce the complexity of life. Some of their uses in the context of research on ageing, a process that can be seen as the corruption of life, are discussed below.

2.1 *Functional genomics*

The goal of functional genomics is to understand the function of genes, in particular the function of the roughly 25 000 genes that make up the human genome. Functional genomics focuses mostly on analyses of the expression of genes and/or proteins and their interactions (i.e. protein-protein interactions and protein-DNA interactions).

Clearly, one of the most powerful genomic technologies is the microarray, which is normally used to study gene expression. Microarrays allow researchers to quantify the mRNA levels of thousands of genes. Typically, gene expression studies of ageing aim to identify genes whose expression levels change with age. Briefly, many ageing tissues have been studied in animals and, although ageing changes in gene expression are often tissue-specific, a few common patterns have emerged. For instance, some pathways appear to change with age in different model organisms, such as genes linked to cellular detoxification (McElwee *et al.*, 2007). A complementary approach is to identify genes differentially expressed in long-lived strains (e.g. in long-lived knock-outs) or conditions (e.g. under caloric restriction). In an interesting work, Murphy *et al.* (2003) used microarray analysis to identify genes differentially expressed in long-lived *C. elegans* mutant strains, many of which they later showed to influence ageing.

One problem with gene expression studies of ageing is that genes differentially expressed may not necessarily be related to the ageing process but rather to specific age-related diseases. They may also reflect programmed responses to the loss of function and tissue homeostasis that normally represents ageing, rather than an underlying mechanism of ageing (de Magalhães and Toussaint, 2004a). As such, interpreting ageing microarray data is not as straightforward as when studying other processes, like the cell cycle.

Microarrays have numerous pharmacological applications, including for age-related diseases. Briefly, one example comes from the Connectivity Map, a collection of gene-expression profiles obtained from cultured human cells treated with 164 small molecules. The Connectivity Map can be used to find connections among small molecules, diseases and physiological processes and has been used to identify candidate small molecules that might be used to help treat Alzheimer's disease (Lamb *et al.*, 2006). Another area in biogerontology in which microarrays may be valuable is in finding biomarkers of ageing. Although the search has so far been largely fruitless and even somewhat controversial, being able to quantify how aged a given organism or patient is would be extremely useful to basic and applied research (Miller, 2001). Gene expression studies – alone or integrated with other parameters – may help identify biomarkers of ageing that allow us to quantify ageing (Kriete, 2006). For example, genes differentially expressed with age whose changes are attenuated in long-lived strains are potentially good candidates. Interestingly, it has been proposed that a cheaper and quicker approach to screen candidate life-extending therapies in mice is to employ gene expression biomarkers at earlier ages rather than conduct full longevity studies (Spindler and Mote, 2007).

Because of their ability to generate large amounts of data in a relatively short period of time, gene expression studies have thus far been more used than protein studies, even though protein modifications can alter protein function and go undetected by gene expression studies. With proteomics, however, it is possible not only to analyse protein levels but inquire about post-translational modifications (Gafni, 2004). Although proteomic studies are less prevalent in ageing research, and arguably in biomedical research in general, there have been some successes. The most widely used rationale for applying proteomic analyses to ageing is the same for doing gene expression studies, that is, to identify protein differences with age or between short- and long-lived strains (Gafni, 2004). Recently, Dong *et al.* (2007) employed mass spectrometry to identify 86 proteins more or less abundant in a long-lived *daf-2* mutant *C. elegans* strain. They also showed that some of these proteins act in processes regulated by DAF-2, including development and ageing. In another study, Sowell *et al.* (2007) recently reported a mass-spectrometry-based survey of the *Drosophila* proteome at nine time points across adulthood. Overall, while not as high-throughput as gene expression studies, and in spite of advances in protein microarrays to assess protein abundance and function (LaBaer and Ramachandran, 2005), proteomics offers a complementary set of tools capable of providing unique insights.

Ultimately, one goal of functional genomics is to obtain gene expression maps for whole organisms: in other words, to know the expression patterns of nearly all genes in different cells or tissues and at different stages of the lifespan. Deriving such maps for lower organisms appears within reach, in particular for *C. elegans* since it only has 959 adult somatic cells and its cell lineage has been described. To achieve this objective, Dupuy *et al.* (2004) generated a 'Promoterome', a genome-wide resource of about 6000 *C. elegans* promoters which can be used to drive the expression of protein-encoding genes, such as localization markers like the green fluorescent protein (GFP). In mice, one fascinating resource is the Allen Brain Atlas (http://www.brain-map.org/), a digital atlas of gene expression patterns of approximately 20 000 genes in the adult mouse brain. Although it does not presently contain different time points, the Allen Brain Atlas can serve as a reference point for comparisons with ageing and/or disease states (Lein *et al.*, 2007).

Another area of intense research is the study of transcriptional regulation and regulatory networks. This is a major challenge of the post-genome era because, even knowing the building blocks of life, genes can regulate their own expression and/or the expression of other genes in complex networks and regulatory loops. Phenotypes often arise from these multiple components and their complex interactions. One powerful technology to study transcriptional regulation is the ChIP-microarray or ChIP-chip where ChIP stands for chromatin immunoprecipitation. Succinctly, to identify targets of a given transcription factor, a whole-genome microarray can be used to find genomic regions to which the transcription factor is bound (Buck and Lieb, 2004). For example, Meier *et al.* (2007) used ChIP-chip technology to study mammalian DNA-damage response factors in damaged telomeres. Other studies of ageing have also taken into account gene regulatory regions. For example, one study in the ageing human brain reported that genes with reduced expression with age had increased DNA damage in their promoters, and in cell culture these same promoters were more susceptible to damage by oxidative stress (Lu *et al.*, 2004).

Understanding transcriptional regulation is further complicated, however, because gene expression can be regulated at many levels and our knowledge of these mechanisms is still incomplete. For example, the recent discovery of microRNAs and their capacity to regulate gene expression has added a new dimension to the problem and it is likely that other unknown mechanisms will still surprise researchers. Interestingly, Boehm and Slack (2005) showed that a microRNA regulates lifespan in *C. elegans*, so studies of these mechanisms have a huge potential to provide insights about ageing.

Although the above technologies are powerful by themselves, they can of course be used in combination with each other or with other techniques. For example, Hubner *et al.* (2005) focused on two rat tissues relevant to the pathogenesis of the metabolic syndrome and integrated gene expression profiles with linkage analysis to identify 73 candidate genes for hypertension. Studies combining genetic screens with microarray analysis have also been proposed to identify longevity genes in mice (de Haan and Williams, 2005). Another interesting example combining basic biology in a model organism with genome-wide approaches comes from the *SLC24A5* gene. A mutation in its zebrafish homolog was found to be responsible for pigmentation. Because it was known that the human gene showed a strong difference between populations, the researchers then demonstrated a strong association between an allele in *SLC24A5* and variation in pigmentation between European and African populations (Lamason *et al.*, 2005). Lastly, in *Drosophila*, Lai *et al.* (2007) studied gene expression in two strains for which quantitative trait loci (QTLs) affecting lifespan had been previously identified. Combining information on genes differentially expressed with age and between the two strains, they obtained a set of 49 candidate genes.

To cope with the large amounts of data being generated, including the growing number of genes being associated with ageing, several computational resources have been developed to assist researchers in navigating the seas of information (see *Table 1*). Indeed, we developed the Human Ageing Genomic Resources (de Magalhães *et al.*,

Table 1. *Major databases of relevance to biogerontologists.*

Name	URL	Type of data
AGEMAP	http://server1-kimlab.stanford.edu/cgi-bin/index.cgi?Home	Gene expression database for ageing in mice
Aging Gene Nexus	http://gan.usc.edu/	Microarray gene expression data
AGINGDB	http://aging.pharm.pusan.ac.kr/	Oxidative stress and calorie restriction in the study of ageing
AnAge	http://genomics.senescence.info/species/	Life history traits, longevity records, and ageing-related observations in organisms
Baltimore Longitudinal Study of Aging	http://blswww.grc.nia.nih.gov/	Repository for most of the longitudinal data collected during this study
GenAge	http://genomics.senescence.info/genes/	Ageing- and longevity-associated genes in model organisms and in humans
Human Mortality Database	http://www.mortality.org/	Mortality and population data
SAGE KE Database	http://uwaging.org/genesdb/	Genes and interventions that have been studied with respect to their effects on life-span or age-related neurological diseases

2005). One of its key components is GenAge (http://genomics.senescence.info/genes/), the first curated database of genes possibly related to human ageing. Recently, we also added a database of genes associated with ageing and/or longevity in model organisms as well as a list of genes analysed for their possible association with human longevity. Our aim in developing GenAge is to collect, manage and analyse the data. Not only do we want to summarize basic information about genes related to ageing, but we also want GenAge to be useful for researchers in many types of studies, including genetic association studies, or even for exploiting clinical interventions in human ageing. Another database of ageing- and longevity-associated genes is the SAGE KE database, which is now being hosted at the University of Washington (http://uwaging.org/genesdb/index.php). Interestingly, Lunetta *et al.* (2007) recently reported a genome-wide association study of longevity genes and performed an analysis focusing on candidate genes obtained from GenAge and from the SAGE KE database. Finally, another potentially useful database in biogerontology is the Gene Aging Nexus, a database of ageing gene-expression data (Pan *et al.*, 2007).

2.2 *Comparative and evolutionary genomics*

The human genome can be seen as the digital code from which proteins, the molecular machines of life, emerge (Hood and Galas, 2003). Comparative genomics, which is the analysis and comparison of different genomes (either from different species or from different individuals of the same species), provides opportunities to decipher that code by allowing researchers to associate DNA variation with phenotypic variation.

Ageing and longevity are quantitative traits. Both are complex in the sense that they are modulated by two or more genes and their interactions with each other and with the environment. Associating complex diseases with DNA variants is challenging but researchers are now benefiting from the availability of large numbers of genetic markers and new genotyping technologies (Cardon and Bell, 2001). One common type of genetic variation is the single-nucleotide polymorphism (SNP). SNPs are now widely used genetic markers in genetic association studies and one major resource is HapMap, a public database with several million SNPs (International HapMap Consortium, 2005). Thanks to this and other projects, massive genome-wide association studies are ongoing to search for disease genes. One example is the Wellcome Trust Case Control Consortium (2007), which examined about 500 000 SNPs in 14 000 individuals affected by major diseases and in 3000 controls, yielding evidence for the involvement of many loci in the diseases.

Such studies promise to help bridge the gap between genotype and phenotype for hundreds of complex diseases. By focusing on longevity, genome-wide studies have a huge potential to help unravel the genetic basis of longevity, age-related diseases, and possibly of the ageing process itself. Of course, genes associated with longevity may or may not be related to the basic ageing process. For example, one of the first genes associated with longevity was *APOE*, with one of its alleles known to promote premature atherosclerosis reported to be less frequent in centenarians than in controls (Schachter *et al.*, 1994). Whether *APOE* is involved in the basic process of ageing remains to be determined. Nonetheless, associating genes with longevity and age-related diseases has a remarkable potential to improve human health and in some cases may provide mechanistic clues about ageing (Barzilai and Shuldiner, 2001). Moreover, with the dropping costs of resequencing the human genome, knowing the genetic determinants of human longevity is crucial for developing personalized medicine (Hood, 2003).

Certainly, an exciting area of research is identifying the genetic basis for individual variance in longevity and susceptibility to age-related diseases. An even greater phenotypic variation than that observed between individuals, however, is of course observed between species, and this is also reflected in the biology of ageing. Even similar species, like primates, can age at remarkably different rates (*Figure 1*). As in population studies, the rationale behind evolutionary genomics is that by studying different phenotypes of the process under study we can gather clues about its underlying genetics and mechanisms. It is clear that the genomic information determines the features of each species' ageing process to a large extent (Austad, 2005; de Magalhães, 2003). Multiple sequenced genomes provide data-mining opportunities to at least derive testable hypotheses. Identifying some of the genes involved in species differences in longevity would help understand the cellular and molecular basis for differences in the ageing process and in diseases like cancer, type II diabetes and neurodegenerative diseases. Eventually, it would inform about human susceptibility to such diseases.

Due to its much smaller size, sequencing the mitochondrial DNA (mtDNA) is considerably cheaper and quicker than sequencing the nuclear genome. As such, the mtDNA of hundreds of vertebrates has now been sequenced. Because mitochondria

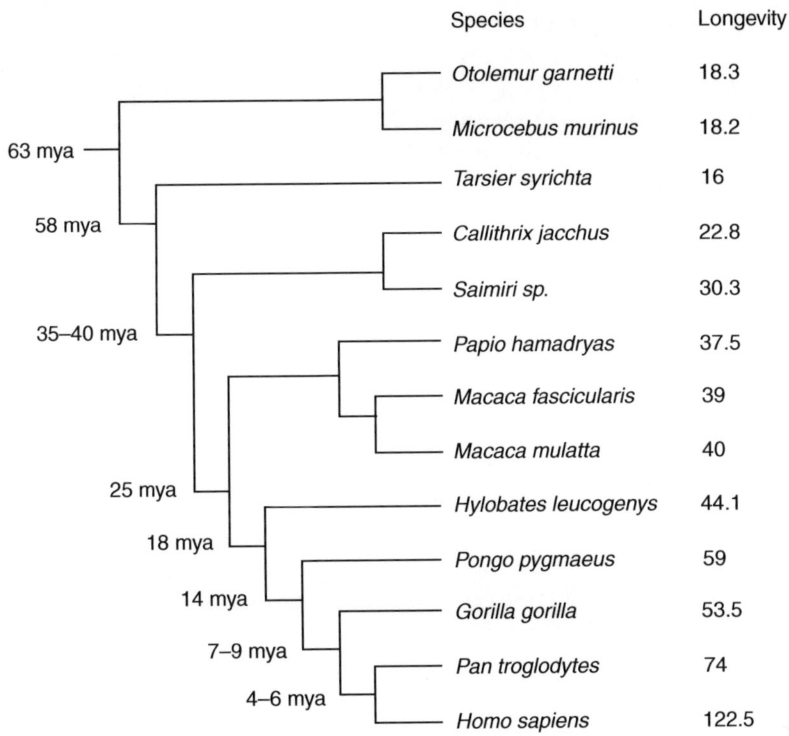

Figure 1. *Phylogeny and longevity of primate species approved for sequencing at the time of writing (http://www.genome.gov/10002154). The strategy behind evolutionary genomics of ageing, which can be seen as a form of comparative genomics, is to take advantage of the variation in longevity between similar species and of the growing number of sequenced genomes to understand not only the evolutionary pressures shaping longevity but obtain biologically significant clues about the genetics of ageing and age-related diseases. Longevity refers to maximum longevity and was obtained from build 9 of the AnAge database (de Magalhães* et al., *2007). Because it is not yet possible to quantify rate of ageing, maximum longevity under protected conditions is often used as an approximation to compare differences in ageing between species (de Magalhães, 2006). Human longevity, of course, may be overestimated due to the much larger sample size. Phylogeny from Goodman, 1999. Branch sizes are not to scale. mya, million years ago.*

have been associated with ageing, and hypothesized to play a causal role in the human ageing process, some studies have been conducted trying to associate some feature of the mtDNA with the longevity of different species (de Magalhães, 2005a; Khaidakov *et al.*, 2006; Lehmann *et al.*, 2006; Rottenberg, 2007). For example, in one intriguing study Samuels (2005) demonstrated a relation between the free energy of mitochondrial DNA, which can be seen as a factor in its susceptibility to mutation, and longevity. Although there are alternative explanations for all these studies, they could suggest a general pattern of selection on the mitochondrion that is associated with the evolution of longevity.

When studying the evolution of the much-larger nuclear genome to gather clues about ageing, one major drawback is the fact that ageing is a largely mysterious process. The genetic basis of longevity is also mostly unknown, which means researchers are virtually searching needles in a haystack. An example may be useful. Because humans tend to live longer and exhibit signs of ageing at later ages than chimpanzees, we recently studied the evolutionary forces acting on human-chimpanzee orthologous gene pairs. With only two species, however, our signal-to-noise ratio was still low and hence our analyses were limited to protein-coding sequences and to pathways and mechanisms previously associated with ageing. Surprisingly, we found that ageing-associated genes tend to evolve under stronger-than-average functional constraints. Albeit interesting from an evolutionary perspective, however, we failed to obtain significant insights regarding the genetic basis for differences in ageing and longevity between these two species (de Magalhães and Church, 2007).

Many different theoretical frameworks seeking to explain species differences in ageing have been put forward. One of the most popular explanations is the free radical theory of ageing, which succinctly argues that species with higher generation of reactive oxygen species (ROS) accumulate damage faster and thus age faster. Because mitochondria, the cells' powerhouses, are one major source of ROS, the free radical theory offered an elegant explanation to the long-held view that short-lived mammals have higher metabolic rates (Harman, 1981). Recently, however, this explanation has been put into question by the demonstration that metabolic rates, when statistically corrected for the effects of body mass and phylogeny, do not correlate with longevity in placental mammals or in birds (de Magalhães *et al.*, 2007a).

In addition to ageing being largely mysterious, another problem in evolutionary genomics (of ageing and otherwise) is that we have few clues regarding which genomic features are involved in species differences and in adaptation. New genes can obviously give rise to new phenotypes. For example, transgenic mice expressing a CPD-photolyase, enzymes involved in repair of UV-induced damage that were evolutionarily lost in placental mammals, were protected from sunlight-induced tumorigenesis (Jans *et al.*, 2005). Probably, though, more subtle adaptive changes in gene products or in their regulatory regions underlie the augmented human longevity. Considering that humans are thought to have essentially the same set of genes as chimpanzees, it is unlikely that differences in ageing are due to the loss or evolution of new genes, at least among such closely related species. It has long been thought that differences between similar species may be due to transcriptional regulation (King and Wilson, 1975), though probably a combination of changes in regulatory and protein-coding regions are involved (Hoekstra and Coyne, 2007). The challenge is to identify such changes from the background of neutral mutations.

In spite of these difficulties, it appears plausible that, as the genomes of more organisms are sequenced, it will be possible to associate selection patterns in one or more genes with the evolution of longevity in mammalian and/or primate lineages. In that context, sequencing the genome of long-lived mammals would be a tremendous help (de Magalhães *et al.*, 2007b). New genomic methods also promise to aid in these efforts. Because differences in gene expression might play a role in species differences in ageing, gene expression studies can be used in combination with genome comparisons to study regulatory networks. Using a multispecies cDNA microarray, Gilad *et al.* (2006) identified genes with expression levels significantly elevated or reduced in the human lineage when compared to other primates. Transcription factors were particularly prevalent among the identified genes. Overall, I am convinced that increases in the signal-to-noise ratio through better methods and/or more data will make it possible to identify candidate human longevity genes. Eventually, with the dropping costs of resequencing the human genome (Church, 2006), such genes can be further studied in genetic association studies in human populations and may provide important clues about the basic cellular and molecular mechanisms of ageing.

3 Systems biology

The aforementioned technologies have allowed researchers to survey biological systems and their molecular components on a global scale. Systems biology is not only the study of these components but how they interact to give rise to the system under study (Hood, 2003; Kriete *et al.*, 2006). In biogerontology, many genes have been related to ageing in model systems but it is now necessary to study how these genes interact and how they exert their influence as an aggregate to modulate the ageing process. As such, the ageing process is suitable for a system-level approach (Kirkwood *et al.*, 2003; Vijg and Suh, 2003). GenAge, in fact, was developed as a tool for systems biology by providing a core dataset that allows researchers to employ data-mining algorithms and computational approaches to study ageing from a genomic, proteomic and evolutionary perspective.

One simple way to study a complex system is to observe the effects of disrupting each of its components. Indeed, based on GenAge, I developed a system-level interpretation of genetic interventions that alter the ageing process in model systems to help comprehend the mechanisms of ageing. One of my conclusions was that DNA alterations appear to be more relevant to ageing than other forms of molecular damage (de Magalhães, 2005b).

As aforementioned, a key aspect of systems biology involves the study of how the system's components – typically genes and/or proteins, but they could be metabolites or cells – interact. Some potentially useful network studies of ageing have already been conducted (Bergman *et al.*, 2007; Promislow, 2004). Of particular interest, a number of researchers have studied networks of ageing-associated proteins in order to identify new candidate genes or proteins that may play a role in ageing (Budovsky *et al.*, 2007; de Magalhães and Toussaint, 2004b; Ferrarini *et al.*, 2005). In this 'guilt-by-association' method, the principle is that proteins that tend to interact with proteins previously associated with ageing have a higher probability of also being involved in ageing than what would be expected by chance, and are thus promising targets for further experiments (see *Figure 2*).

The future of systems biology and of biological research in general is in integrating different sources of information to understand complex processes. Indeed systems biology is not limited to a few technologies or approaches. For example, as discussed above, evolutionary biology provides a unique perspective on gene action, and evolutionary studies can be integrated into systems biology. In fact, evolutionary systems biology is a major emerging field whose broad goal is to study how gene networks evolve (Medina, 2005). Understanding regulatory networks also benefits from the integration of different types of data, such as combining microarray data with literature mining, which can be seen as a form of systems biology (Werner, 2007). Likewise, some

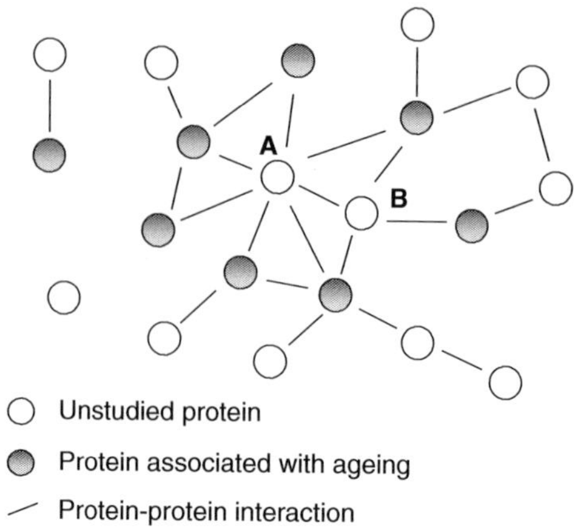

Figure 2. *'Guilt-by-association' method in a simple hypothetical network of protein-protein interactions. Each circle represents a protein with grey circles representing proteins previously associated with ageing and white circles representing proteins not previously studied in the context of ageing. Just from visualizing the network, it is clear that proteins A and B are the best targets for future experiments because they interact with a large number of proteins already linked to ageing, which makes A and B more likely to play a role in ageing. Typically, though, protein networks tend to be more complex with more proteins and interactions. It is also possible to define the type of the interactions and take into consideration regulatory interactions, such as activation or repression, which require more advanced algorithms to infer the best targets for future experiments.*

of the integrative genomic approaches described above can be classified as systems biology approaches.

Although promising, applying systems biology to ageing research still faces some significant hurdles. For instance, ageing research suffers from a lack of appropriate experimental models (de Magalhães and Toussaint, 2004a). In cancer biology, applying the principles of systems biology is much easier because cancer researchers have at their disposal multiple cancer cell lines in which to conduct a battery of large-scale analyses using '-omic' technologies, which can then be integrated to identify cancer markers that make it possible to develop personalized cancer therapies (Weinstein, 2004). In fact, one exciting project is the Cancer Genome Atlas, a comprehensive effort to explore the spectrum of genomic changes associated with cancer (http://cancergenome.nih.gov). Systems biology is also appropriate to study infectious diseases, which perturb the network structure of a system (e.g. the immune system), and are themselves caused by agents that are made of interacting components and whose genome can be (or has been) sequenced. Although cellular models of ageing are suitable for systems biology, applying systems biology to complex organisms like mammals still suffers from a lack of experimental paradigms. Other applications and problems of systems biology in the study of ageing are discussed below.

3.1 *Modelling the ageing process*

One goal of systems biology is to develop computer models of the whole system from its parts and their relationships. *In silico* models are easier to interpret and may allow novel insights into the process under study. The ultimate aim of modelling is to predict phenotypes that may help guide experiments and may even have clinical applications. For example, as mentioned above, quantifying different aspects of ageing may reveal a fingerprint of ageing that allows researchers to quantify the whole ageing process (Kriete, 2006).

Models can be constructed at different levels, focusing on different components of the system, and with varying degrees of mathematical complexity (de Jong, 2002). The type of model depends on what is sought from the model and on the data available, with better data of course allowing more complete models. Modelling can follow a top-down or a bottom-up approach (or a combination of the two). A top-down model begins with an overview of the system. Its parts are then refined until a level of detail is reached that provides suitable predictions. In bottom-up modelling, the components of the system are first identified and characterized and their relationships are then put together until the entire system is modelled.

Probably the main problem in modelling ageing is that we still do not understand the essence of the ageing process (de Magalhães, 2005b). Some researchers have even argued that there is no basic ageing process and that what we call ageing is in fact the product of different age-related diseases and degenerative processes occurring approximately in parallel (Holliday, 1995). Recent breakthroughs in the genetics of ageing showing it is possible to modulate the entire ageing process by manipulating single genes, however, have cast doubts on this view (de Magalhães, 2003).

Others view ageing as the result of one or more forms of molecular damage accumulating with age, in particular damage caused by chemical reactions such as oxidative damage or DNA mutations. But even if ageing is caused by the accumulation of molecular damage and inefficient repair systems, there are different mechanisms by which this may occur and these need to be modelled in different ways. Bell (1998) argued that there could be conceptually different types of repair. DNA, for example, can be repaired based on an original blueprint. Another way of eliminating errors is to have a large number of DNA copies, for example in different cells, and then subject those cells to a given test that eliminates cells with a number of DNA mutations above a certain threshold (in this case, the outcome might be apoptosis). To put it another way, evolution may increase longevity by optimizing repair mechanisms – and even a slight increase in DNA repair will have major consequences over the large number of cell divisions necessary to produce an adult mammal – or evolution might operate by optimizing cell selection (e.g. acting on processes like apoptosis and mitosis).

The point has also been made that degeneration is not caused by damage to the components of a system but rather by the inability to replace those components (Williams, 1957). Ageing can be seen as a result of systems being redundant in irreplaceable elements (Gavrilov and Gavrilova, 2001), and some authors have suggested a 'network theory of ageing' integrating different forms of molecular damage (Franceschi *et al.*, 2000; Kowald and Kirkwood, 1996). Alternatively, a completely different hypothesis is that some developmental mechanisms shaped by evolution to optimize reproduction have an impact on ageing and age-related diseases (de Magalhães and Church, 2005). The free radical theory of ageing can, in fact, be interpreted in light of the antagonistic roles of ROS in development/growth and in ageing (de Magalhães and Church, 2006). It appears that at least some forms of damage may be caused by genetic programmes (e.g. as a form of antagonistic pleiotropy) rather than inefficient repair systems.

Overall, there is no consensus on what the driving force of ageing is or what type of components play a role in it, which makes it nearly impossible to model ageing in higher organisms with precision. For now, modelling ageing has been restricted to lower organisms or simple pathways. A model of yeast ageing has been proposed (Gillespie *et al.*, 2004), as has a model of the role of chaperones in ageing (Proctor *et al.*, 2005). Ultimately, however, the goal of systems biology is to reconstruct the causal structure of the human ageing process at different levels and understand its genetic network.

4 Future prospects

'By the year 2030, we will have (1) developed a complete model of all human cell types, obviating the need for many laboratory experiments [by doing computer simulations instead]; (2) lowered the cost of doing a complete genomic sequence for an human individual to less than $1,000 each; and (3) catalogued all the genes involved in ageing. Therefore, human clinical trials to extend lifespan could already be underway by this date.' – Francis Collins

It is my belief that understanding ageing will require what is known in reverse engineering as 'a grey box approach': a combination of white and black box. A white box is the source code, the binary code being decoded: in this case, the genome. The black box approach involves observing results of different inputs in the system being analysed to understand its functioning: in this case, genetic manipulation experiments followed by observations from gross phenotypes to changes in genes and proteins. In this context, microarrays and proteomics provide new high-throughput ways to generate massive amounts of detailed observations of the system under study, while a systems-biology thinking allows researchers to fit these immense amounts of data into coherent models.

The lack of a theoretical framework that explains ageing (or the large numbers of frameworks that may or may not be correct) is a difficulty in applying genomic technologies to ageing, from interpreting gene expression results to modelling ageing, passing by evolutionary genomics. On the other hand, genomic technologies have so far been mostly used for hypothesis generation rather than hypothesis testing and may even provide new conceptual frameworks that have so far eluded biogerontologists.

Biomedical discovery will continue to benefit from the extraordinary progress of technology and provide researchers with new horizons. For example, as mentioned above, comparative genomics is an emerging field and it will soon be possible to resequence the genome of thousands of patients. As resequencing costs continue to drop, however, even more ambitious projects become feasible. One endeavour that may allow researchers to address the role of mutations in ageing would be to produce a phylogenetic tree of cells. The idea would be to catalogue somatic mutations in the genomes of individual cells during development and ageing of mammals, to sequence the genome of many cells in a mouse. This would allow researchers to infer the history of cell divisions and obtain a mammalian cell-fate map describing cellular lineages during the entire lifespan (Salipante and Horwitz, 2007).

Ultimately, biomedical research will gradually transform biologists and physicians from observers to architects. At present, there are inherited limitations to radically engineering biological systems and not many success stories. In non-biological fields, engineers have an in-depth understanding that permits computer models and an abstraction of the task at hand. The situation is quite different in biology, at least for now (Heinemann and Panke, 2006). Progress in the emergent field of synthetic biology, however, coupled with a more complete understanding of biological processes made possible by genomic technologies, promises gradually to augment our capacity to engineer biology. De Grey (2003), for example, has argued that it will be possible within a foreseeable future to engineer human ageing to the point of reversing it.

In conclusion, this is an exciting time for research in the biomedical sciences with unprecedented opportunities for discovery. When compared to other biomedical fields, ageing research is still in its infancy. Biogerontologists have now high-throughput technologies with which to prod the mechanisms of ageing, and the computational capacity to build models of the ageing process with an increasing ability to predict phenotypes. In the next decade or two, I anticipate we will learn more about the basic ageing process and its genetic, cellular and molecular underpinnings than ever before.

Acknowledgements

I would like to thank everyone at the Annual Scientific Meeting of the Society for Experimental Biology in Glasgow, March–April 2007, for excellent discussions on these topics. Further thanks to the participants of the Workshop on Complexities of Aging in Biological Systems in Santa Fe, March 2007, for useful suggestions and ideas particularly on applying systems biology to ageing. My work is supported by a NIH-NHGRI CEGS grant to George Church.

References

Aebersold, R. and Mann, M. (2003) Mass spectrometry-based proteomics. *Nature* **422:** 198–207.

Austad, S.N. (2005) Diverse aging rates in metazoans: targets for functional genomics. *Mech. Ageing Dev.* **126:** 43–49.

Barzilai, N. and Shuldiner, A.R. (2001) Searching for human longevity genes: the future history of gerontology in the post-genomic era. *J. Gerontol. A Biol. Sci. Med. Sci.* **56:** M83–87.

Bell, G. (1988) *Sex and Death in Protozoa: the History of Obsession*. Cambridge University Press, Cambridge.

Bergman, A., Atzmon, G., Ye, K., MacCarthy, T. and Barzilai, N. (2007) Buffering mechanisms in aging: a systems approach toward uncovering the genetic component of aging. *PLoS Comput. Biol.* **3:** e170.

Boehm, M. and Slack, F. (2005) A developmental timing microRNA and its target regulate life span in *C. elegans*. *Science* **310:** 1954–1957.

Brown, P.O. and Botstein, D. (1999) Exploring the new world of the genome with DNA microarrays. *Nat. Genet.* **21:** 33–37.

Buck, M.J. and Lieb, J.D. (2004) ChIP-chip: considerations for the design, analysis, and application of genome-wide chromatin immunoprecipitation experiments. *Genomics* **83:** 349–360.

Budovsky, A., Abramovich, A., Cohen, R., Chalifa-Caspi, V. and Fraifeld, V. (2007) Longevity network: construction and implications. *Mech. Ageing Dev.* **128:** 117–124.

Cardon, L.R. and Bell, J.I. (2001) Association study designs for complex diseases. *Nat. Rev. Genet.* **2:** 91–99.

Church, G.M. (2006) Genomes for all. *Sci. Am.* **294:** 46–54.

de Grey, A.D. (2003) An engineer's approach to the development of real anti-aging medicine. *Sci. Aging Knowledge Environ.* **2003:** VP1.

de Haan, G. and Williams, R.W. (2005) A genetic and genomic approach to identify longevity genes in mice. *Mech. Ageing Dev.* **126:** 133–138.

de Jong, H. (2002) Modeling and simulation of genetic regulatory systems: a literature review. *J. Comput. Biol.* **9:** 67–103.

de Magalhães, J.P. (2003) Is mammalian aging genetically controlled? *Biogerontology* **4:** 119–120.

de Magalhães, J.P. (2005a) Human disease-associated mitochondrial mutations fixed in nonhuman primates. *J. Mol. Evol.* **61:** 491–497.

de Magalhães, J.P. (2005b) Open-minded scepticism: inferring the causal mechanisms of human ageing from genetic perturbations. *Ageing Res. Rev.* **4:** 1–22.

de Magalhães, J.P. (2006) Species selection in comparative studies of aging and anti-aging research. In: *Handbook of Models for Human Aging* (ed. P.M. Conn). Elsevier Academic Press, Burlington, MA, pp. 9–20.

de Magalhães, J.P. and Church, G.M. (2005) Genomes optimize reproduction: aging as a consequence of the developmental program. *Physiology (Bethesda)* **20:** 252–259.

de Magalhães, J.P. and Church, G.M. (2006) Cells discover fire: employing reactive oxygen species in development and consequences for aging. *Exp. Gerontol.* **41:** 1–10.

de Magalhães, J.P. and Church, G.M. (2007) Analyses of human-chimpanzee orthologous gene pairs to explore evolutionary hypotheses of aging. *Mech. Ageing Dev.* **128:** 355–364.

de Magalhães, J.P. and Toussaint, O. (2004a) How bioinformatics can help reverse engineer human aging. *Ageing Res. Rev.* **3:** 125–141.

de Magalhães, J.P. and Toussaint, O. (2004b) GenAge: a genomic and proteomic network map of human ageing. *FEBS Lett.* **571:** 243–247.

de Magalhães, J.P., Costa, J. and Toussaint, O. (2005) HAGR: the Human Ageing Genomic Resources. *Nucleic Acids Res.* **33 Database Issue:** D537–543.

de Magalhães, J.P., Costa, J. and Church, G.M. (2007a) An analysis of the relationship between metabolism, developmental schedules, and longevity using phylogenetic independent contrasts. *J. Gerontol. A Biol. Sci. Med. Sci.* **62:** 149–160.

de Magalhães, J.P., Sedivy, J.M., Finch, C.E., Austad, S.N. and Church, G.M. (2007b) A proposal to sequence genomes of unique interest for research on aging. *J. Gerontol. A Biol. Sci. Med. Sci.* **62:** 583–584.

Dierick, J.F., Dieu, M., Remacle, J., Raes, M., Roepstorff, P. and Toussaint, O. (2002) Proteomics in experimental gerontology. *Exp. Gerontol.* **37:** 721–734.

Dong, M.Q., Venable, J.D., Au, N., Xu, T., Park, S.K., Cociorva, D., Johnson, J.R., Dillin, A. and Yates, J.R. 3rd (2007) Quantitative mass spectrometry identifies insulin signaling targets in *C. elegans*. *Science* **317:** 660–663.

Dupuy, D., Li, Q.R., Deplancke, B., *et al.* (2004) A first version of the *Caenorhabditis elegans* Promoterome. *Genome Res.* **14:** 2169–2175.

Eriksson, M., Brown, W.T., Gordon, L.B., *et al.* (2003) Recurrent de novo point mutations in lamin A cause Hutchinson-Gilford progeria syndrome. *Nature* **423:** 293–298.

Ferrarini, L., Bertelli, L., Feala, J., McCulloch, A.D. and Paternostro, G. (2005) A more efficient search strategy for aging genes based on connectivity. *Bioinformatics* **21:** 338–348.

Franceschi, C., Valensin, S., Bonafe, M., Paolisso, G., Yashin, A.I., Monti, D. and De Benedictis, G. (2000) The network and the remodeling theories of aging: historical background and new perspectives. *Exp. Gerontol.* **35:** 879–896.

Gafni, A. (2004) Proteomics in aging-related research. *Sci. Aging Knowledge Environ.* **2004:** pe41.

Gavrilov, L.A. and Gavrilova, N.S. (2001) The reliability theory of aging and longevity. *J. Theor. Biol.* **213:** 527–545.

Gilad, Y., Oshlack, A., Smyth, G.K., Speed, T.P. and White, K.P. (2006) Expression profiling in primates reveals a rapid evolution of human transcription factors. *Nature* **440:** 242–245.

Gillespie, C.S., Proctor, C.J., Boys, R.J., Shanley, D.P., Wilkinson, D.J. and Kirkwood, T.B. (2004) A mathematical model of ageing in yeast. *J. Theor. Biol.* **229:** 189–196.

Goodman, M. (1999) The genomic record of Humankind's evolutionary roots. *Am. J. Hum. Genet.* **64:** 31–39.

Hamilton, B., Dong, Y., Shindo, M., Liu, W., Odell, I., Ruvkun, G. and Lee, S.S. (2005) A systematic RNAi screen for longevity genes in *C. elegans*. *Genes Dev* **19:** 1544–1555.

Hannon, G.J. (2002) RNA interference. *Nature* **418:** 244–251.

Harman, D. (1981) The aging process. *Proc. Natl Acad. Sci. USA* **78:** 7124–7128.

Heinemann, M. and Panke, S. (2006) Synthetic biology – putting engineering into biology. *Bioinformatics* **22:** 2790–2799.

Hoekstra, H.E. and Coyne, J.A. (2007) The locus of evolution: evo devo and the genetics of adaptation. *Evol. Int. J. Org. Evol.* **61:** 995–1016.

Holliday, R. (1995) *Understanding Ageing*. Cambridge University Press, Cambridge.

Hood, L. (2003) Systems biology: integrating technology, biology, and computation. *Mech. Ageing Dev.* **124:** 9–16.

Hood, L. and Galas, D. (2003) The digital code of DNA. *Nature* **421:** 444–448.

Hubner, N., Wallace, C.A., Zimdahl, H., *et al.* (2005) Integrated transcriptional profiling and linkage analysis for identification of genes underlying disease. *Nat. Genet.* **37:** 243–253.

International HapMap Consortium (2005) A haplotype map of the human genome. *Nature* **437:** 1299–1320.

Jans, J., Schul, W., Sert, Y.G., *et al.* (2005) Powerful skin cancer protection by a CPD-photolyase transgene. *Curr. Biol.* **15:** 105–115.

Kaeberlein, M. (2004) Aging-related research in the '-omics' age. *Sci. Aging Knowledge Environ.* **2004:** pe39.

Khaidakov, M., Siegel, E.R. and Shmookler Reis, R.J. (2006) Direct repeats in mitochondrial DNA and mammalian lifespan. *Mech. Ageing Dev.* **127:** 808–812.

King, M.C. and Wilson, A.C. (1975) Evolution at two levels in humans and chimpanzees. *Science* **188:** 107–116.

Kirkwood, T.B., Boys, R.J., Gillespie, C.S., Proctor, C.J., Shanley, D.P. and Wilkinson, D.J. (2003) Towards an e-biology of ageing: integrating theory and data. *Nat. Rev. Mol. Cell Biol.* **4:** 243–249.

Kowald, A. and Kirkwood, T.B. (1996) A network theory of ageing: the interactions of defective mitochondria, aberrant proteins, free radicals and scavengers in the ageing process. *Mutat. Res.* **316:** 209–236.

Kriete, A. (2006) Biomarkers of aging: combinatorial or systems model? *Sci. Aging Knowledge Environ.* **2006:** pe1.

Kriete, A., Sokhansanj, B.A., Coppock, D.L. and West, G.B. (2006) Systems approaches to the networks of aging. *Ageing Res. Rev.* **5:** 434-448.

LaBaer, J. and Ramachandran, N. (2005) Protein microarrays as tools for functional proteomics. *Curr. Opin. Chem. Biol.* **9:** 14–19.

Lai, C.Q., Parnell, L.D., Lyman, R.F., Ordovas, J.M. and Mackay, T.F. (2007) Candidate genes affecting *Drosophila* life span identified by integrating microarray gene expression analysis and QTL mapping. *Mech. Ageing Dev.* **128:** 237–249.

Lamason, R.L., Mohideen, M.A., Mest, J.R., *et al.* (2005) SLC24A5, a putative cation exchanger, affects pigmentation in zebrafish and humans. *Science* **310:** 1782–1786.

Lamb, J., Crawford, E.D., Peck, D., *et al.* (2006) The Connectivity Map: using gene-expression signatures to connect small molecules, genes, and disease. *Science* **313:** 1929–1935.

Lee, S.S., Lee, R.Y., Fraser, A.G., Kamath, R.S., Ahringer, J. and Ruvkun, G. (2003) A systematic RNAi screen identifies a critical role for mitochondria in *C. elegans* longevity. *Nat. Genet.* **33:** 40–48.

Lehmann, G., Budovsky, A., Muradian, K.K. and Fraifeld, V.E. (2006) Mitochondrial genome anatomy and species-specific lifespan. *Rejuvenat. Res.* **9:** 223–226.

Lein, E.S., Hawrylycz, M.J., Ao, N., *et al.* (2007) Genome-wide atlas of gene expression in the adult mouse brain. *Nature* **445:** 168–176.

Lim, C.A. and Ng, H.H. (2007) Application of advanced technologies in ageing research. *Mech. Ageing Dev.* **128:** 149–160.

Lu, T., Pan, Y., Kao, S.Y., Li, C., Kohane, I., Chan, J. and Yankner, B.A. (2004) Gene regulation and DNA damage in the ageing human brain. *Nature* **429:** 883–891.

Lunetta, K.L., D'Agostino, R.B. Sr., Karasik, D., *et al.* (2007) Genetic correlates of longevity and selected age-related phenotypes: a genome-wide association study in the Framingham Study. *BMC Med. Genet.* **8 Suppl 1:** S13.

McElwee, J.J., Schuster, E., Blanc, E., *et al.* (2007) Evolutionary conservation of regulated longevity assurance mechanisms. *Genome Biol.* **8:** R132.

Medina, M. (2005) Genomes, phylogeny, and evolutionary systems biology. *Proc. Natl Acad. Sci. USA* **102 Suppl 1:** 6630–6635.

Meier, A., Fiegler, H., Munoz, P., Ellis, P., Rigler, D., Langford, C., Blasco, M.A., Carter, N. and Jackson, S.P. (2007) Spreading of mammalian DNA-damage response factors studied by ChIP-chip at damaged telomeres. *EMBO J.* **26:** 2707–2718.

Miller, R.A. (2001) Biomarkers of aging. *Sci. Aging Knowledge Environ.* **2001:** pe2.

Murphy, C.T., McCarroll, S.A., Bargmann, C.I., Fraser, A., Kamath, R.S., Ahringer, J., Li, H. and Kenyon, C. (2003) Genes that act downstream of DAF-16 to influence the lifespan of *Caenorhabditis elegans*. *Nature* **424:** 277–283.

Pan, F., Chiu, C.H., Pulapura, S., *et al.* (2007) Gene Aging Nexus: a web database and data mining platform for microarray data on aging. *Nucleic Acids Res.* **35:** D756–759.

Proctor, C.J., Soti, C., Boys, R.J., Gillespie, C.S., Shanley, D.P., Wilkinson, D.J. and Kirkwood, T.B. (2005) Modelling the actions of chaperones and their role in ageing. *Mech. Ageing Dev.* **126:** 119–131.

Promislow, D.E. (2004) Protein networks, pleiotropy and the evolution of senescence. *Proc. Biol. Sci.* **271:** 1225–1234.

Raghothama, C., Harsha, H.C., Prasad, C.K. and Pandey, A. (2005) Bioinformatics and proteomics approaches for aging research. *Biogerontology* **6:** 227–232.

Rottenberg, H. (2007) Coevolution of exceptional longevity, exceptionally high metabolic rates, and mitochondrial DNA-coded proteins in mammals. *Exp. Gerontol.* **42:** 364–373.

Salipante, S.J. and Horwitz, M.S. (2007) A phylogenetic approach to mapping cell fate. *Curr. Top. Dev. Biol.* **79:** 157–184.

Samuels, D.C. (2005) Life span is related to the free energy of mitochondrial DNA. *Mech. Ageing Dev.* **126:** 1123–1129.

Schachter, F., Faure-Delanef, L., Guenot, F., Rouger, H., Froguel, P., Lesueur-Ginot, L. and Cohen, D. (1994) Genetic associations with human longevity at the APOE and ACE loci. *Nat. Genet.* **6:** 29–32.

Sowell, R.A., Hersberger, K.E., Kaufman, T.C. and Clemmer, D.E. (2007) Examining the proteome of *Drosophila* across organism lifespan. *J. Proteome Res.* **6:** 3637–3647.

Spindler, S.R. and Mote, P.L. (2007) Screening candidate longevity therapeutics using gene-expression arrays. *Gerontology* **53:** 306–321.

Vijg, J. and Suh, Y. (2003) Functional genomics of ageing. *Mech. Ageing Dev.* **124:** 3–8.

Weinstein, J.N. (2004) Integromic analysis of the NCI-60 cancer cell lines. *Breast Dis.* **19:** 11–22.

Wellcome Trust Case Control Consortium (2007) Genome-wide association study of 14,000 cases of seven common diseases and 3,000 shared controls. *Nature* **447:** 661–678.

Werner, T. (2007) Regulatory networks: linking microarray data to systems biology. *Mech. Ageing Dev.* **128:** 168–172.

Williams, G.C. (1957) Pleiotropy, natural selection, and the evolution of senescence. *Evolution* **11:** 398–411.

7

Telomeres, ageing and oxidation

Kelly J. Lamb and Paul G. Shiels

1 Introduction

Oxidation processes are considered to be significant contributory factors to cellular and organ senescence and age-related diseases. Typically, these processes effect reduced physiological capacity and decreased longevity as a consequence of accelerated biological ageing. As a consequence, many of the diseases associated with increasing chronological age are influenced by the cumulative burden of acquired oxidative damage.

Oxidant stress has been implicated in a growing number of age-related pathologies, including cardiovascular disease (Brouilette *et al.*, 2003; Samani *et al.*, 2001), neurodegeneration (Shiels and Davies, 2003), transplant dysfunction (Joosten *et al.*, 2003; Lamb *et al.*, 2006; Shiels and Jardine, 2003), reduced immune function, psychological disorders (Epel *et al.*, 2004) and mortality associated with chronic kidney disease (Carrero *et al.*, 2008). Determining suitable characteristics and measuring the efficacy of antioxidants designed to tackle clinically such pathologies is not straightforward. There are still equivocal reports as to the exact nature of how cells respond to acute oxidant challenge. Central to this is how telomere biology is linked to such damage responses.

2 Telomeres

Telomeres are nucleo-protein complexes at the end of eukaryotic chromosomes, comprising an array of tandem TTAGGG repeats bound to a variety of proteins, with roles in chromosomal protection, nuclear attachment and replication. The proteins maintaining the structure of the telomere also function in sensing, signalling and repair of DNA damage (Shiels and Davies, 2003). Telomeres function to cap the chromosome, preventing chromosomal fusions and the recognition of the chromosome end as a DNA break. Telomeres facilitate chromosomal attachment within the correct subcellular compartment and have a critical role in DNA replication.

3 Telomere structure

Though most of the telomere comprises double-stranded DNA, the distal end contains a single strand 3′-overhang called the G-tail (Kanoh and Ishikawa, 2003). This is depicted in the schematic representation of a mammalian telomere outlined in *Figure 1*. This G-tail is folded back upon itself to form a lasso-like structure, known as the

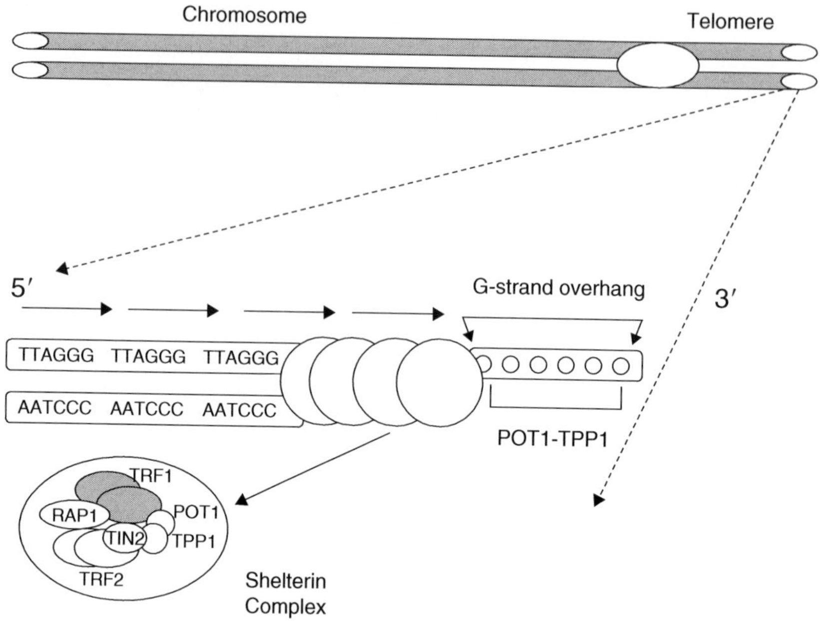

Figure 1. *Schematic representation of the mammalian telomere. Telomeres are specialized DNA nucleoprotein complexes which act as a protective cap on the ends of eukaryotic chromosomes. They are comprised of stretches of a* $(TTAGGG)_n$ *repeat, associated with a variety of proteins. The telomere DNA has a single-stranded G-rich overhang at the 3′ end called the G-tail. Protection of the G-tail, distinguishing it from a normal DNA strand break, is achieved through creation of a double loop structure, held in place by telomeric proteins.*

telomere loop (T-loop). It has been proposed that this loop invades the homologous strand of the same telomere within the double-stranded telomeric tract, forming a displacement loop (D-loop; Griffith *et al.*, 1999). The arrangement of this loop, and the associated proteins, with the double-stranded component of the telomere is thought to confer functional stability on the telomere. This acts as a protective cap for the chromosomal DNA against enzymatic degradation and from DNA-end fusion events (Kanoh and Ishikawa, 2003; Kurz *et al.*, 2004).

This unique structure is pertinent to the critical role telomeres have in facilitating chromosomal replication. During cell division, telomeres gather in a 'bouquet form', acting as chromosomal anchors to the inner nuclear membrane in early meiosis, and are involved in the pairing and recombination of homologous chromosomes (Maddar *et al.*, 2001; Siderakis and Tarsounas, 2007). Typically, in larger mammals, the telomeric DNA component shortens with successive cell divisions. This arises from what has been termed the end replication problem (Olovnikov, 1971).

The end replication problem occurs due to the fact that replication of the lagging strand of DNA, at the replication fork, must occur discontinuously, resulting in a loss of telomeric DNA from one strand every replication cycle. DNA polymerases synthesize, moving 5′ to 3′, from a starting point that is within the chromosome and not at its end. Synthesis on the leading strand proceeds 5′-3′ without any difficulty, but the lagging strand is problematic. This requires short RNA sequences (Miller and Collins, 2002) to prime lagging strand synthesis. However, when the end of the chromosome is reached, there is no DNA strand in front of the RNA primer to enable complete replication. Consequently, the telomeric DNA shortens at the 5′ end of both strands with each round of replication.

This gradual attrition in telomere length reaches a critical point, where genomic stability is compromised and the cell enters replicative senescence, a permanent state of growth arrest. This point is also termed the Hayflick limit and was first described in seminal observations by Hayflick and Moorhead (1961). Telomere attrition is, therefore, thought to represent a molecular clock, at least at the cellular level, where it has been hypothesized to act as an anti-neoplastic mechanism (Campisi and d'Adda di Fagagna, 2007). This has been proposed to function by countering the accumulation of mutations over successive cell divisions, which potentially could produce neoplastic cells.

4 Mechanisms of telomere repair

4.1 *Telomerase*

Telomeres can undergo elongation catalysed by the holo-enzyme telomerase. Telomerase is composed of a reverse transcriptase protein (TERT), the telomerase RNA (TERC) and dyskerin (Cohen *et al.*, 2007). Telomerase lengthens the telomere by adding hexanucleotide repeats copied from within its own RNA sequence. Typically, somatic cells display little or no telomerase activity and therefore, undergo progressive telomere attrition with cell division (Levy *et al.*, 1992). Stem cells, germ-line cells and most cancer cells express telomerase (Cheung and Deng, 2008). Telomerase is also expressed at high levels in T and B cells (Hodes *et al.*, 2002), where it facilitates expansion in response to immune challenge, or immune cell re-population following bone marrow transplantation. Introduction of telomerase into telomerase-negative cells extends their replicative lifespan (Bodnar *et al.*, 1998; Vaziri and Benchimol, 1998), while telomerase dysfunction, due to mutation of one of its component parts, can result in diseases such as dyskeratosis congentia, aplastic anaemia and idiopathic pulmonary fibrosis (Garcia *et al.*, 2007; Xin *et al.*, 2007).

4.2 *Alternate pathway of telomere extension*

Telomere length may also be increased by a further pathway, termed the alternative lengthening of telomeres (ALT; Weinert and Timiras, 2003). ALT is proposed to occur when the ends of the telomere, normally hidden in the T-loop, become exposed and processed by homologous recombination-dependent repair mechanisms, leading to elongation of the telomere (Verdun and Karlseder, 2006). Elongation of the telomere by either telomerase or ALT allows mutated cells to bypass normal regulatory checkpoints and to proliferate freely. This can potentially lead to tumour formation, as immortalized cells may gain additional mutations allowing for malignant transformation (Cheung and Deng, 2008).

5 Other telomere-associated proteins

A key feature from pathologies associated with aberrant function of telomere-associated proteins is that they have a role in the sensing, signalling and repair of oxidative damage to DNA (Shiels and Davies, 2003; Sklavounou *et al.*, 2006). Key players in this process are discussed below.

5.1 *Shelterin*

An important component of these processes is shelterin (see *Figure 1*), a protein complex whose location and function is specific to the telomere (de Lange, 2005). Shelterin comprises six proteins: TRF1, TRF2, RAP1, TPP1, TIN2 and POT1. Functionally, this complex shapes and protects human telomeres from degradation. Shelterin is inhibited by telomere attrition (Denchi and de Lange, 2007). In a murine fibroblast model, short telomeres have been shown to mediate DNA damage responses by the activation of the ATM (Ataxia Telangiectasia Mutated protein) and ATR (ATM and Rad 3-related) signalling pathways (Azzalin and Lingner, 2007). Normally, components of the shelterin complex act to suppress the activation of ATM and ATR to protect the telomere, which may be recognized as damaged DNA (Azzalin and Lingner, 2007) as it contains similar structural components. ATM is repressed by TRF2, while ATR is acted upon by POT1 (Denchi and de Lange, 2007). ATM and ATR are protein kinases that act as a key regulator of signalling responses to DNA double-strand breaks (DSBs).

5.2 *Tankyrase*

Tankyrase is a modular protein initially identified in HeLa cells (Smith *et al.*, 1998) involved in telomere length regulation and sister chromatid separation. Two forms of tankyrase exist: tankyrase 1 and its homolog, tankyrase 2. Tankyrase 1 binds to TRF-1, a component of the shelterin complex, via its ANK domain (Sbodio *et al.*, 2002) and is transported to the telomere. TRF-1 acts as a negative regulator of telomere length by acting as an inhibitor of telomerase at telomere ends. Recombinant tankyrase has been shown to have poly-ADP ribosylation activity, which acts on TRF-1 diminishing its ability to bind to telomeric DNA and therefore controlling telomere length (Smith *et al.*, 1998).

5.3 *Ku complex proteins*

The Ku heterodimeric autoantigen complex comprises the two highly conserved nuclear proteins G22P1 (Ku70) and XRCC5 (Ku 80) which, along with DNA-PK*cs*, promote DNA repair as part of the non-homologous end-joining pathway (NHEJ; Stellwagen *et al.*, 2003; Walker *et al.*, 2001). XRCC5 also has a protective role in maintaining the structural integrity of the chromosome (Difilippantonio *et al.*, 2000; Stellwagen *et al.*, 2003) by regulating the precise structure of telomeric termini (Tuteja and Tuteja, 2000). Knock out of XRCC5 has resulted in accelerated ageing in mice, highlighting the importance of these structures to cellular and organ integrity. G22P1 may also play a role in mitochondria-dependent apoptosis as a suppressor of Bax (Sawada *et al.*, 2003). Consequently, these genes are key components of DNA damage responses and as such mediators of the cell's response to oxidative damage.

5.4 *Sirtuins*

Sirtuins are members of a widely conserved gene family that regulate cell stress and damage responses and are implicated in apoptosis and senescence (Ashraf *et al.*, 2006; Brunet *et al.*, 2004; Chua *et al.*, 2005; Kobayashi *et al.*, 2005; Langley *et al.*, 2002; Luo *et al.*, 2001). They are orthologues of the yeast SIR2 gene (Frye, 1999). In budding yeast (*Saccharomyces cerevisiae*), a complex composed of SIRT2/3/4 is recruited to sites of DNA damage, along with XRCC5 (Ku 80) and G22P1 (Ku 70), where it is involved in the repair of DSBs via NHEJ (Kanoh and Ishikawa, 2003). In addition to their protein deacetylase activity, sirtuins are also NAD-dependent ADP-ribosylases. These activities suggest that sirtuins may act in a sensory capacity responsive to changes in the redox and metabolic state of the cell and regulate its function accordingly (Ashraf *et al.*, 2006; Shi *et al.*, 2005; Shiels and Davies, 2003). These are highlighted in *Figure 2.*

In humans, seven sirtuins, SIRT1–7, have been identified. SIRT1 negatively regulates p53 in response to oxidative stress (Luo *et al.*, 2001; Vaziri *et al.*, 2001), while SIRT2 participates in a cell-cycle checkpoint, causing growth arrest in the face of chromosomal abnormalities (Dryden *et al.*, 2003; North *et al.*, 2003). SIRT3 is implicated in fuel utilisation and energy production in times of stress (Hallows *et al.*, 2006) and SIRT7 is associated with growth and proliferation (Ford *et al.*, 2006). SIRT6 has also been shown to promote resistance to DNA damage and suppress genomic instability by participating in DNA repair (Mostoslavsky *et al.*, 2006).

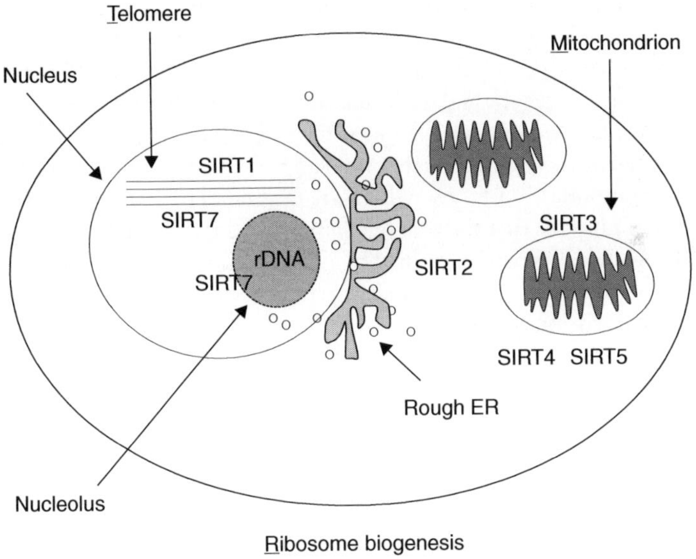

Figure 2. *Sirtuins and the MTR trinity. Sirtuins are members of a widely conserved gene family implicated in cell-cycle control, epigenetic regulation, apoptosis and senescence. Sirtuins have been hypothesized to link the functions of the mitochondrion and the telomere nucleoprotein complex with ribosome production, the MTR trinity (Shiels and Davies, 2003; Shiels* et al., *1999).*

Sirtuins have been hypothesized to integrate the functions of the mitochondrion and the telomere nucleo-protein complex with ribosome production (the MTR trinity; Shiels and Davies, 2003; Shiels *et al.*, 1999). Sirtuin activity, within the components of the MTR, allows the cell to monitor oxidant load, facilitate oxidative damage repair and regulate protein biosynthesis.

A notable success of this theory is predicting the involvement of sirtuins in cancer, a disease of ageing with dysregulated cellular proliferation. Notably, SIRTS 3 (mitochondrial) and 7 (nucleolar-rDNA) have all been implicated in disease state progression based on telomere maintenance and its knock-on effects on protein biosynthesis and energy requirements in the cancerous cell (Ashraf *et al.*, 2006).

6 Telomeres, oxidation and ageing

The sirtuins provide a pertinent link between telomere biology and oxidative damage, which centres on the ability of the cell to detect, assess, signal and repair the damage, if and when appropriate. The physical consequence of such damage manifests in the phenotype of ageing and a finite lifespan in mammals. How this is accomplished is the subject of much debate. Several theories of ageing involving oxidative damage have been proposed, though their testing *in vivo* has not always been straightforward. Nor has their relationship with telomere biology been fully explored.

7 Ageing

Ageing is a complex, multi-factorial process, originally described as an exponential increase in the likelihood of mortality with increasing age (Gompertz, 1825). More recently, this has been formulated as 'the process of growing old and/or the sum of all changes, physiological, genetic and molecular that occur with the passage of time from fertilization until death' (Weinert and Timiras, 2003). Multiple theories have been proposed to explain why ageing occurs. These theories can be broadly categorized into those which have a genetic component and those which are facilitated by 'wear and tear' phenomena. Genetic theories propose that the rate of ageing, as marked by a decline in cognitive and physiological processes, is predetermined by our genetic code and that this 'planned obsolescence' may be an inheritable trait. The genetic theory of ageing, however, is also compatible with the inclusion of telomere biology, as first postulated by Olovnikov (1973).

Several theories of ageing exist which can be very broadly categorized under the heading of 'wear and tear'. In this context of this chapter, 'wear and tear' is defined as the process of declining function or integrity caused by constant use as part of normal existence and by extraneous damage. This grouping encompasses several hypotheses including 'the free radical theory of ageing', 'the mitochondrial theory of ageing', 'the glycation theory of ageing' and 'the neuro-endocrine theory of ageing'. In-depth discussion of these theories is outwith the scope of this chapter. However, short descriptions are outlined below.

7.1 *The glycation theory of ageing*

'The glycation theory of ageing' involves the cross-linkage of two, or more, molecules by covalent bonding, thus impairing their function. Biologically, the most important example of this type of non-enzymatic bonding, the Maillard reaction, results in the formation of cross-linkages between aldehydes, ketones and reducing sugars, and amino acids, peptides and proteins (Robert and Robert, 2007). An elegant and comprehensive review of the role of glycation and other protein modifications in ageing can be found within Chapter 12 of this book (Xue *et al.*, 2008). In brief, however, protein modification may impact upon telomere biology via its effects on the MTR trinity.

The proteosome is the major site of cellular protein degradation, via its multicatalytic enzyme degradation complex (Gonzales-Dosal *et al.*, 2006). However, the proteinase activities of the proteosome have been observed to decline with age, resulting in an accumulation of modified proteins, such as the 'wear and tear' marker, lipofuscin (Terman and Brunk, 2004). Studies using human skin and lung fibroblasts reported age-related increases in markers of protein glycation in senescent cells, which were not present in young cells, or telomerase-immortalized bone marrow cells (hTERT-BMCs; Ahmed *et al.*, 2007; Gonzales-Dosal *et al.*, 2006). In addition, the accumulation of free adduct glycation oxidation and nitration products can inhibit proteosome function by the formation of large aggregates, which cannot be degraded and clog up the proteosome (Hernebring *et al.*, 2006). Consequently, increased intracellular stress will impact on the MTR and have a direct effect on cellular senescence.

Glycation also impacts on caloric restriction (CR). CR has been shown to increase longevity in a wide range of species from *Drosophila* to non-human primates (Mair *et al.*, 2004; Nikolich-Žugich and Messaoudi, 2005). One of the main mechanisms by which CR is proposed to reduce ageing is via a reduction in reactive oxygen species (ROS) generated by the mitochondria. The rate of mitochondrial production of ROS is thought to be one of the major contributory factors in the rate of ageing. Reducing the levels of dietary methionine content has been shown to decrease mitROS in rat liver, oxidative damage to mitDNA, and thus increase longevity (Pamploma and Barja, 2006). Furthermore, these observations are in keeping with the MTR theory and are congruent with a scenario whereby increases in antioxidant enzyme levels, associated with CR, may no longer *per se* be viewed as a direct cause of the longevity, but of CR driving carbon into the tricarboxylic acid cycle (TCA), thus increasing respiration.

7.2 *The neuro-endocrine theory of ageing*

This theory was first proposed by Dilman (1971). It proposes that with increasing age, hypothalamic and peripheral endocrine receptors lose their sensitivity to negative feedback inhibition by hormones, neurotransmitters and other chemical signals. This leads to a disruption in endocrine homeostasis, leading to pathological conditions and death.

7.3 *Metabolic rate, free radical and mitochondrial theories of ageing*

In 1908, Rubner discovered a relationship between metabolic rate, body size and longevity, which proposed that since larger animals expend fewer calories per gram of body mass than smaller, shorter-lived species, as a result of their comparatively slow rate of metabolism, they will live longer. Later work by Pearl (1928) put forward the 'rate of living theory' in which metabolic rate dictated the rate of ageing – increasing the metabolic rate would result in an increased ageing process.

These theories were combined in the 'free radical theory of ageing' with the concept of free radicals as toxic agents (Gerschman *et al*., 1954; Harman, 1956). Harman's theory was based on the premise that macromolecules, such as DNA, are subject to damage by reactions with free radicals and oxidants, collectively known as reactive oxygen species (ROS), generated thorough normal aerobic metabolism. The amount of ROS produced correlated with metabolic rate and antioxidant capacity, resulting in cumulative oxidative damage that contributed to a loss of functionality, leading to increased pathology and death. Therefore, increasing the rate of metabolism would promote ageing and decrease lifespan. Free radicals, in this context, are generally molecules that have an unpaired electron making them highly reactive but short-lived species. A fuller explanation of free radicals and oxidants is outlined below.

In 1972, Harman again proposed a role for ROS in ageing. In an extension of his previous theory, the 'mitochondrial theory of ageing' implicates the production of an excess of ROS by increased metabolism in the mitochondria, which causes damage to mtDNA and membrane leakage, leading to a loss of structural integrity and function. The associated decline in essential mitochondrial energy production in essential organ repair and maintenance was proposed to lead to organ failure and eventually death.

These theories are not mutually exclusive. Indeed, they display an intuitive degree of biological overlap. A distinction between chronological age and biological age, however, must be considered at this juncture.

8 Telomeres and biological ageing

The telomere theory of cellular ageing is based upon the seminal observations of Hayflick and Moorhead that indicated that human diploid cells in culture had a finite lifespan defined by a discrete number of cell divisions (Hayflick, 1965; Hayflick and Moorhead, 1961). The mechanism for this mitotic clock is considered to be governed by telomere length, which is proposed to act as a cellular replicometer. Olovnikov (1973) first proposed a telomeric explanation for Hayflick's theory, based around the end under-replication problem. The resultant loss of DNA with successive cell divisions eventually led to the loss of essential genetic information that acted as a trigger for cellular senescence. It should be noted that the telomere-based theory is mutually compatible with pre-existing theories, as loss of cells with short telomeres would result in decreasing tissue function and integrity over time, which impacts on disease susceptibility and mortality (Carrero *et al*., 2008)

9 Nuclear transfer, telomeres and the MTR trinity

Support for a telomere-based theory of ageing has emerged on a number of fronts. These include the study of progeroid syndromes and nuclear transfer (NT) experimentation, as well as rodent transgenics.

9.1 *Telomeres and progeria*

Telomeres provide a number of functions within the capacity of cellular defences against oxidative damage. Apart from the capacity to prevent chromosomal fusions, telomeric proteins aid in signalling and the repair of such DNA damage. Dysregulation of telomere-associated protein function can lead to accelerated biological ageing and disease. A number of progerias have been linked to specific proteins, some examples of which are described below.

One example is the Werner's protein (WRN), the key player in Werner's syndrome, an autosomal recessive disorder characterized by the rapid onset of ageing characteristics at puberty. WRN is a member of the recQ family of DNA helicases (Gray *et al.*, 1997) and is necessary for efficient replication of telomeric DNA, preventing telomere dysfunction and consequent genomic instability (Crabbe *et al.*, 2004). In studies with fibroblasts derived from patients with Werner's syndrome, the forced expression of telomerase extended cellular lifespan (Wyllie *et al.*, 2000), suggesting that Werner's syndrome hastens the rate of normal telomere-driven replicative senescence.

Another progeroid syndrome associated with mutation of a recQ helicase is Bloom's syndrome. Bloom's syndrome is an autosomal recessive condition first described in the 1950s (Bloom, 1954). Bloom's syndrome is caused by a mutation in the BLM protein causing a high frequency of chromosomal breaks and rearrangement. This leads to the high prevalence of early-onset cancers. This may also be influenced by an abnormal p53 response to DNA damage observed in cells from Bloom's patients (Lu and Lane, 1993).

Hutchinson-Gilford progeria syndrome (HGPS) is an extremely rare form of progeria characterized by an extreme precocious ageing phenotype. It is caused by a *de novo* point mutation on chromosome 1q of the lamin A/C (LMNA) gene, which results in the substitution of a cytosine nucleotide with a thymine nucleotide (Cao *et al.*, 2007). This substitution leads to synthesis of a mutant form of prelamin A, known as progerin (Moulson *et al.*, 2007). Over time, progerin accumulates in the cytoplasm and nuclear envelope, damaging the structure and function of the nucleus and contributing to premature cell death. Additional alterations on chromosome 11, as well as complex chromosomal rearrangements, also implicate chromosomal instability in HGPS (Corso *et al.*, 2005). Reduced levels of the antioxidant enzymes manganese-containing superoxide distmutase, catalase and glutathione peroxidase have been shown in skin fibroblasts from a HGPS patient. As these enzymes are involved in scavenging ROS their loss will result in elevated oxidative stress, which may contribute to accelerated ageing in progeria (Yan *et al.*, 1999). As such, they provide a compelling link between telomere structure and function, and the control of redox state in the cell and biological ageing and disease.

One final example is Cockayne's syndrome. This is an extremely rare form of progeria characterized by short stature and a premature ageing phenotype. It results from a mutation in the *ERCC6* and *ERCC8* genes, which code for the excision repair cross-complementing proteins 6 and 8, involved in DNA damage repair. Cockayne's patients are abnormally sensitive to UV light damage (Jaeken *et al.*, 1989) and some chemical carcinogens. Additionally, they have an abnormal DNA and RNA repair mechanism (Mayne and Lehmann, 1982).

9.2 *Telomeres and nuclear transfer*

In NT experiments, germ-line reset of telomere length is bypassed. Analysis of cloned animals produced by NT has resulted in a number of equivocal reports on the role of telomeres in biological ageing, as healthy animals that are phenotypically identical to aged matched controls are produced with widely differing telomere lengths. These have been reported as longer (Jeon *et al.*, 2005; Lanza *et al.*, 2000), the same (Betts *et al.*, 2001; Kubota *et al.*, 2004; Tian *et al.*, 2000) or shorter than controls (Alexander *et al.*, 2007; Miyashita *et al.*, 2002; Shiels and Jardine, 2003; Shiels *et al.*, 1999), for a number of different cell types and species. The reasons for this are unclear. Factors affecting the length of the telomeres in NT-derived animals include:

(i) pre-existing telomere length in the donor tissue;
(ii) the length of time in culture, as *in vitro* growth results in substantial elevation in oxidant stress levels;
(iii) differential telomerase activity in the donor cell type and species;
(iv) the composition and distribution of the nucleo-proteins on the telomere nucleo-protein complex.

Nuclear transplantation experiments have indicated that telomere biology *per se* is not the sole driver of biological ageing. Despite telomere length reset in many animals derived by this procedure, cryptic oxidative damage from cell culture and epigenetic changes have resulted in what has been termed the 'survival of the least damaged' (Humphreys *et al.*, 2002; Ogonuki *et al.*, 2002). This is indicative of more-global damage being a key driving event in biological ageing.

9.3 *The MTR trinity*

The lack of an advanced ageing phenotype in animals derived from NT, however, indicates that the mitochondria do have a large part to play in maintenance of an age-normal phenotype (Campbell *et al.*, 1996; Shiels *et al.*, 1999; Wilmut *et al.*, 1997). This has led to the development of a theory that combines separate, non-mutually exclusive components of both ageing and damage detection and responses. This revolves around a functional trinity, comprising the <u>m</u>itochondrion, <u>t</u>elomere nucleo-protein complexes and the control of <u>r</u>ibosome biosynthesis, the MTR (Shiels and Davies, 2003). Their interactions integrate energy production, protein synthesis and DNA integrity. The telomere nucleo-protein complexes form part of the cell's DNA damage-sensing system, while the mitochondrion generates energy and is a key component of apoptotic cell-death mechanisms. Ribosome biosynthesis, hence energy utilization, is modulated via proteins associated with the rDNA cluster, which are in balance with the telomere for protein complex binding.

One outcome of such a regulatory system is that if one part responds it will be reflected in the status of the others. For example, though oxidative damage can result in accelerated telomere erosion, which leads to premature cell death, rDNA hypersynthesis can produce a sink for the telomere nucleo-protein complex, exposing the telomere with the same consequence. As defects in any of these components are hypothesized to accelerate ageing, they can be construed as contributing to the ageing process. Many models of ageing have already been proposed with one or other of the trinity at centre stage, as discussed above.

A further key consideration arising from the MTR is whether mitochondrial dysfunction and associated free-radical generation is the primary source of age-related damage. Data pertaining to caloric restriction indicate that, counterintuitive to classical mitochondrial theories of ageing, increased mitochondrial electron transport activity is actually beneficial and increases longevity (Lin *et al.*, 2002).

When viewed from the perspective of the MTR, however, these considerations must be offset by the fact that it is *damaged mitochondria* that are central to apoptotic processes, which may be a critical feature of organ/tissue ageing.

10 Integrating reactive oxygen species generation and telomeres

Reactive oxygen species are mostly small molecules whose unpaired electrons make them highly reactive. ROS include hydrogen peroxide, the hypochlorite ion, the hydroxyl radical and the superoxide anion. These radicals may be generated by a range of biological processes, which can impact on telomere biology. The generation of these entities is discussed below.

10.1 *Mitochondrial ROS*

During normal cellular metabolism in the mitochondria, leakage of electrons from the electron transport chain during oxidative phosphorylation can result in the formation of superoxide radicals through a reaction with molecular oxygen. These superoxide radicals and their reaction products, that is the hydroxyl ion, can cause damage to mitochondrial DNA, leading to mitochondrial dysfunction. Mitochondrial dysfunction has been implicated in increasing levels of oxidative stress, as well as in disrupting cellular energy production and calcium homeostasis (Anandatheerthavarada and Devi, 2007; Passos *et al.*, 2007). Mouse embryos treated with the compound carbonyl cyanide p-trifluoromethoxyphenylhydrazone (FCCP) were found to have chromosomal instability and telomere attrition, in keeping with this postulate. FCCP is a protonophore compound that uncouples the mitochondrial electron transport system and disrupts mitochondrial function. Telomere attrition has been postulated to be caused by the ROS generated by mitochondrial dysfunction (Liu *et al.*, 2002).

10.2 *Peroxisomal ROS*

Peroxisomes are ubiquitous organelles found in great numbers within virtually all eukaryotic cells. Peroxisomes contain at least 50 different enzymes that are involved in various biochemical pathways. Three of these enzymes, D-amino acid oxidase, ureate oxidase and catalase, are involved in the breakdown of toxic substances, especially in the liver and kidney. This is accomplished via an oxidation reaction involving other compounds, resulting in the formation of hydrogen peroxide that is converted to H_2O and oxygen by the action of catalase. Both the highly reactive oxygen radicals generated and hydrogen peroxide can contribute to oxidative stress (del Río *et al.*, 1992).

10.3 *Cytochrome system*

Another detoxification process, which may in itself produce ROS, is the cytochrome P450 system. The cytochrome P450 family (CYP450), found in bacteria, plants and animals, comprises a large number of hemeproteins whose major role is in the detoxification of endogenous and exogenous compounds. In humans, the CYP450 system is associated mainly with phase I detoxification reactions in the liver. As the CYP450 system is part of an electron transport system, singlet oxygen radicals are routinely formed in the metabolism of other compounds. If these ROS are formed in excess of antioxidant capacity this will result in oxidative stress (Barouki and Morel, 2001).

10.4 *Immune-generated ROS*

ROS may also be formed as part of the innate immune response to bacteria or fungi. During the process of phagocytosis, neutrophils and macrophages engulf invading cells and undergo a 'respiratory burst'. This requires an increase in oxygen consumption catalysed by the action of NADPH-oxidase, generating the superoxide anion and hydrogen peroxide, via the action of SOD. This burst effectively kills invading cells but also has a detrimental effect on surrounding tissues and other immune cells (Ramaiah and Jaeschke, 2007).

11 ROS-induced oxidative damage in DNA

ROS damage to DNA and/or its subsequent repair may result in DNA-protein cross linkage, DNA base and sugar lesions, base-free sites or strand breaks. Telomeres are particularly sensitive to damage by oxidative insult and oxidative DNA damage is repaired less well in the telomere than in the rest of the chromosome (von Zglinicki, 2002). Cumulative oxidative damage is considered to be a significant contributory factor to cellular senescence, organ senescence and age-related diseases (Shiels and Jardine, 2003). Typically, this results in reduced organ function and decreased longevity as a consequence of accelerated biological ageing.

In addition to free radicals produced by normal metabolism, cells may also show a rapid growth arrest – termed 'stress- and stimulation-induced senescence' (STASIS) – in response to various stressors that cause single strand breaks in the DNA (Petersen *et al.*, 1998). STASIS has also been reported to be independent of telomere length (Rubio *et al.*, 2004). Both replicative senescence and STASIS may have an important biological role in limiting the incidence of cancer, by limiting the replicative potential of cells that have damaged DNA. Consequently, disruption of the telomere also may lead to other pathologies, including osteoporosis (Pignolo *et al.*, 2008), cardiovascular disease (Brouilette *et al.*, 2003; Samani *et al.*, 2001), neurodegeneration (Shiels and Davies, 2003), transplant dysfunction (Shiels and Jardine, 2003), impaired renal function (Carrero *et al.*, 2008) and psychological stress (Epel *et al.*, 2004).

12 Telomeres and endogenous antioxidant defences

Endogenous antioxidant defences can go some way to decrease the deleterious effects of ROS generated either via normal metabolism, or via the immune response. However, antioxidants cannot completely eliminate these molecules, as ROS are required for important physiological processes (Loft and Poulsen, 1996), such as ion transport, cell signalling, phagocytic killing, and regulation (Halliwell, 1999). Endogenous antioxidant defences include enzymes such as catalase, superoxide dismutase and peroxidases, and compounds such as vitamin C and E and glutathione. However, levels of endogenous antioxidants, such as glutathione, may vary greatly between tissues of the same donor. As telomere length in humans has been shown to be affected by oxidative stress, it is intuitive to hypothesize that telomere length would differ between differing tissues dependent on their antioxidant profile and, therefore, the level of biological ageing between tissues of the same individual may vary. Tissue-dependent variation of telomere length was found in a study of elderly individuals between inter-individual samples of skin, leukocytes and synovial tissue (Freidrich *et al.*, 2000).

Exogenous antioxidants may also have a role in the protection of telomeres from oxidative challenge. When human foetal lung fibroblasts and human MRC-5 fibroblasts were incubated with the antioxidant compound l-carnosine and the free radical spin trap PBN (α-phenyl-t-butyl-nitrone), respectively, they displayed a decreased rate of telomere attrition and increased longevity when compared to controls. This was thought to be due to their protective effects on telomeric DNA (von Zglinicki *et al.*, 2000).

The implications from such observations are wide ranging, as the availability of exogenous dietary antioxidants may be a key driver in the rate of biological ageing between differing social groups. Groups in areas of high deprivation may have a less-balanced diet and as a consequence be subject to increased oxidant stress, resulting in accelerated biological ageing, earlier onset of disease and increased likelihood of mortality at an earlier chronological age (Adams and White, 2004; Bartley *et al.*, 2004). Measurement of peripheral blood leucocyte (PBL) telomere lengths in groups under elevated levels of psycho-social stress supports the hypothesis (Epel *et al.*, 2004), while a link with social status is still debated (Adams *et al.*, 2007; Cherkas *et al.*, 2006) as differing methodologies to measure telomere lengths have led to equivocal results. What is not debated is that telomere length reflects earlier onset of disease and shorter telomeres are associated with mortality in disease cohorts (Brouilette *et al.*, 2003; Carrero *et al.*, 2008; Harris et al 2006; Samani *et al.*, 2001; Starr *et al.*, 2007). The influence of psycho-social (housing, deprivation) and lifestyle factors (smoking, alcohol, drug intake, diet) is intuitive, but their influence on telomere biology has yet to be fully elucidated.

13 Using telomere biology and the MTR to refine chemistry

The innate cellular responses to stress, mediated through telomere biology, can be exploited for use as a biological tool with which to direct chemistry. This is especially pertinent to the development of novel antioxidants, as oxidation drives cellular ageing. In practice, however, this can be applied to any chemical entity.

Changes in primary biological damage responses, as manifested in rapid transcriptional changes in cell stress and telomere-associated gene expression, has already proven successful in identifying superior antioxidants (Sklavounou *et al.*, 2006). The use of endogenous molecular monitors of cell stress gives a direct read out, not only of efficacy but also of biological safety. Significantly, these changes in gene expression are a cellular front-line response to potential genotoxic agents, designed to assess, signal and repair genomic damage. They supersede cytoplasmic enzyme systems, evolved to act as redox protectants, in that they are specific genome (DNA) protectants. Notably, this system is bypassed in immortalized human cells where telomere maintenance occurs, thus negating the function of an important part of the system. The use of primary human cells to exploit this biology is thus crucial.

Differences in such gene expression have an application in the screening of new chemical entities (NCEs) for cytotoxic effects and for the identification of novel antioxidant compounds. Further work within our group has shown that these genetic markers are also applicable to certain pathological states, such as late renal transplant dysfunction and cancer (Ashraf *et al.*, 2006; Lamb *et al.*; 2006).

We have exploited telomere-associated protein expression to evaluate how renal allograft function is affected by the oxidative insult of ischaemia and reperfusion associated with transplantation (Joosten *et al.*, 2003). Using peri-transplant biopsies as a starting material, such analyses can predict graft function at 6 months and a year post transplant (Lamb *et al.*, 2006). Consequently, telomere biology evaluated, the MTR may provide a useful tool in developing tailored therapeutics for these conditions.

An extension of such analyses is the identification of fetuin A as a PBL telomere-dependent factor in mortality, associated with chronic kidney disease (Carrero *et al.*, 2008). Fetuin A is a serum calcium-binding protein with a role in maintaining redox homeostasis in the blood. Once again, this highlights how redox control is associated with telomere biology and the importance of this system in human health. Significantly, it provides a link between redox control and telomere biology in the blood system, which can act as a surrogate for predicting biological ageing in solid organs and hence predisposition to disease.

14 Conclusions

This chapter has explored the role of telomere biology as an integral part of biological ageing at the cellular level. It has described the nature of telomere nucleo-protein complexes and how their dysfunction can result in disease, when cellular effects are extrapolated to the level of tissues and organs.

The crucial role telomere-associated proteins have in DNA damage responses makes them key respondents to redox changes within the cell. As such, they are part of the cell's primary defence mechanisms. The interplay between the individual components of these defences remains to be fully understood in order to appreciate their impact on oxidation processes and, subsequently, biological ageing. What is clear, however, is that their study is opening up new pathways for investigating disease and providing significant new targets for therapeutic intervention. Importantly, they are a rich harvesting ground for both evaluating and improving the efficacy of any novel therapeutics.

Acknowledgements

P.S. is supported by the Cunningham Trust and Darlinda's Charitable Trust. The authors thank Dr L. McGlynn and Dr S. Zino for critical reading of the manuscript.

References

Adams, J.M. and White, M. (2004) Biological ageing: a fundamental, biological link between socio-economic status and health? *Eur. J. Public Health* **14:** 331–334.

Adams, J., Martin-Ruiz, C., Pearce, M.S., White, M., Parker, L., von Zglinicki, T. (2007) No association between socio-economic status and white blood cell telomere length. *Ageing Cell* **6:** 125–128.

Ahmed, E.K., Picot, C.R., Bulteau, A.L. and Friguet, B. (2007) Protein oxidative modifications and replicative senescence of WI-38 human embryonic fibroblasts. *Ann. N. Y. Acad. Sci.* **1119:** 88–96.

Alexander, B., Coppola, G., Perrault, S.D., Peura, T.T., Betts, D.H. and King, W.A. (2007) Telomere length status of somatic cell sheep clones and their offspring. *Mol. Reprod. Dev.* **74:** 1525–1537.

Anandatheerthavarada, H.K. and Devi, L. (2007) Amyloid precursor protein and mitochondrial dysfunction in Alzheimer's disease. *Neuroscientist* **13:** 626–638.

Ashraf, N., Zino, S., Macintyre, A., Kingsmore, D., Payne, A.P., George, W.D. and Shiels, P.G. (2006) Altered sirtuin expression is associated with node-positive breast cancer. *Br. J. Cancer* **95:** 1056–1061.

Azzalin, C.M. and Lingner, J. (2007) 'Damage control.' *Nature* **448:** 1001–1002.

Barouki, R. and Morel, Y. (2001) Repression of cytochrome P450 1A1 gene expression by oxidative stress: mechanisms and biological implications. *Biochem. Pharmacol.* **61:** 511–516.

Bartley, M., Martikainen, P., Shipley, M. and Marmot, M. (2004) Gender differences in the relationship of partner's social class to behavioural risk factors and social support in the Whitehall II study. *Soc. Sci. Med.* **59:** 1925–1936.

Betts, D., Bordignon, V., Hill, J., Winger, Q., Westhusin, M., Smith, L. and King, W. (2001) Reprogramming of telomerase activity and rebuilding of telomere length in cloned cattle. *Proc. Natl Acad. Sci. USA* **98:** 1077–1082.

Bloom, D. (1954) Congenital telangiectatic erythema resembling lupus erythematosus in dwarfs; probably a syndrome entity. *Am. J. Dis. Child.* **88:** 754–758.

Bodnar, A.G., Ouellette, M., Frolkis, M., *et al.* (1998) Extension of life-span by introduction of telomerase into normal human cells. *Science* **279:** 349–352.

Brouilette, S., Singh, R.K., Thompson, J.R., Goodall, A.H. and Samani, N.J. (2003) White cell telomere length and risk of premature myocardial infarction. *Arterioscler. Thromb. Vasc. Biol.* **23:** 842–846.

Brunet, A., Sweeney, L.B., Sturgill, J.F., *et al.* (2004) Stress-dependent regulation of FOXO transcription factors by the SIRT1 deacetylase. *Science* **303:** 2011–2015.

Campbell, K.H., McWhir, J., Ritchie W.A. and Wilmut, I. (1996) Sheep cloned by nuclear transfer from a cultured cell line. *Nature* **380:** 64–66.

Campisi, J. and d'Adda di Fagagna, F. (2007) Cellular senescence: when bad things happen to good cells. *Nat. Rev. Mol. Cell Biol.* **8:** 729–740.

Cao, K., Capell, B.C., Erdos, M.R., Djabali, K. and Collins, F.S. (2007) A lamin A protein isoform overexpressed in Hutchinson-Gilford progeria syndrome interferes with mitosis in progeria and normal cells. *Proc. Natl Acad. Sci. USA* **104:** 4949–4954.

Carrero, J.J., Stenvinkel, P., Fellström, B., *et al.* (2008) Telomere attrition is associated with inflammation, low fetuin-A levels and high mortality in prevalent haemodialysis patients *J. Intern. Med.* **263:** 302–312.

Cherkas, L.F., Aviv, A., Valdes, A.M., Hunkin, J.L., Gardner, J.P., Surdulescu, G.L., Kimura, M. and Spector, T.D. (2006) The effects of social status on biological ageing as measured by white-blood-cell telomere length. *Ageing Cell.* **5:** 361–365.

Cheung, A.L. and Deng, W. (2008) Telomere dysfunction, genome instability and cancer. *Front. Biosci.* **13:** 2075–2090.

Chua, K.F., Mostoslavsky, R., Lombard, D.B., *et al.* (2005) Mammalian SIRT1 limits replicative life span in response to chronic genotoxic stress. *Cell Metab.* **2:** 67–76.

Cohen, S.B., Graham, M.E., Lovrecz, G.O., Bache, N., Robinson, P.J. and Reddel, R.R. (2007) Protein composition of catalytically active human telomerase from immortal cells. *Science* **315:** 1850–1853.

Corso, C., Parry, E.M., Faragher, R.G., Seager, A., Green, M.H. and Parry, J.M. (2005) Molecular cytogenetic insights into the ageing syndrome Hutchinson-Gilford Progeria (HGPS). *Cytogenet. Genome Res.* 111: 27–33.

Crabbe, L., Verdun, R.E., Haggblom, C.I. and Karlseder, J. (2004) Defective telomere lagging strand synthesis in cells lacking WRN helicase activity. *Science* **306:** 1951–1953.

de Lange, T. (2005) Shelterin: the protein complex that shapes and safeguards human telomeres. *Genes Dev.***19:** 2100-2110.

del Río, L.A., Sandalio, L.M., Palma, J.M., Bueno, P. and Corpas, F.J. (1992) Metabolism of oxygen radicals in peroxisomes and cellular implications. *Free Radic. Biol. Med.* **13:** 557–580.

Denchi, E.L. and de Lange, T. (2007) Protection of telomeres through independent control of ATM and ATR by TRF2 and POT1. *Nature* **448:** 1068–1071.

Difilippantonio, M.J., Zhu, J., Chen, H.T., Meffre, E., Nussenzweig, M.C., Max, E.E., Ried, T. and Nussenzweig, A. (2000) DNA repair protein Ku80 suppresses chromosomal aberrations and malignant transformation. *Nature* **404:** 510–514.

Dilman, V.M. (1971) Age-associated elevation of hypothalamic, threshold to feedback control, and its role in development, ageing, and disease. *Lancet* **1:** 1211–1219.

Dryden, S.C., Nahhas, F.A., Nowak, J.E., Goustin, A.S. and Tainsky, M.A. (2003) Role for human SIRT2 NAD-dependent deacetylase activity in control of mitotic exit in the cell cycle. *Mol. Cell Biol.* **23:** 3173–3185.

Epel, E.S., Blackburn, E.H., Lin, J., Dhabhar, F.S., Adler, N.E., Morrow, J.D. and **Cawthon, R.M.** (2004) Accelerated telomere shortening in response to life stress. *Proc. Natl Acad. Sci. USA* **101:** 17312–17315.

Ford, E., Voit, R., Liszt, G., Magin, C., Grummt, I. and Guarente, L. (2006) Mammalian Sir2 homolog SIRT7 is an activator of RNA polymerase I transcription. *Genes Dev.* **20:** 1075–1080.

Friedrich, U., Griese, E., Schwab, M., Fritz, P., Thon, K. and Klotz, U. (2000) Telomere length in different tissues of elderly patients. *Mech. Ageing Dev.* **119:** 89–99.

Frye, R.A. (1999) Characterization of five human cDNAs with homology to the yeast SIR2 gene: Sir2-like proteins (sirtuins) metabolize NAD and may have protein ADP-ribosyltransferase activity. *Biochem. Biophys. Res. Commun.* **260:** 273–279.

Garcia, C.K., Wright, W.E. and Shay, J.W. (2007) Human diseases of telomerase dysfunction: insights into tissue ageing. *Nucleic Acids Res.* **35:** 7406–7416.

Gerschman, R., Gilbert, D.L., Nye, S.W., Dwyer, P. and Fenn, W.O. (1954) Oxygen poisoning and X-irradiation: a mechanism in common. *Science* **119:** 623–626.

Gompertz, B. (1825) On the nature of the function expressive of the law of human mortality, and on a new mode of determining the value of life contingencies. *Philos. Trans R. Soc. Lond.* **115:** 513–585.

Gonzales-Dosal, R., Sørenson, M.D., Clark, B.F., Rattan, S.I. and Kristensen, P. (2006) Phage-displaying antibodies for the detection of glycated proteasome in ageing cells. *Ann. NY Acad. Sci.* **1067:** 474–478.

Gray, M.D., Shen, J.C., Kamath-Loeb, A.S., Blank, A., Sopher, B.L., Martin, G.M., Oshima, J. and Loeb, L.A. (1997) The Werner syndrome protein is a DNA helicase. *Nat. Genet.* **17:** 100–103.

Griffith, J.D., Comeau, L., Rosenfield, S., Stansel, R.M., Bianchi, A., Moss, H. and de Lange, T. (1999) Mammalian telomeres end in a large duplex loop. *Cell* **97:** 503–514.

Halliwell, B. (1999) Antioxidant defence mechanisms: from the beginning to the end (of the beginning). *Free Radic. Res.* **31:** 261–272.

Hallows, W.C., Lee, S. and Denu, J.M. (2006) Sirtuins deacetylate and activate mammalian acetyl-CoA synthetases. *Proc. Natl Acad. Sci. USA* **103:** 10230–10235.

Harman, D. (1956) Ageing: a theory based on free radical and radiation chemistry. *J. Gerontol.* **11:** 298–300.

Harman, D. (1972) The biological clock: the mitochondria? *J. Am. Geriatr. Soc.* **20:** 145–147.

Harris, S.E., Deary, I.J., MacIntyre, A., Lamb, K.J., Radhakrishnan, K., Starr, J.M., Whalley, L.J. and Shiels, P.G. (2006) The association between telomere length, physical health, cognitive ageing, and mortality in non-demented older people. *Neurosci. Lett.* **406:** 260–264.

Hayflick, L. (1965) The limited *in vitro* lifetime of human diploid cell strains. *Exp. Cell Res.* **37:** 614–636.

Hayflick, L. and Moorhead, P.S. (1961) The serial cultivation of human diploid cell strains. *Exp. Cell Res.* **25:** 585–621.

Hernebring, M., Brolén, G., Aquilaniu, H., Semb, H. and Nyström, T. (2006) Elimination of damaged proteins during differentiation of embryonic stem cells. *Proc. Natl Acad. Sci. USA* **103:** 7700–7705.

Hodes, R.J., Hathcock, K.S. and Weng, N.P. (2002) Telomeres in T and B cells. *Nat. Rev. Immunol.* **2:** 699–706.

Humphreys, D., Eggan, K., Akutsu, H., Friedman, A., Hochedlinger, K., Yanagimachi, R., Lander, E.S., Golub, T.R. and Jaenisch, R. (2002) Abnormal gene expression in cloned mice derived from embryonic stem cell and cumulus cell nuclei. *Proc. Natl Acad. Sci. USA* **99:** 12889–12894.

Jaeken, J., Klocker, H., Schwaiger, H., Bellmann, R., Hirsch-Kauffmann, M. and Schweiger, M. (1989) Clinical and biochemical studies in three patients with severe early infantile Cockayne syndrome. *Hum. Genet.* **83:** 339–346.

Jeon, H.Y., Hyun, S.H., Lee, G.S., *et al.* (2005) The analysis of telomere length and telomerase activity in cloned pigs and cows. *Mol. Reprod. Dev.* **71:** 315–320.

Joosten, S.A., van Ham, V., Nolan, C.E., Borrias, M.C., Jardine, A.G., Shiels, P.G., van Kooten, C. and Paul, L.C. (2003) Telomere shortening and cellular senescence in a model of chronic renal allograft rejection. *Am. J. Pathol.* **162:** 1305–1312.

Kanoh, J. and Ishikawa, F. (2003) Composition and conservation of the telomeric complex. *Cell Mol. Life Sci.* **60:** 2295–2302.

Kobayashi, Y., Furukawa-Hibi, Y., Chen, C., Horio, Y., Isobe, K. Ikeda, K. and Motoyama, N. (2005) SIRT1 is critical regulator of FOXO-mediated transcription in response to oxidative stress. *Int. J. Mol. Med.* **16:** 237–243.

Kubota, C., Tian, X.C. and Yang, X. (2004) Serial bull cloning by somatic cell nuclear transfer. *Nat. Biotechnol.* **22:** 693–694.

Kurz, D.J., Decary, S., Hong, Y., Trivier, E., Akhmedov, A. and Erusalimsky, J.D. (2004) Chronic oxidative stress compromises telomere integrity and accelerates the onset of senescence in human endothelial cells. *J. Cell Sci.* **117:** 2417–2426.

Lamb, K., Stevenson, K.S., Zino, S., MacIntyre, A., Kingsmore, D., Jardine, A.G., Marson, L., Forsyth, J.L.R. and Shiels, P.G. (2006) Telomere binding protein expression correlates with cold ischemic time and late graft dysfunction in human renal allografts, as a result of accelerated organ senescence. *Nephrol. Dial. Transplant* **21(supp):** 259–260.

Langley, E., Pearson, M., Faretta, M., Bauer, U.M., Frye, R.A. Minucci, S., Pelicci, P.G. and Kouzarides, T. (2002) Human SIR2 deacetylates p53 and antagonizes PML/p53-induced cellular senescence. *EMBO J.* **21:** 2383–2396.

Lanza, R.P., Cibelli, J.B., Blackwell, C., Cristofalo, V.J., Francis, M.K., Baerlocher, G.M., Mak, J., Schertzer, M., Chavez, E.A., Sawyer, N., Lansdorp, P.M., West, M.D. (2000) Extension of cell life-span and telomere length in animals cloned from senescent somatic cells. *Science.* **288:** 665–669.

Levy, M.Z., Allsopp, R.C., Futcher, A.B., Greider, C.W. and Harley, C.B. (1992) Telomere end-replication problem and cell ageing. *J. Mol. Biol.* **225:** 951–960.

Liu, J., Atamna, H., Kuratsune, H. and Ames, B.N. (2002) Delaying brain mitochondrial decay and ageing with mitochondrial antioxidants and metabolites. *Ann. NY Acad. Sci.* **959:** 133–166.

Lin, S.J., Kaeberlein, M., Andalis, A.A., Sturtz. L.A., Defossez, P.A., Culotta, V.C., Fink, G.R., Guarente, L. (2002) Calorie restriction extends Saccharomyces cerevisiae lifespan by increasing respiration. *Nature.* **418:** 344–348.

Loft, S. and Poulsen, H.E. (1996) Cancer risk and oxidative DNA damage in man. *J. Mol. Med.* **74:** 297–312.

Lu, X. and Lane, D.P. (1993) Differential induction of transcriptionally active p53 following UV or ionizing radiation: defects in chromosome instability syndromes? *Cell* **75:** 765–778.

Luo, J., Nikolaev, A.Y., Imai, S., Chen, D., Su, F., Shiloh, A., Guarente, L. and Gu, W. (2001) Negative control of p53 by Sir2alpha promotes cell survival under stress. *Cell* **107:** 137–148.

Maddar, H., Ratzkovsky, N. and Krauskopf, A. (2001) Role for telomere cap structure in meiosis. *Mol. Biol. Cell* **12:** 3191–3203.

Mair, W., Sgró, C.M., Johnson, A.P., Chapman, T. and Partridge, L. (2004) Lifespan extension by dietary restriction in female *Drosophila melanogaster* is not caused by a reduction in vitellogenesis or ovarian activity. *Exp. Gerontol.* **39:** 1011–1019.

Mayne, L.V. and Lehmann, A.R. (1982) Failure of RNA synthesis to recover after UV irradiation: an early defect in cells from individuals with Cockayne's syndrome and xeroderma pigmentosum. *Cancer Res.* **42:** 1473–1478

Miller, M.C. and Collins, K. (2002) Telomerase recognizes its template by using an adjacent RNA motif. *Proc. Natl Acad. Sci. USA* **99:** 6585–6590.

Miyashita, N., Shiga, K., Yonai, M., *et al.* (2002) Remarkable differences in telomere lengths among cloned cattle derived from different cell types. *Biol. Reprod.* **66:** 1649–1655.

Mostoslavsky, R., Chua, K.F., Lombard, D.B., *et al.* (2006) Genomic instability and ageing-like phenotype in the absence of mammalian SIRT6. *Cell* **124:** 315–329.

Moulson, C.L., Fong, L.G., Gardner, J.M., Farber, E.A., Go, G., Passariello, A., Grange, D.K., Young, S.G. and Miner, J.H. (2007) Increased progerin expression associated with unusual LMNA mutations causes severe progeroid syndromes. *Hum. Mutat.* **28:** 882–889.

Nikolich-Žugich, J. and Messaoudi, I. (2005) Mice and flies and monkeys too: caloric restriction rejuvenates the aging immune system of non-human primates. *Exp. Gerontol.* **40:** 884–893.

Nolan, C.E., Melin, J., Marson, L., *et al.* (2006) Insulin protects from accelerated senescence secondary to ischaemia/reperfusion injury in STZ diabetic rats. *Nephrol. Dial. Transplant* **21(suppl)4:** iv4 (SO004).

North, B.J., Marshall, B.L., Borra, M.T., Denu, J.M. and Verdin, E. (2003) The human Sir2 ortholog, SIRT2, is an NAD^+-dependent tubulin deacetylase. *Mol. Cell* **11:** 437-444.

Ogonuki, N., Inoue, K., Yamamoto, Y., *et al.* (2002) Early death of mice cloned from somatic cells. *Nat. Genet.* **30:** 253–254.

Olovnikov, A.M. (1971) Principle of marginotomy in template synthesis of polynucleotides. *Dokl. Akad. Nauk. SSSR* **201:** 1496–1499.

Olovnikov, A.M. (1973) A theory of marginotomy: the incomplete copying of template margin in enzymic synthesis of polynucleotides and biological significance of the phenomenon. *J. Theor. Biol.* **41:** 181–190.

Pamplona, R. and Barja, G. (2006) Minotchondrial oxidative stress, ageing and caloric restriction: the protein and methionine connection. *Biochem. Biophys. Acta* **1757:** 496–508.

Passos, J.F., Saretzki, G., Ahmed, S., *et al.* (2007) Mitochondrial dysfunction accounts for the stochastic heterogeneity in telomere-dependent senescence. *PLoS Biol.* **5:** e11.

Pearl, R. (1928) *The Rate of Living.* Knopf, New York, NY.

Petersen, S., Saretzki, G. and von Zglinicki, T. (1998) Preferential accumulation of single-stranded regions in telomeres of human fibroblasts. *Exp. Cell Res.* **239:** 152–160.

Pignolo, R.J., Suda, R.K., McMillan, E.A., Shen, J., Lee, S.H., Choi, Y., Wright, A.C. and Johnson, F.B. (2008) Defects in telomere maintenance molecules impair osteoblast differentiation and promote osteoporosis. *Ageing Cell* **7:** 23–31.

Ramaiah, S.K. and Jaeschke, H. (2007) Role of neutrophils in the pathogenesis of acute inflammatory liver injury. *Toxicol. Pathol.* **35:** 757–766.

Robert, L. and Robert, A.M. (2007) The Maillard reaction: physiopathological role and pharmacological approach. *J. Soc. Biol.* **201:** 167–174.

Rubio, M.A., Davalos, A.R. and Campisi, J. (2004) Telomere length mediates the effects of telomerase on the cellular response to genotoxic stress. *Exp. Cell Res.* **298:** 17–27.

Rubner, M. (1908) *Das Problem det Lebensdaur und seiner beziehunger zum Wachstum und Ernarnhung.* Oldenberg, Munich.

Samani, N.J., Boultby, R., Butler, R., Thompson, J.R. and Goodall A.H. (2001) Telomere shortening in atherosclerosis. *Lancet* **358:** 472–473.

Sawada, M., Sun, W., Hayes, P., Leskov, K., Boothman, D.A. and Matsuyama, S. (2003) Ku70 suppresses the apoptotic translocation of Bax to mitochondria. *Nat. Cell Biol.* **5:** 320–329.

Sbodio, J.I., Lodish, H.F. and Chi, N.W. (2002) Tankyrase-2 oligomerizes with tankyrase-1 and binds to both TRF1 (telomere-repeat-binding factor 1) and IRAP (insulin-responsive aminopeptidase). *Biochem. J.* **361:** 451–459.

Shi, T., Wang, F., Stieren, E. and Tong, Q. (2005) SIRT3, a mitochondrial sirtuin deacetylase, regulates mitochondrial function and thermogenesis in brown adipocytes. *J. Biol. Chem.* **280:** 13560–13567.

Shiels, P.G. and Davies, R.W. (2003) Ageing and death in neurones. In: *The Molecular Biology of the Neurone*, 2nd Edn (eds Davies and Morris). Oxford University Press, Oxford, pp. 435–464.

Shiels, P.G. and Jardine, A.G. (2003) Dolly, no longer the exception: telomeres and implications for transplantation. *Cloning Stem Cells* **5:** 157–160.

Shiels, P.G., Kind, A.J., Campbell, K.H., Waddington, D., Wilmut, I., Colman, A. and Schnieke, A.E. (1999) Analysis of telomere lengths in cloned sheep. *Nature* **399:** 316–317.

Siderakis, M. and Tarsounas, M. (2007) Telomere regulation and function during meiosis. *Chromosome Res.* **15:** 667–679.

Sklavounou, E., Hay, A., Ashraf, N., Lamb, K., Brown, E., MacIntyre, A., George, W.D., Hartley, R.C. and Shiels, P.G. (2006) The use of telomere biology to identify and develop superior nitrone based anti-oxidants. *Biochem. Biophys. Res. Commun.* **347:** 420–427.

Smith, S., Giriat, I., Schmitt, A. and de Lange, T. (1998) Tankyrase, a poly (ADP-ribose) polymerase at human telomeres. *Science* **282:** 1484–1487.

Starr, J.M., McGurn, B., Harris, S.E., Whalley, L.J., Deary, I.J. and Shiels, P.G. (2007) Association between telomere length and heart disease in a narrow age cohort of older people. *Exp. Gerontol.* **42:** 571–573.

Stellwagen, A.E., Haimberger, Z.W., Veatch, J.R. and Gottschling, D.E. (2003) Ku interacts with telomerase RNA to promote telomere addition at native and broken chromosome ends. *Genes Dev.* **17:** 2384–2395.

Terman, A. and Brunk, U.T. (2004) Molecules in focus: lipofuscin. *Int. J. Biol. Cell Biol.* **36:** 1400–1404.

Tian, X.C., Xu, J. and Yang, X..(2000) Normal telomere lengths found in cloned cattle. *Nat. Genet.* **26:** 272–273.

Tuteja, R. and Tuteja, N. (2000) Ku autoantigen: a multifunctional DNA-binding protein. *Crit. Rev. Biochem. Mol. Biol.* **35:** 1–33.

Vaziri, H. and Benchimol, S. (1998) Reconstitution of telomerase activity in normal human cells leads to elongation of telomeres and extended replicative life span. *Curr. Biol.* **8:** 279–282.

Vaziri, H., Dessain, S.K., Ng Eaton, E., Imai, S.I., Frye, R.A., Pandita, T.K., Guarente, L. and Weinberg, R.A. (2001) hSIR2(SIRT1) functions as an NAD-dependent p53 deacetylase. *Cell* **107:** 149–159.

Verdun, R.E. and Karlseder, J. (2006) The DNA damage machinery and homologous recombination pathway act consecutively to protect human telomeres. *Cell* **127:** 709–720.

von Zglinicki, T. (2002) Oxidative stress shortens telomeres. *Trends Biochem. Sci.* **27:** 339–344.

von Zglinicki, T., Pilger, R. and Sitte, N. (2000) Accumulation of single-strand breaks is the major cause of telomere shortening in human fibroblasts. *Free Radic. Biol. Med.* **28:** 64–74

Walker, J.R., Corpina, R.A. and Goldberg, J. (2001) Structure of the Ku heterodimer bound to DNA and its implications for double-strand break repair. *Nature* **412:** 607–614.

Weinert, B.T. and Timiras, P.S. (2003) Invited review: Theories of ageing. *J. Appl. Physiol.* **95:** 1706–1716.

Wilmut, I., Schnieke, A.E., McWhir, J., Kind, A.J. and Campbell, K.H. (1997) Viable offspring derived from fetal and adult mammalian cells. *Nature* **385:** 810–813.

Wyllie, F.S., Jones, C.J., Skinner, J.W., Haughton, M.F., Wallis, C., Wynford-Thomas, D., Faragher, R.G. and Kipling, D. (2000) Telomerase prevents the accelerated cell ageing of Werner syndrome fibroblasts. *Nat. Genet.* **24:** 16–17.

Xin, Z.T., Beauchamp, A.D., Calado, R.T., *et al.* (2007) Functional characterization of natural telomerase mutations found in patients with hematologic disorders. *Blood* **109:** 524–532.

Xue, M.Z., Sunil, A.A., Rabbani, N. and Thornalley, P.J. (2008) Chapter 15. In: *Ageing and Oxidants in Plants and Animals*, SEB Vol. 62 (eds C. Foyer, P.J. Thornalley, and R. Faragher). Taylor and Francis, Abingdon.

Yan, T., Li, S., Jiang, X. and Oberley, L.W. (1999) Altered levels of primary antioxidant enzymes in progeria skin fibroblasts. *Biochem. Biophys. Res. Commun.* **257:** 163–167.

8

A-type lamins, disease and ageing: A stress-induced relationship?

Vanja Pekovic and Christopher J. Hutchison

1 Introduction

Ageing is associated with a number of degenerative mesenchymal disorders and inflammatory conditions including sarcopenia, osteoarthritis and osteopenia, lipoatrophy with insulin resistance and diabetes, and cardiovascular conditions such as hypertension, atherosclerosis and cardiomyopathy (Pekovic and Hutchison, 2008; see *Figure 3*). These disorders are believed to reflect an altered proliferative homeostasis between cell loss and renewal and include a number of processes ranging from tissue fibrosis and sclerosis to hyperplasias and atrophy (Martin, 2007). Fibrotic lesions are a hallmark of ageing in various organs, including heart, muscle and kidney (Pekovic and Hutchison, 2008). In aged tissues such as muscle and liver there is a massive accumulation of lipid droplets. Muscle fibres, heart tissue and bone matrix are often replaced with an adipose-like connective tissue (Kirkland *et al.*, 2002), whilst an inappropriate ossification of cell types such as the interstitial fibroblasts occurs in ligaments, tendons, muscles and heart valves. One model that explains the functional decline of various organ systems with increasing age argues that the cells essential for tissue function and regeneration reach a finite lifespan termed 'cellular senescence' and as such accumulate in the body (Hayflick and Moorhead, 1961; Wright and Shay, 2001). Senescent cells arise as a result of a stress response to a number of inefficient intrinsic cellular mechanisms including errors in DNA repair, failures in protection against oxidative damage and protein cross-linking, accumulation of defective mitochondria and failures in protein or cell turnover and tumour protection mechanisms (von Zglinicki, 2002; Wright and Shay, 2002). Senescent cells accumulate at many tissue sites associated with age-related pathologies such as atherosclerotic lesions, skin ulcers, arthritic joints and hyperproliferative regions in the prostate and liver (Faragher and Kipling, 1998). Recently, it was demonstrated that *in vivo* senescent cells account for over 15% of the cell populations in mitotic tissues of ageing primates (Herbig *et al.*, 2006; Jeyapalan *et al.*, 2007). Moreover, the altered functions of senescent cells are cell-type specific and often include the expression of genes that have long-range, pleiotropic effects including degradative enzymes, growth factors and inflammatory cytokines, which alter the tissue microenvironment and compromise tissue repair and renewal (Campisi, 2005). Premature senescence is also a hallmark of the so-called premature ageing syndromes or segmental progerias which show an early manifestation of one or a few aspects of ageing, such as Hutchinson Gilford progeria syndrome and Werner's syndrome (Bridger and Kill, 2004; Kipling *et al.*, 2004). At least two converging mechanisms regulate growth arrest in senescent human cells (see *Figure 1* in the colour plate section). One is known as replicative senescence, which is a cell intrinsic mechanism triggered by critically short dysfunctional telomeres that are sensed as double-stranded DNA breaks (d'Adda di Fagagna *et al.*, 2003). The second mechanism that triggers a permanent proliferation arrest in primary cells is reminiscent of replicative senescence and is referred to as 'stress-induced premature senescence'(SIPS; Toussaint *et al.*, 2000; von Zglinicki *et al.*, 1995). These senescence-inducing stresses include non-telomeric DNA damage and telomore uncapping, oxidative stress, epigenetic alterations, dysfunctional mitochondria, mitotic misregulation and oncogenic signalling (Karlseder *et al.*, 2002).

According to 'disposable soma theory', the strongest candidates for longevity genes are those regulating somatic maintenance and repair, including the cellular responses to stress (Kirkwood, 1977). Recently, the lamin A gene (*LMNA*) has been linked to longevity and was proposed to be a guardian of somatic cells during their lifetime (Hutchison and Worman, 2004). Mutations in the *LMNA* gene cause a spectrum of 20 human age-related diseases and progeroid syndromes, termed laminopathies, which affect the maintenance of one or more tissues of mesenchymal origin including skeletal muscle, tendons, adipose, skin, bone, peripheral neurones, myocardium and vasculature (Pekovic and Hutchison, 2008; see *Figure 3* in the colour plate section). Tissues affected by *LMNA* mutations often show degenerative changes accompanied with an increased fibrosis and/or adipose-like tissue replacements (Hutchison and Worman, 2004). The compromised tissue functions in laminopathies are proposed to be a consequence of decreased cellular proliferation (Pekovic *et al.*, 2007), a failure to maintain a differentiated state (Markiewicz *et al.*, 2005) and a loss of tissue repair during regeneration (Bakay *et al.*, 2006). Mouse models of lmna knock-out (KO), mutated lmna knock-in (KI) or transgenic mice have been produced that manifest muscular dystrophy (Arimura *et al.*, 2005; Sullivan *et al.*, 1999), premature ageing (Mounkes *et al.*, 2003; Varga *et al.*, 2006) and dilated cardiomyopathy (Mounkes *et al.*, 2005), respectively. These mouse models have a shortened lifespan and die prematurely between 2 and 13 months of age. Moreover, mouse models null in the lamin A processing enzyme Zmpste24 also have shortened lifespan and show progeria-like pathologies of bone and muscle (Bergo *et al.*, 2002; Pendas *et al.*, 2002), whilst down-regulation of Ce-lamin in *Caenorhabditis elegans* leads to a 16% decrease in lifespan (Haithcock *et al.*, 2005).

In this chapter, we will address the role of A-type lamins as a stress-resistant filamentous network in cells and discuss some of the latest findings implicating their involvement in a number of signalling pathways critical to regulation of growth and adaptive responses to stress. Different tissues in the body experience distinct stresses, which elicit distinct adaptive signalling pathways. Such stresses may occur simultaneously or in isolation, and depending on the success of the elicited response, they will have a cumulative ageing effect on the body as a whole leading to a critical threshold of damage that results in an activation of a common senescent response. Since cellular senescence can be viewed as a 'last-resort'cellular response to stress, we propose that dysfunction in A-type lamins leads to a failure in adaptive stress responses necessary for post-natal growth, development and long-term tissue maintenance, which in turn triggers senescence-associated stress signalling and drives organismal ageing.

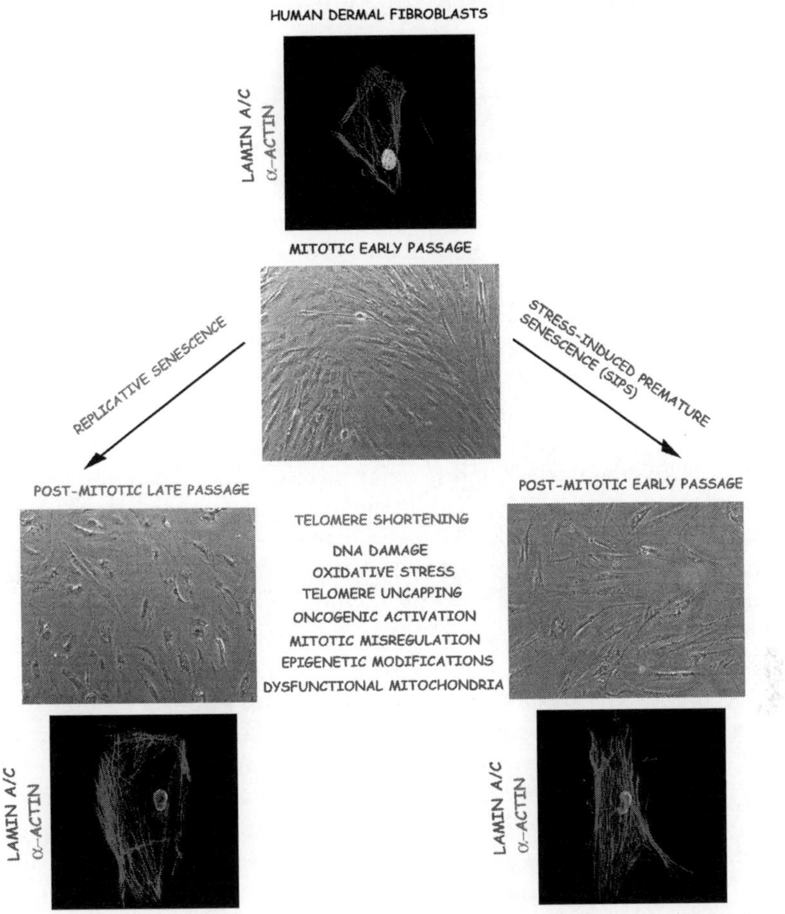

Figure 1. *Human cells have two modes of regulating permanent growth arrest through either replicative senescence or stress-induced premature senescence (SIPS). A number of mechanisms leading to either replicative (marked in red) or premature (marked in green) senescence are outlined. Early mitotic passages of human dermal fibroblasts (marked in blue) were aged through serial passage (Pekovic et al., 2007) until they reached a post-mitotic stage through replicative senescence (marked in red). Alternatively, early mitotic fibroblasts were induced to senesce prematurely by two treatments with an oxidizing agent hydrogen peroxide (H_2O_2) using 300μM final concentration for 1 hr after which they became post-mitotic within the next 7 days (marked in green). Phase contrast images of cells were collected by a Nicon Diaphot 300 fluorescence microscope. Green/red colour merged images were collected by Zeiss confocal microscope following immuno-staining of cells with Sigma anti-lamin A rabbit antibody (in green) and anti-α-actin mouse antibody (in red). Senescent cultures display flat appearance and an increased cell size. As a result they have a lower number of cells per unit area devoid of cell-cell contacts as opposed to spindle-shaped morphology of early passage cultures in which any given cell is encircled by a number of neighbouring cells. Senescent cells display increased levels of senescence-associated β-galactosidase (SA-β-gal) activity, α-actin stress fibres in the cytoplasm and dysmorphic nuclei.*

2 The nuclear lamina as a bridge connecting the cytoplasm to the nucleus

The main feature of eukaryotic cells is the nucleus, which encircles the chromosomes, and is the site of many diverse and fundamental cellular processes including DNA replication, RNA transcription and processing and ribosome assembly. Moreover, the nuclear envelope (NE) is not simply a periphery of the nucleus, but a complex organelle that forms a meeting point between the nuclear and cytoplasmic compartments during interphase. It is involved in the maintenance of proper nuclear morphology, nuclear growth and in the organization of nuclear domains and diverse processes that occur within the nucleus (Broers *et al.*, 2006; Goldberg and Allen, 1995). The NE is composed of the inner and outer nuclear membranes (INM and ONM, respectively), which are separated by the luminal space continuous with the endoplasmic reticular (ER) lumen (see *Figure 2* in the colour plate section). Communication between the nucleoplasm and cytoplasm takes place through many thousands of large supra-molecular structures, called the nuclear pores, which span the NE, where the INM and ONM join (Goldberg and Allen, 1995). Within these pores are anchored the nuclear pore complexes (NPCs), which mediate a bi-directional nucleo-cytoplasmic transport.

Underneath the INM is the nuclear lamina, a term used to describe a meshwork of nuclear intermediate filament proteins, the nuclear lamins, and a growing number of INM and peripheral membrane (PM) proteins (see *Figure 2*; Broers *et al.*, 2006; Mounkes and Stewart, 2004; Pekovic and Hutchison, 2008; Smith *et al.*, 2005). The INM proteins are type II integral membrane proteins and their binding to the nuclear lamins ensures their efficient retention at the NE. Lamins are not restricted to the nuclear periphery but exist throughout the nuclear interior (Bridger *et al.*, 1993; Kennedy *et al.*, 2000; Zastrow *et al.*, 2004). Photo-bleaching of intranuclear GFP-lamin A argues for the existence of both immobile lamin A structures and a free lamin A in the nucleoplasm (Broers *et al.*, 1999). In addition, internal lamins are also found in nuclear bodies called nuclear speckles, which are components of the nuclear matrix (Jagatheesan *et al.*, 1999). Close to the INM is the peripheral chromatin, a large proportion of which is the heterochromatin. The distribution of lamins coincides with the peripherally located chromatin (Belmont *et al.*, 1993) and polymerized lamins can act as anchoring sites for heterochromatin domains at the nuclear periphery (Zhao *et al.*, 1996). Recently, a family of giant cytoskeletal linker proteins, termed nesprins, have been identified which localize at both the ONM and INM (Padmakumar *et al.*, 2005; Zhen *et al.*, 2002). They contain an N-terminal α-actinin type actin-binding domain (ABD), a central coiled-coil rod domain and a C-terminal transmembrane domain, and are proposed to structurally link the nuclear lamina and nucleoskeleton to the actin cytoskeleton in the cytoplasm.

Historically, it has been assumed that the nuclear lamina is an inert scaffolding structure that essentially functions to provide structural support to the nucleus and to protect the peripheral chromatin (Hutchison, 2002). Today, the nuclear lamina is thought to be directly or indirectly involved in a number of fundamental molecular processes, including DNA replication (Moir *et al.*, 2000; Spann *et al.*, 1997), RNA transcription and splicing (Muralikrishna *et al.*, 2001; Spann *et al.*, 2002), genome integrity and DNA repair (Liu *et al.*, 2005; Manju *et al.*, 2006), nucleo-cytoplasmic transport and signalling (Lammerding *et al.*, 2004; Markiewicz *et al.*, 2006; Muchir *et al.*, 2007; van Berlo *et al.*,

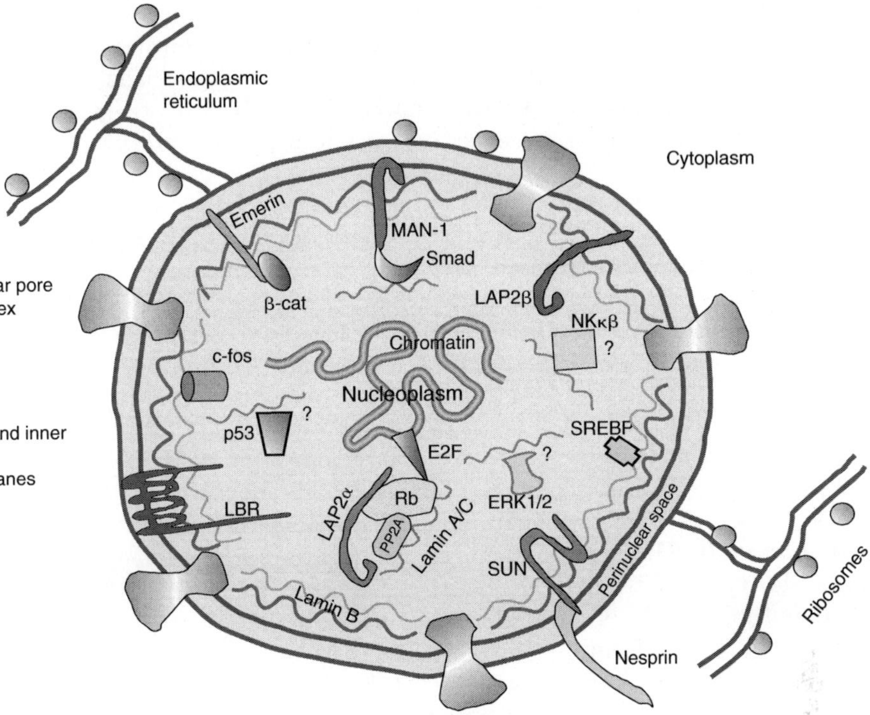

Figure 2. *A schematic diagram depicting an overview of the major structural components of the nuclear envelope (NE) and their interactions and organization in the nucleus. These include the outer and inner nuclear membranes (ONM and INM; the former of which is continuous with the endoplasmic reticulum lumen lined with ribosomes), the nuclear pore complexes, and the nuclear lamina (NL) composed of A-type and B-type lamin oligomers and INM and ONM proteins. Some selected INM proteins include: lamin B receptor (LBR), lamina-associated protein 2β (LAP2β), MAN1, emerin and SUN1. SUN1 acts as a tether for Nesprin 2 in the ONM through interactions which span the perinuclear space and link cytoskeleton to the nuclear membrane. Chromatin domains are also anchored to the nuclear lamina. A-type lamins and their binding partners (emerin, MAN1, LAP2α, SUN1) form functional complexes at the nuclear periphery and in the nucleoplasm, which have a role in gene regulation and signalling pathways. Not depicted are complexes involving B-type lamin binding partners such as LAP2β and LBR. Some selected nucleoplasmic partners include LAP2α, PP2A, Rb, E2F, β-catenin (β-cat), c-fos, Smads and SREBP1. Question marks indicate suggested but not yet proven A-type lamin interactions with p53, NF-κβ and ERK1/2.*

2005; Varela *et al.*, 2005), nucleo-cytoskeletal integrity (Broers *et al.*, 2004; Nikolova *et al.*, 2004; Salpingidou *et al.*, 2007), epigenetic silencing and genome organization (Reddy *et al.*, 2008), spatial organization and anchorage of the NPCs and NE proteins (Liu *et al.*, 2000; Smythe *et al.*, 2000), cell cycle regulation (Dorner *et al.*, 2006; Johnson *et al.*, 2004; Pekovic *et al.*, 2007), and chromosome segregation and cytokinesis (Cao *et al.*, 2007; Dechat *et al.*, 2007; Liu *et al.*, 2000).

How may proteins of the nuclear lamina control these distinct molecular and cellular processes? The nuclear lamina proteins interact with each other as well as a plethora of transcriptional regulators, chromatin-binding proteins and signalling molecules such as kinases and phosphatases (Pekovic and Hutchison, 2008; Zastrow *et al.*, 2004). A growing number of transcription factors, most of which are repressors, localize at the NE. The octamer-binding transcription factor (Oct-1), for example, co-localizes with the lamins and actively represses the ageing-associated collagenase gene when present at the NE (Imai *et al.*, 1997). In ageing cells, translocation of Oct-1 from the NE coincides with a loss of transcriptional silencing by Oct-1 and an increased collagenase activity, which contributes to the pathogenesis of many tissue-degenerative processes. Some of the binding partners of A-type lamins include: the NE proteins such as B-type lamins (Schirmer *et al.*, 2001), emerin (Holaska and Wilson, 2007; Lee *et al.*, 2001; Vaughan *et al.*, 2001) and MAN1 (Mansharamani and Wilson, 2005); cytoskeletal linker proteins such as nesprins (Libotte *et al.*, 2005; Mislow *et al.*, 2002; Zhang *et al.*, 2005) and SUN domain (Sad/UNC-84) proteins (Haque *et al.*, 2006; Padmakumar *et al.*, 2005), and gene-regulatory proteins such as retinoblastoma protein Rb (Mancini *et al.*, 1994; Ozaki *et al.*, 1994), LAP2α (lamina-associated polypeptide 2α; Dechat *et al.*, 2000), Smads (van Berlo *et al.*, 2005), SREBP1 (sterol-regulatory element-binding protein; Lloyd *et al.*, 2002) and c-fos (Ivorra *et al.*, 2006; see *Figure* 2). In the absence of A-type lamin, many of its nuclear binding partners mislocalize to either the ER (emerin), the proteasome/splicing speckles (Rb) or to the one pole of the nucleus (LAP2α; Johnson *et al.*, 2004; Pekovic *et al.*, 2007; Sullivan *et al.*, 1999), or accumulate in the nucleus in an untimely manner, such as transcription factors Smads and c-fos (Ivorra *et al.*, 2006; van Berlo *et al.*, 2005).

3 A-type lamins in post-natal growth, development and tissue maintenance

The complexity of the lamin proteins has notably increased during evolution (Cohen *et al.*, 2001). Vertebrates have three lamin genes: *LMNA*, *LMNB1* and *LMNB2* (Hutchison, 2002). The human *LMNA* gene, located on chromosome 1q21.2-3, encodes four alternatively spliced A-type lamin isoforms: somatic lamins A and C, lamin AΔ10, found in somatic cells and tumours, and male germ-cell-specific lamin C2. The human *LMNB1* gene on chromosome 5q23.2-31.3 encodes lamin B1, and the *LMNB2* gene on chromosome 19p13.3 encodes lamins B2 and male germ-cell-specific lamin B3. In addition, the *Drosophila melanogaster* genome only encodes one A-type (lamin C) and one B-type lamin (Dm0) genes, whilst in *C. elegans* only one B-type lamin (with A-type lamin features) has been found (Ce-lamin). B-type lamins are dynamically expressed throughout all stages of embryonic development and in all cells of the adult organism (Broers *et al.*, 1997). At the cellular level, lamins B1 and B2 are essential for cell growth and survival (Harborth *et al.*, 2001). *Lmnb1* null mice show abnormal lung and bone development and die at birth (Vergnes *et al.*, 2004). In humans, duplications of *LMNB1* lead to a neurodegenerative disease, autosomal dominant leukodystrophy (Padiath *et al.*, 2006), whilst mutations in *LMNB2* cause acquired partial lypodystrophy (Hegele *et al.*, 2006).

In contrast, the A-type lamin expression is developmentally regulated. Some of the transcriptional regulators that control the promoter activity of the *LMNA* gene include

AP1, Sp1/3 (Mularikrishna and Parnaik, 2001), hepatocyte nuclear factor-3β (HNF-3-β), retinoic X receptor β (RXRβ) (Arora *et al.*, 2004) and p53 (Rahman-Roblick *et al.*, 2007). A-type lamins are not detected in embryonic stem cells and their early progenitors, but they become expressed in a tissue-specific manner during mid stages of embryonic development and post-natally, after which they continue to be expressed in most differentiated adult cells except for undifferentiated lymphoid cells of the spleen, thymus, blood and bone marrow (Broers *et al.*, 1997; Stewart and Burke, 1987). Consistent with these findings, loss of A-type lamin expression is associated with dedifferentiation in some tumour cells (Broers *et al.*, 1997; Venables *et al.*, 2001). During mouse embryonic development, lamin A appears first during myogenesis of the eye and trunk muscles followed by appearance during epidermis stratification, whereas the expression in skeletal muscle, heart, liver, lung and brain is post-natal (Rober *et al.*, 1989; Stewart and Burke, 1987). Human homozygous *LMNA* mutation, which leads to a premature stop codon and thus completely abrogates the expression of functional A-type lamins, is incompatible with late embryonic development and shows early foetal lethality (Muchir *et al.*, 2003).

Interestingly, primary embryonic fibroblasts from lmna KO mice models show no differences in proliferation patterns when compared to wild-type cells across their lifespan (Sullivan *et al.*, 1999), whilst fibroblasts from early post-natal lmna KO mice show severe proliferation defects and rapidly die (Mounkes and Stewart, 2004). This phenomenon resembles the early post-natal growth retardation seen in a number of progeroid laminopathies. Intriguingly, most lamin A-associated disorders become evident in patients during the early post-natal period or at the puberty stage when individuals fail to cope with the increased stress levels associated with a rapid post-natal growth and development. Moreover, in HGPS patients and other lamin A-associated progeroid conditions (mandibulo-acral dysplasia and atypical Werner's syndrome), which do not show an expected growth spurt in these periods, growth retardation may manifest itself as a consequence of the conserved stress response attempting to reduce the stress levels in the body rather than as a cause of disease *per se*, as has recently been observed in other short-lived progeroid mice and humans (Niedernhofer *et al.*, 2006; van der Pluijm *et al.*, 2006). Taken together, it becomes apparent that A-type lamins are essential for both post-natal growth and development as well as for long-term tissue maintenance. Since the rate of ageing of more complex tissues depends on adequate tissue maintenance and regeneration, A-type lamins may directly influence longevity by affecting the onset and/or the rate of adult stem cell ageing (Pekovic and Hutchison, 2008).

4 The nuclear lamina as a stress-resistant filamentous network

With respect to sequence homology, lamins belong to the type V large family of intermediate filament (IF) proteins but, unlike types I–IV which reside in the cytoplasm, they contain the nuclear localization sequence (NLS) and hence localize to the nucleus (Hutchison, 2002). Like all IF proteins, lamin monomers have a tripartite structure consisting of a small N-terminal head, a coiled-coil rod domain and a globular C-terminal tail. Lamins form α-helical coiled-coil dimers, which are the building blocks for further assembly (Stuurman *et al.*, 1998). Lamin dimers associate longitudinally to form polar head-to-tail tetramers, which associate laterally in an anti-parallel half-staggered fashion to form IF-like assemblies.

The C-terminal ends of all lamins, except mammalian and *Drosophila* lamin C, contain a sequence motif CAAX box (C, cysteine; a, any aliphatic amino acid; X, any amino acid). This motif represents a signal for the modification of proteins by an isoprenoid group, and is found in many prenylated proteins, including fungal mating factors, Ras, Ras-related proteins and G-protein subunits (Rusinol and Sinensky, 2006). This motif undergoes a sequence of modifications including farnesylation (attachment of the farnesyl moiety) of the C-terminal cysteine residue, a cleavage of the final three C-terminal residues (-AAX) and methyl esterification of the newly exposed carboxyl group on the cysteine. The attached hydrophobic prenyl moiety can target and anchor lamins to the NE (Rusinol and Sinensky, 2006). Whilst B-type lamins remain farnesylated throughout their lifetime, lamin A is processed further, from a pre-lamin A form to a mature lamin A. The C-terminal 15 residues of lamin A and the prenyl tail are removed by a proteolytic cleavage to yield a mature lamin A (Sasseville and Raymond, 1995). This property accounts for the distinct behaviour of lamin A during mitosis. Following mitotic disassembly, lamin A is soluble during mitosis in the form of dimers and/or tetramers, whereas B-type lamins remain attached to nuclear membrane vesicles (Hutchison, 2002). In contrast, the A-type lamin isoform, lamin C, lacks the C-terminal CAAX motif and is thought to become attached to the INM via interaction with prelamin A (Hutchison *et al.*, 2001).

It has been proposed that the lamin network has properties of both mechanical stability and flexibility, which facilitates transmission of mechanical signals across the NE (Dahl *et al.*, 2004). Recent reports suggest the existence of closely juxtaposed A-type and B-type lamin homo-polymers, which are unevenly distributed at the NE (Delbarre *et al.*, 2006). As such, the lamin network is thought to assume a disordered pattern in the nucleus that can easily dissipate mechanical stress by the reversible rearrangements of its filaments (Dahl *et al.*, 2006). In HGPS cells, mutant A-type lamins assume a strikingly ordered pattern of interactions forming hetero-polymers with B-type lamins (Delbarre *et al.*, 2006), which determines their less-flexible and rigid behaviour upon mechanical stress and leads to nuclear 'wrinkling', a prominent characteristic of these nuclei.

5 A-type lamins and a spectrum of laminopathies

The indispensable role of lamins in diverse and fundamental cellular processes ranging from DNA replication to DNA repair may account for a wide spectrum of human disorders (at least 20) being ascribed to mutations at the *LMNA* locus (Worman and Bonne, 2007). Although the nuclear lamina has an essential structural function in the nuclei of all cell types, the emerging evidence from human laminopathies suggests that the lamina proteins also have highly specialized as well as tissue-specific functions. One of the hallmarks of laminopathy cells are dramatic deformations of the nuclear morphology, ranging from dysmorphic nuclear shapes to envelope raffling, blebbing, alterations of peripheral heterochromatin and INM proteins and nuclear fragility upon stress (Goldman *et al.*, 2004; Hutchison, 2002). In addition, abnormal anaphase chromatin bridges between nuclei in single cells were observed after RNAi knock-down of Ce-lamin in *C. elegans* (Liu *et al.*, 2000) and in human cells lacking prelamin A processing protease FACE1 (Gruber *et al.*, 2005). The most severe nuclear abnormalities, including extensively lobulated nuclei and nuclear fragmentation with a subsequent formation of micronuclei, have only been seen in HGPS cells (Bridger and Kill, 2004; Goldman *et al.*, 2004) and in cells null for the prelamin-A-processing protease, either mouse Zmpste24 or human FACE1 (Gruber *et al.*, 2005; Navarro *et al.*, 2004). Mutations in A-type lamins not only lead to the NE fragility but also to nucleoplasmic lamin disorganization (Broers *et al.*, 2005; Wiesel *et al.*, 2008), and can affect both the structural organization of lamin dimers and filaments as well as protein-protein and protein-DNA interactions (Dhe Paganon *et al.*, 2002; Krimm *et al.*, 2002).

Laminopathies can be loosely divided into four categories including those affecting primarily one specific tissue, such as striated muscle, adipose tissue or peripheral neurones; and those affecting several tissues in a systemic way, such as the premature ageing syndromes and the related progeroid-like disorders (Broers *et al.*, 2006; Pekovic and Hutchison, 2008; Worman and Bonne, 2007). However, it is important to note that even diseases affecting predominantly one or two tissues will have variable involvement of other tissues, which often leads to overlapping clinical phenotypes (see *Figure 3*).

Diseases of striated muscle caused by mutations in the *LMNA* gene include: the autosomal dominant and recessive forms of Emery-Dreyfuss muscular dystrophy (AD- and AR-EDMD; Bonne *et al.*, 1999; Raffaele di Barletta *et al.*, 2000); limb girdle muscular dystrophy with atrioventricular conduction disturbances type 1B

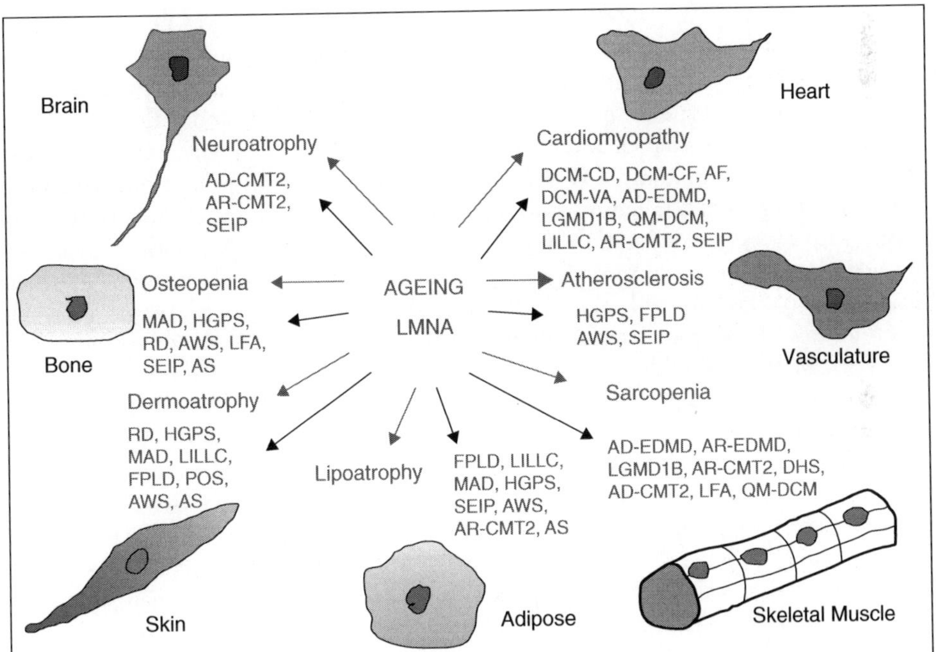

Figure 3. *A schematic diagram drawing parallels between tissues affected in a number of age-associated degenerative disorders (labelled in red) and human diseases associated with mutations in the LMNA gene (labelled in blue). The full names of LMNA-associated diseases in the diagram are explained in the text (see Sections 5 and 6). Please see the colour plate section for colour coded labels noted in legend.*

(LGMD1B; Muchir *et al.*, 2000) and dilated cardiomyopathy with conduction system disease (DCM-CD; Fatkin *et al.*, 1999). Other atypical striated muscle conditions include: dilated cardiomyopathy with early-onset cardiac fibrosis (DCM-CF; van Tintelen *et al.*, 2007); dilated cardiomyopathy with apical left ventricular aneurysm (DCM-VA; Forissier *et al.*, 2003); early-onset atrial fibrillation (AF; Sebillon *et al.*, 2003); amyotrophic quadricipital myopathy with dilated cardiomyopathy (QM-DCM; Charniot *et al.*, 2003); and 'dropped head' syndrome (DHS; D'Amico *et al.*, 2005).

Generally, EDMD symptoms are first detected at 4–5 years of age and the disease progresses slowly (Mercuri and Muntoni, 2001). Patients exhibit tendon contractures of the heels, elbows and the neck, a rigid spine, wasting and weakness of the musculature of the pelvic girdle, biceps and triceps, the disappearance of deep tendon reflexes, cardiac conduction problems and facial muscle weakness. Cardiomyopathy leading to ventricular dysfunction and often requiring a heart transplant is a common feature of AD-EDMD. In contrast, LGMD1B patients show proximal muscle weakness and wasting, no rigid spine, minimal tendon contractures and sometimes calf hypertrophy.

A single autosomal recessive mutation and a dominant mutation have been implicated in axonal Charcot-Marie-Tooth syndrome type 2B1 (AR- and AD-CMT2B1), a peripheral neuropathy with reduced axon density, demyelinated axons and peripheral muscle weakness and wasting (Benedetti *et al.*, 2005; de Sandre-Giovanolli *et al.*, 2002; Goizet *et al.*, 2004). Dunnigan's familial lipodystrophy (FPLD; Cao and Hegele, 2000; Shackleton *et al.*, 2000) and mandibulo-acral dysplasia (MAD; Novelli *et al.*, 2002) primarily result in the loss and redistribution of white adipose tissue, hyperlipidemia, insulin resistance and diabetes. Other FPLD-related disorders include polycystic ovaries with type-A insulin resistance syndrome (POS; Young *et al.*, 2005) and FPLD with variable skeletal and heart involvement (Garg *et al.*, 2002; van der Kooi *et al.*, 2002; Vantyghem *et al.*, 2004). FPLD is not evident until puberty at which time there is a loss of subcutaneous fat in regions such as the extremities and gluteal areas along with excess fat deposition in the neck, back and face. MAD patients also show bone defects including craniofacial abnormalities, osteolysis of the terminal digits and hypoplasia of clavicles and mandibula, as well as post-natal growth retardation, joint contractures, alopecia and cutaneous atrophy (Novelli *et al.*, 2002). Interestingly, mutations in ZMPSTE24, the prelamin-A-processing protease, which lead to the accumulation of unprocessed prelamin A precursors, also cause MAD (Agarwal *et al.*, 2003).

Mutations in the *LMNA* gene include missense point mutations, frame-shift mutations, deletions and non-sense mutations (Goldman *et al.*, 2002). In addition, the position of the mutation within the *LMNA* gene often determines the cell type that becomes affected. Mutations in AD-EDMD are present in all domains of the lamin A protein gene; DCM and LGMD1B mutations are mainly present in the rod and carboxy-terminal domains; CMT2B1 mutations are present in the rod domain; and FPLD and MAD mutations have mainly been found in the tail domain. Krimm *et al.*, (2002) have found that AD-EDMD mutations in the C-terminal domain of lamins A/C destabilize the three-dimensional structure, which leads to altered head-to-tail association and later stages of filament assembly (Strelkov *et al.*, 2004). On the other hand, FPLD mutations in the same region lead to the suppression of positively charged character sites, which decreases protein-protein interactions or protein-DNA interactions (Krimm *et al.*, 2002; Stierle *et al.*, 2003). These findings suggested that the particular lamin A/C domains have a high degree of functional specificity, and opened the possibility that domain-specific effects could underlie the phenotypic spectrum of diseases.

Surprisingly, however, the clinical expression of the same *LMNA* mutation often varies in severity among members of the same family (Worman and Bonne, 2007). Moreover, the clinical symptoms of laminopathies often overlap and patients with the most severe forms of AD-EDMD, for example, frequently have symptoms of other laminopathies such as FPLD. Since lamin A is an abundant nuclear protein in all tissues affected by *LMNA* mutations, the question arose as to whether laminopathies represent variable expressions of the same disease rather than separate diseases (Hutchison *et al.*, 2001; Novelli and d'Apice, 2003). In addition, novel missense mutations in the head and rod domains of lamins A/C have been reported in patients with both FPLD and DCM-CD and thus raised the possibility of a multi-system dystrophy syndrome (Garg *et al.*, 2002). Consequently, it is believed that laminopathies represent a spectrum of related disorders rather than separate diseases (Bonne and Levy, 2003) with the same underlying cause modulated by genetic and/or environmental factors (Stewart *et al.*, 2007).

6 Progeria and progeroid laminopathies

Hutchinson-Gilford progeria syndrome (HGPS) is a rare disease, undetectable at birth, recently demonstrated to be caused by the splicing mutation in the *LMNA* gene (Cao and Hegele 2003; de Sandre-Giovanolli *et al.*, 2003; Eriksson *et al.*, 2003). Since it was first described in 1886, more than 100 cases have been reported worldwide. Patients show symptoms of premature ageing (they appear to age 10 times faster than normal individuals), including severe postnatal growth retardation, midface hypoplasia, micrognathia, hypogonadism, wrinkled skin, loss of subcutaneous fat, alopecia, generalized osteodysplasia with osteolysis, poor muscle development and premature atherosclerosis (Broers *et al.*, 2006). The average age at death in patients is 12–15 years, usually caused by myocardial infarction due to coronary artery disease or stroke. Genome-scale expression profiling of HGPS cells revealed widespread transcriptional misregulation at the developmental level leading to mesodermal/mesenchymal defects and accelerated atherosclerosis (Csoka *et al.*, 2004a). Since patients with HGPS show a low incidence of certain features associated with ageing, including an increase in tumour susceptibility, osteoporosis, diabetes, cataract formation or cognitive degeneration, HGPS is thought to only partially reproduce the ageing process and is thus classified as a segmental progeroid syndrome (Martin and Oshima, 2000). Recently, however, it was reported that there is an association of non-classical HGPS with the mesenchymal neoplasm, osteosarcoma (Shalev *et al.*, 2007).

A single base substitution from C to T in the *LMNA* gene resulting in a silent Gly to Gly change at the codon 608 within exon 11 (G608G) has been found in a subset of individuals with HGPS (de Sandre-Giovanolli *et al.*, 2003). This silent mutation reveals a cryptic splice site in the lamin A RNA, resulting in a deletion of 50 amino acids from the C-terminal domain of lamin A but not affecting the extreme C-terminal end containing the CAAX motif (de Sandre-Giovannoli *et al.*, 2003; Eriksson *et al.*, 2003). This shortened form of lamin A has been named progerin. The absence of 50 amino acids from the C-terminus of lamin A deletes an endoproteolytic cleavage site required for the final processing step of prelamin A to its mature form. The mutated prelamin

A thus remains permanently farnesylated and/or methylated on its extreme C-terminal cysteine (Cao *et al.*, 2007; Dechat *et al.*, 2007). HGPS cells produce both normal and truncated lamin A transcripts by the same mutated allele, indicating that the mutation inhibits transcription of the normal allele, acting as a dominant negative mutation (de Sandre-Giovanolli *et al.*, 2003). Patients typically express only 25% of normal lamin A. Mutant progerin and normal unprocessed prelamin A significantly increase with age at the NE of HGPS cells (Goldman *et al.*, 2004), presumably due to mutated progerin interfering with the normal processing, turnover and assembly state of lamin A in the lamina. Recently, it was demonstrated that the introduction of modified oligonucleotides targeted to the aberrantly activated cryptic splice site in the lamin A RNA reverses the cellular HGPS phenotype (Scaffidi and Misteli, 2005). In addition, there are other rare single-base mutations within exon 11 or other exons of the *LMNA* gene that have been reported to cause atypical HGPS (Rankin and Ellard, 2006).

Another premature ageing syndrome is Werner's Syndrome (WS). In the majority of patients (83%), WS is inherited as an autosomal recessive disease due to mutations in *WRN* gene which encodes a 3′-5′ RecQ DNA helicase-exonuclease that unwinds DNA and cleaves nucleotides from DNA termini (Martin and Oshima, 2000). Atypical cases of Werner's Syndrome (AWS) do not carry detectable mutations in the *WRN* gene. A recent report revealed that 13% of these atypical cases had missense mutations in the *LMNA* gene resulting in amino acid substitution in the N-terminal or rod domains of lamin A (Chen *et al.*, 2003). Unlike typical WS patients who have a mean age of diagnosis at 39 years, AWS patients have a more severe phenotype with a mean age of initial symptoms being 13 years (Chen *et al.*, 2003). The clinical pathology of AWS includes muscular atrophy, lipodystrophy, insulin-resistant diabetes mellitus, cardiovascular pathology, a short stature, atrophic skin and grey or sparse hair.

The most recently described five progeroid-like laminopathies include a systemic condition in a patient showing generalized lipoatrophy, insulin-resistant diabetes, liver steatosis, disseminated leukomelanodermic papules and hypertrophic cardiomyopathy with valvular involvement (LILLC) caused by the same mutation involved in AWS (Caux *et al.*, 2003). The second condition is an autosomal recessive arthropathy syndrome (AS) affecting mainly the distal femora and proximal tibia in the knee with tendinous calcification and progeroid features such as micrognathia, alopecia, generalized lypodystrophy and sclerodermatous skin (van Esch *et al.*, 2005). Another progeroid condition is restrictive dermopathy (RD), which is one of the most deleterious laminopathies so far identified in humans. RD or tight skin contracture'syndrome is characterized by intrauterine growth retardation, tight and rigid skin with erosions and epidermal hyperkeratosis, multiple joint contractures, progeroid facial features, mineralization defects of the skull, absent eyelashes and eyebrows, pulmonary hypoplasia and early neonatal death (Navarro *et al.*, 2004). RD is caused by either a heterozygous splicing mutation in the *LMNA* gene found in HGPS, or heterozygous non-sense mutations in the prelamin-A-processing enzyme encoded by *ZMPSTE-24/FACE* 1 gene (Navarro *et al.*, 2004). Lethal foetal akinesis (LFA) is another devastating condition described in a newborn deceased child with a homozygous non-sense *LMNA* mutation which results in complete functional loss of A-type lamins (Muchir *et al.*, 2003). The phenotype showed severe generalized muscular dystrophy, absence of inter-costal muscles, long-bone fractures, severe contractures of fingers and toes and facial dysmorphism with retrognathia. The fifth condition is a congenital Seip-

Berardinelli syndrome (Seip) characterized by a generalized lipodystrophy, cardiomyopathy, osteosklerosis, retinopathy, anginal pectoris, pancreatic amyloidosis and a variable degree of cognitive impairment (Csoka *et al.*, 2004b).

7 Laminopathies and disease models

A number of distinct but complementary disease models have been proposed to explain how mutations in the *LMNA* gene cause such diverse tissue-specific diseases. These include a structural/mechanical stress model (Hutchison *et al.*, 2001; Lammerding *et al.*, 2004, 2005), an epigenetic/transcriptional de-regulation model (Cohen *et al.*, 2001; Wilson, 2000), an ER protein misfolding model (Novelli and d'Apice, 2003) and an impaired cell-cycle model (Dorner *et al.*, 2006; Gotzmann and Foisner, 2006; Pekovic *et al.*, 2007). In addition, the toxic effects of defectively processed mutant lamin A in progeroid syndromes has been explained by an altered DNA repair model (Liu *et al.*, 2005; Manju *et al.*, 2006) and a mitotic misregulation model (Cao *et al.*, 2007; Dechat *et al.*, 2007).

7.1 *Structural/mechanical stress hypothesis*

Mutant lamin A protein has been proposed to weaken the lamin polymer, resulting in fragile nuclei that do not provide adequate structural integrity and instead exhibit a greater susceptibility to mechanical stress (Hutchison *et al.*, 2001). In cardiac and skeletal muscles and tendons, the contractile forces generated from muscle contraction may deform or rupture these fragile nuclei, resulting in cell death and muscle wasting by inducing apoptosis (Sullivan *et al.*, 1999). Direct analysis of the mechanical properties of lmna KO mouse fibroblasts in response to physical stretching revealed that their nuclei were less rigid than those of the wild type and lmna KO cells were more prone to apoptosis and necrosis in response to repetitive mechanical strain (Broers *et al.*, 2004; Lammerding *et al.*, 2004). Moreover, lmna KO mouse cells showed disturbed interactions of all cytoskeletal structures including actin, vimentin and tubulin filaments in the perinuclear region, an indication of the defective nucleo-cytoskeletal integrity (Broers *et al.*, 2004). In addition, lmna KO cells showed diminished activation of NF-κβ mechano-transduction signalling in response to mechanical stress (Lammerding 2005). However, although this hypothesis addresses the cause of muscle tissue-associated laminopathies, it does not explain how adipose tissue or peripheral neurons become affected in laminopathies such as FPLD and CMT2, respectively.

7.2 *Epigenetic/transcriptional regulation hypothesis*

Ultra-structural studies of lmna KO mouse cells revealed that heterochromatin was not adjacent to the NE in the blebbed regions of dysmorphic nuclei (Sullivan *et al.*, 1999). Such changes in heterochromatin organization and/or alterations in the lamin A-DNA binding needed to establish particular patterns of gene expression may play an important role in the pathophysiology of some laminopathies due to a loss of gene silencing (Stierle *et al.*, 2003). Mutant A-type lamins may affect tissue-specific gene expression by altering the assembly and/or localization of components of transcriptional complexes and thus alter transcriptional regulation (Wilson, 2000). These altered interactions may involve general transcriptional regulators such as Rb and Smads (Markiewicz *et al.*, 2005; van Berlo *et al.*, 2005) or tissue-specific partners such as the RNA splicing components in skeletal muscle (Wilkinson *et al.*, 2003). For example, in fibroblasts from FPLD, MAD and AWS patients, which accumulate prelamin A at the NE, the adipocyte differentiation factor SREBP1 is sequestered at the NE. This inhibits SREBP1 localisation in the nucleus and subsequently impairs adipocyte differentiation (Capanni *et al.*, 2005).

7.3 ER protein misfolding hypothesis

This hypothesis suggests that mutations in the *LMNA* gene bring about changes in the NE, which in turn may lead to perturbation of the peripheral ER due to an altered distribution of INM proteins, such as emerin, which mislocalize to the ER (Mounkes and Stewart 2004; Novelli and d'Apice, 2003). The ER is a main site of cholesterol and fatty acid synthesis and abnormal accumulation of INM proteins in the ER could alter lipogenesis or lipogenic signalling in lamin A mutant cells, leading to aberrant adipocyte development and lipodystrophic disease. On the other hand, in skeletal and cardiac muscles, Ca^{2+} release from the sarcoplasmic reticulum (SR) may be compromised during muscle contractions by alterations in the SR, which is contiguous with the ER in muscle tissues. Alternatively, a generalized ER stress resulting from the inappropriate accumulation of INM proteins in the ER could promote aberrant ER/nucleus signalling related to misfolding or ER overload with consequent downstream effects on gene expression and cell viability. Although very attractive, this hypothesis is still waiting to be tested.

7.4 Cell cycle-proliferation-ageing hypothesis

This model is concerned with the A-type lamins' regulation of the transcriptional factor Rb, which is crucial in controlling the balance between adult stem-cell proliferation and differentiation (Gotzmann and Foisner, 2006). A-type lamins and their binding partner LAP2α bind to and tether Rb/E2F transcriptional complexes in the nucleus during the G1 phase of the cell cycle (Dorner *et al.*, 2006; Markiewicz *et al.*, 2002). This is essential for Rb's role in repressing activation of E2F-dependent S-phase genes and inhibiting growth (Mittnacht and Weinberg 1991). In lmna-null cells, Rb has lost its proliferative control, which leads to a rapid growth phenotype (Johnson *et al.*, 2004; van Berlo *et al.*, 2005) and the eventual exhaustion of their proliferative capacity (Pekovic *et al.*, 2007). Interestingly, HGPS fibroblasts exhibit a hyper-proliferative phenotype at earlier passages but then experience a dramatic decline in their proliferative ability and enter a senescent arrest at later passages (Bridger and Kill, 2004). Similarly, human fibroblasts from a patient functionally null for A-type lamins display a poor index of growth and early senescence (Muchir *et al.*, 2003; Pekovic *et al.*, 2007), and human cells lacking FACE1 undergo abrupt cell cycle arrest (Gruber *et al.*, 2005).

7.5 Altered DNA repair hypothesis

Recently, HGPS human cells and Zmpste24-null mouse cells have been shown to accumulate chromosomal abnormalities due to increased DNA damage and altered

DNA repair (Liu *et al.*, 2005), which classifies them as genome instability syndromes along with the other known progerias (Martin and Oshima, 2000). These results are supported by the earlier findings of the shortened telomeres and diminished DNA repair in HGPS fibroblasts (Allsopp *et al.*, 1992; Wang *et al.*, 1991). Recent studies have shown that farnesyl transferase inhibitor (FTI) treatments, which inhibit farnesylation of many cellular proteins including lamins, can correct the nuclear morphology defects in HGPS cells by displacing progerin from the NE to the nucleoplasm (Mallampalli *et al.*, 2005; Toth *et al.*, 2005). However, FTI treatments were not successful in ameliorating some other cellular features of HGPS cells such as DNA damage accumulation (Liu *et al.*, 2006). It is still not known, however, whether these DNA lesions result from oxidative damage, telomere shortening/uncapping or inefficient repair pathways. It is speculated that due to the absence of functional A-type lamins from the nucleoplasmic complexes in cells, chromatin may be more vulnerable to DNA damage and/or cells may be unable to efficiently activate repair pathways once damage has occurred.

7.6 Mitotic misregulation hypothesis

Recently, it has been shown that the toxic effects of accumulated progerin in HGPS cells may result from its inappropriate behaviour during both interphase and mitosis, which resembles that of the B-type lamin (Cao *et al.*, 2007; Dechat *et al.*, 2007). Progerin preferentially forms hetero-polymers with B-type rather than A-type lamins across the NE during interphase (Delbarre *et al.*, 2006) and remains in the B-type-containing mitotic vesicles and insoluble aggregates following nuclear disassembly (Cao *et al.*, 2007; Dechat *et al.*, 2007). As a result, HGPS cells show chromosome segregation defects and subsequent blocks in cytokinesis. This, in turn, leads to chromosomal aberrations and abnormal mitotic phenotypes such as bi-nucleated cells with chromosome bridges and giant nuclei. Interestingly, an earlier study also showed mitotic defects and chromosome abnormalities in HGPS cells. Following a microarray analysis of gene expression in fibroblasts from both old individuals and HGPS patients, the same authors reported a significant down-regulation of genes involved in cell-cycle progression and mitotic regulation in those cells (Ly *et al.*, 2000).

7.7 A new disease model: Cellular stress hypothesis

Sensing of cellular concentrations of oxygen is an essential mechanism in many tissues because of the risk of oxidative damage from excess O_2 levels (hyperoxia) and of metabolic demise from insufficient O_2 levels (hypoxia) (Harman, 1956; Semenza, 2001). The oxygen level beyond the 'perceived normoxic range' is a significant stressor leading to growth arrest due to a high level of production of reactive oxygen species (ROS) (Packer and Fuehr, 1977; see *Figure 1*). Many tissues in the body, such as bone and heart, are low-oxygen tension environments and excessive exposure to ROS signals would disturb their normal homeostasis. High levels of ROS have been implicated in the pathogenesis of several human age-related diseases including atherosclerosis, organ fibrosis and neurodegenerative disorders (Haddad, 2004). In response to oxidative stress, mammalian cells mount an adaptive response to prevent further cellular damage by increasing the expression and/or activation of antioxidant enzymes and other protective genes. Interestingly, down-regulation of the antioxidant enzyme catalase has been observed in progeroid Zmpste24-null mice (Varela *et al.*, 2005) and a highly reduced glutathione peroxidase activity has been reported in HGPS fibroblasts (Yan *et al.*, 1999). Moreover, lamin A has been found to be glutathiolated in kidney cells upon oxidative stress during renal ischaemia (Eaton *et al.*, 2003). We therefore propose that A-type lamins protect cells from the damaging effects of ROS by sensing the intracellular redox and in so doing modulate the activity of redox-dependent transcription factors in the nucleus. In some laminopathy cells, mutant A-type lamins may be either over-sensitive or insensitive to redox regulation and thus unable to respond to redox changes, leading to the accumulation of oxidatively modified macromolecules, including DNA, RNA, proteins and lipids. Furthermore, the proposed misregulation in the ROS buffering capacity by mutant A-type lamins could lead to a compromised ability of some laminopathy cells to deal with the additional sources of ROS, such as those associated with a particular genetic predisposition. If this hypothesis is true, dietary supplementation with antioxidants may prove to be useful in the treatment of some laminopathy diseases. Antioxidant therapy has led to the improved function of immune cells in premature ageing mice (Alvarado *et al.*, 2006) and prevented muscle wasting and loss of fat tissue in mouse models of cachexia (Buck and Chojkier, 1996).

8 The nuclear lamins as the signalling receptors of the nuclear envelope

There are at least three signalling pathways central to growth and regeneration which have been linked to the functions of A-type lamins and their binding partners. These include Rb/E2F and Rb/MyoD pathways, the Wnt/β-catenin pathway and the TGF-β/Smad pathway (Pekovic and Hutchison, 2008). These pathways have been particularly studied in the satellite muscle cells and dermal fibroblasts with mutations in either A-type lamins or their binding partners. Mutations in A-type lamin binding partners including emerin, nesprins 1/2, LAP2α and MAN1 have been implicated in X-linked EDMD (Bione *et al.*, 1994), AD-EDMD (Zhang *et al.*, 2007), DCM (Taylor *et al.*, 2005) and disorders of increased bone density, respectively (Osteopoikilosis, Busche-Ollendorf Syndrome and Melorheostosis) (Hellemans *et al.*, 2004). In addition to the pathways associated with tissue regeneration, A-type lamins have been linked to signalling pathways associated with cellular responses to stress, including the MAPK pathways, p53/p21 and Rb/p16 pathways and the NF-κβ pathway, which are either chronically up-regulated or fail to activate upon stress in cells and tissues with *LMNA* mutations (Pekovic and Hutchison, 2008).

As discussed earlier, the regulation of Rb by A-type lamin/LAP2α complexes is essential for Rb's role in repressing the activation of E2F-dependent S-phase genes (Dorner *et al.*, 2006; Markiewicz *et al.*, 2002; see *Figure 2*). This regulation is exerted through either protecting Rb from proteosomal degradation (Johnson *et al.*, 2004) and targeting to subnuclear splicing speckles (Pekovic *et al.*, 2007) or controlling its rate of phosphorylation via protein phosphatase PP2A (van Berlo *et al.*, 2005). Loss of Rb regulation by either LAP2α or A-type lamin sRNAi knock-down in human and mouse cells leads to increased S-phase entry, which is followed by either inadequate growth arrest after withdrawal of mitogens (Dorner *et al.*, 2006) and/or S-phase delay and G2 arrest (Pekovic *et al.*, 2007). Rb also has a well-established role in the myogenesis of satellite myoblasts whereby the interaction of Rb with the myogenic transcription factor MyoD leads to the activation of MyoD-dependent target genes and initiation of muscle differentiation (Galderisi *et al.*, 2006). Recent studies have implicated a de-regulated Rb/MyoD pathway as a cause of decreased differentiation potential of human and mouse satellite cells null for A-type lamins or emerin (Bakay *et al.*, 2006; Favreau *et al.*, 2004; Frock *et al.*, 2006; Markiewicz *et al.*, 2005; Melcon *et al.*, 2006). These myoblasts exhibit a failure to exit the cell cycle and show a decreased expression of proteins critical in muscle differentiation.

The canonical Wnt (Wingless/INT)/β-catenin pathway is one of the most important regulators of mesenchymal tissue proliferation and differentiation (Bowley *et al.*, 2007). The A-type lamin binding partner emerin binds to β-catenin via its C-terminal APC-like (adenomatous polyposis coli) domain. This restricts nuclear accumulation of β-catenin in a CRM1-export-dependent manner (CRM-exportin chromosome region maintenance) and thus represses transcriptional activation of TCF/LEF (T-cell-factor/lymphoid-enhancer-factor) responsive genes (Markiewicz *et al.*, 2006; see *Figure* 2). Indeed, a number of genes associated with Wnt signalling were found to be up-regulated in X-EDMD patient fibroblasts (Markiewicz *et al.*, 2006; Tsukahara *et al.*, 2002). Since cardiac and skeletal muscles accumulate fatty fibrotic tissue in patients with *emerin* and *LMNA* mutations, the aberrant Wnt signalling may lead to this phenotype in EDMD patients through accelerated cell turnover (Markiewicz *et al.*, 2006). Recently, it was reported that the Zmpste24-null mouse model of accelerated ageing shows defective proliferative potential of bulge cells in the hair follicle niche due to a decreased Wnt signalling pathway (Espada *et al.*, 2008).

The canonical tumour growth-factor β (TGF-β)/Smad pathway is another important regulator of mesenchymal tissue homeostasis (Verrecchia *et al.*, 2006). Over-expression of TGF-β signalling is implicated in tissue fibrosis of cardiac and skeletal muscles (Verrecchia *et al.*, 2006). A-type lamins bind to TGF-β-induced Smads 2/3 and repress high basal levels of Smad-dependent gene expression by promoting their dephosphorylation via phosphatase PP2A (van Berlo *et al.*, 2005; see *Figure* 2). In addition, the A-type lamin binding partner MAN1 interacts with receptor R-Smads and antagonizes TGF-β, activin and bone morphogenic protein (BMP) signalling (Lin *et al.*, 2005; Pan *et al.*, 2005). It has been shown that impaired regulation of TGF-β/Smad pathway leads to increased proliferation rate and collagen production in lamin A KO fibroblasts and may therefore have direct implications for tissue fibrosis observed in patients with *LMNA* mutations (van Berlo *et al.*, 2005).

Genome-wide profiling of the hearts in a mouse model of AD-EDMD revealed increases in expression of several genes associated with MAPK signalling, such as *c-jun*, *Elk1*, *pERK1/2* and *pJNK* (Muchir *et al.*, 2007). Activation of MAPK (mitogen-activated protein kinase) signalling either through ERK (extra-cellular signal-regulated kinase) or JNK (jun N-terminal kinase) in the heart has been associated with the development of cardiomyopathy in mouse models (Petrich *et al.*, 2003). The Ras/MAPK pathway over-activation leads to premature senescence in the cells with intact cell-cycle controls (Serrano *et al.*, 1997) through the activation of the stress kinases JNK/SAPK (stress-activated protein kinase) and p38 (Hutter *et al.*, 2002; Iwasa *et al.*, 2003; Wang *et al.*, 2002). Thus, an activation of JNK kinase in the lmna-mutated hearts could lead to cardiomyopathy by inducing premature senescence of either cardiomyocytes or cardiac-resident stem cells.

The p53/p21 and Rb/p16 pathways are essential gatekeepers of senescent arrest in both mouse and human models (Narita *et al.*, 2003; Shay *et al.*, 1991). Hyper-activation of the p53 pathway has recently been reported in Zmpste24 mouse models of accelerated ageing (Varela *et al.*, 2005). The activation of p53 pathway in these mice most likely occurs as a result of increased DNA damage and altered DNA repair (Liu *et al.*, 2005; Varela *et al.*, 2005). Interestingly, both lmna KO and lamin A-processing mutant mouse cells can not arrest in response to either DNA damage (Johnson *et al.*, 2004) or an increased activity of cyclin-dependent kinase (CDK) inhibitor $p16^{ink4}$ (Nitta *et al.*, 2006) due to a defective Rb pathway. In contrast, the cells from patients with lipodystrophy-linked lamin A mutations exhibit prelamin A accumulation, increased oxidative stress levels and p16-dependent premature senescence (Caron *et al.*, 2007). Oxidative stress has also been implicated in the pathogenesis of amyotrophic quadricipital syndrome with cardiac involvement in a large family with the *LMNA* gene mutation (Charniot *et al.*, 2007).

The nuclear factor-κβ (NF-κβ) pathway is involved in the regulation of diverse cellular processes including survival, differentiation and cellular responses to stress (Jones *et al.*, 2003). Following cytokine stimulation or mechanical stress, both lmna and emerin KO mouse fibroblasts show impaired mechano-sensitive transduction pathways involving NF-κβ at a step downstream of its nuclear entry and DNA binding (Lammerding *et al.*, 2004, 2005). Moreover, in response to mechanical loads, lmna KO cells are not able to resist the forces of compression and are more prone to apoptosis (Broers *et al.*, 2004; Lammerding *et al.*, 2004), which is not the case for emerin KO cells. It has been suggested that a lack of dephosphorylation of the NF-κβ subunit relA (reticuloendotheliosis viral oncogene) by A-type lamin-binding phosphatase PP2A may explain its altered activity in lmna or emerin KO cells (van Berlo *et al.*, 2005).

9 Lamin A in physiological ageing

Recently, lamin A has been implicated in both cellular and tissue physiological ageing (Scaffidi and Misteli, 2006; McClintock *et al.*, 2007; Duque and Rivas 2006; Afilalo *et al.*, 2007; Cao *et al.*, 2007; Ukekawa *et al.*, 2007; Candelario *et al.*, 2008; Huang *et al.*, 2008). Interestingly, low levels of progerin mRNA and protein appear to be present in fibroblast cells of young healthy individuals, which surprisingly does not increase in expression with age (McClintock *et al.*, 2007; Scaffidi and Misteli, 2006). Nonetheless, the accumulation of abnormal nuclear shapes accompanied by a loss of peripheral heterochromatin and the aberrant localization of several INM proteins does occur in fibroblasts from old individuals as seen in HGPS cells. Moreover, a subpopulation of fibroblasts that contains progerin show either bi-nucleated cells with chromosome bridges or cells with giant nuclei (Cao *et al.*, 2007), which implies that similar mitotic defects operate in both normal ageing and progeria. Remarkably, inhibition of progeria production in cells of old individuals prevents age-dependent nuclear defects, p53-dependent signalling and age-associated histone modifications (Scaffidi and Misteli, 2006). Interestingly, in progeria cells, there is an age-dependent accumulation of both progerin and normal pre-lamin A (Goldman *et al.*, 2004), whilst lamin A expression shows a dramatic redistribution to the intranuclear sites followed by their complete absence from the nuclei (Bridger and Kill, 2004). In contrast, during normal ageing human cells, reports have shown an increased expression of pre-lamin A as a result of FACE1 down-regulation may be more important for the senescent arrest of fibroblasts than the presence of progerin (Candelario *et al.*, 2008; Ukekawa *et al.*, 2007). In addition, several other studies found a decreased expression of lamin A in osteoblasts and bone marrow cells (Afilalo *et al.*, 2007), as well as in vascular endothelial cells and cardiomyocytes of old mice (Duque and Rivas 2006). Thus, the presence of low levels of progerin in concert with changes to both lamin A and pre-lamin A may be required for the observed nuclear phenotypes in aged cells. Although these reports point out that there is a number of lamin A linked mechanisms leading to cellular senescence during ageing, these may become less ambiguous with further research. Prelamin A accumulation may either act in parallel with low levels of progerin to directly contribute to senescent arrest, or alternatively it may accumulate as a consequence of senescent arrest and thus reinforce its maintenance.

10 Conclusions: A-type lamins as the regulators of growth and responses to stress

Mutations in A-type lamins give rise to a staggering number of age-related degenerative diseases and progeroid syndromes that show many fibrotic lesions and adipose-like tissue replacements normally seen to accumulate in aged individuals. One model that attempts to explain age-dependent tissue degeneration considers the accumulation of senescent cells at many sites in the body associated with age-related pathologies. Senescent cells arise as a last-resort'stress response to many inefficient cellular processes, including those regulating mitogenic signalling and protective mechanisms against oxidative stress. How may A-type lamins regulate maintenance of such diverse tissues and their onset and/or rate of ageing? A-type lamins and their binding partners form a stress-resistant filamentous network at the nuclear envelope and in the nucleoplasm required for post-natal growth and development as well as long-term tissue maintenance. Moreover, the A-type lamin network regulates a number of signalling pathways associated with growth and adaptive responses to stress. Intriguingly, most laminopathy diseases become evident in patients during the early post-natal period or at puberty when individuals may fail to cope with increased levels of ROS normally associated with rapid post-natal growth. High levels of ROS have been implicated in the pathogenesis of several human age-related diseases (Haddad 2004). Interestingly, oxidative stress has also been implicated (Caron *et al.*, 2007; Charniot *et al.*, 2007).

Post-natal growth is associated with an increase in the production of ROS, which can function as intracellular messengers in cellular signalling and modulate responses in physiological conditions (Semenza 2001). These include diverse cellular processes such as the induction of the cell cycle and activation of redox-sensitive transcription factors, which bind to antioxidant response elements (ARE) in the gene promoters and thus regulate gene expression. In response to mild oxidative stress, mammalian cells mount an adaptive response in order to prevent further oxidative damage by increasing the expression and/or activity of antioxidant enzymes and protective genes (Haddad 2004). Interestingly, a down-regulation of antioxidant enzyme catalase has been observed in Zmpste24 KO mice (Varela *et al.*, 2005) whilst highly reduced glutathione peroxidase activity has been reported in progeria cells (Yan *et al.*, 1999).

We therefore propose that A-type lamins protect cells from the damaging effects of growth-related ROS by sensing the intracellular redox and modulating the activity of redox-dependent transcription factors. This process may be important in diverse tissues including many of those affected by *LMNA* mutations, such as skeletal muscle, heart, brain, liver and endothelium. For example, it is known that during mechanical activity in skeletal muscle, the intracellular environment becomes more oxidative, which leads to modulation of protein oxidation states and their conformational flexibility, thus increasing mechanical resistance and survival (Mayans *et al.*, 2001). Such oxidative changes during muscle activity may induce the compaction of A-type lamin filaments and facilitate the survival of cells during oxidative stress by increasing their mechanical resistance. Indeed, the ability of A-type lamins to dynamically reorganize themselves during muscle differentiation seems to be inhibited in myoblasts expressing mutant A-type lamins (Mariappan and Parnaik, 2005; Markiewicz *et al.*, 2005). Thus, mutant A-type lamins may not be able to respond adequately to oxidative changes and may have a decreased ability to defend cells and tissues against ROS. Interestingly, many signalling pathways regulated by A-type lamins or their binding partners, including Wnt, TGF-β, NF-κβ and Ras/MAPK pathways, can be mediated by ROS signals. Wnt signalling has recently been implicated in the abnormal fibrogenic conversion of aged muscle satellite cells, which leads to increased fibrosis during muscle regeneration in response to injury (Brack *et al.*, 2007; Liu *et al.*, 2007). Moreover, increased TGF-β signalling via MAPK/p38 kinase can lead to premature senescence in human fibroblasts (Kim *et al.*, 2004), which is also known to induce β-catenin activation (Sato, 2006). It will be crucial to investigate whether increased fibrosis in hearts and skeletal muscles of patients with mutations in either lamin A or their binding partners results from the induction of ROS-induced premature senescence in these tissues due to the abnormal cross-talk of these pathways. In conclusion, we propose that during physiological ageing and in laminopathy diseases, dysfunction in the A-type lamin signalling network may lead to a loss of cellular protection against growth-related stress and inappropriate tissue regenerative responses. This in turn may switch on chronic stress signalling pathways that serve as a 'last-resort' response to stress intended to maintain the organismal survival but at the same time promoting the senescent phenotype.

Acknowledgements

Our work is supported by the Association for International Cancer Research (AICR), the Muscular Dystrophy Campaign (MDC) and the FP-6 European Framework to C.J.H. We are grateful to Veronika Boczonadi for providing a diagram in Figure 3. We wish to thank Christopher Vaughan for his assistance with the bibliography and for proofreading of the manuscript. This book chapter is dedicated to the memory of Charlotte.

References

Afilalo, J., Sebag, I.A., Chalifour, L.E., *et al.* (2007) Age-related changes in lamin A/C expression in cardiomyocytes. *Am. J. Physiol. Heart Circ. Physiol.* **293(3):** 1451–1456.

Agarwal, A.K., Fryns, J.P., Auchus, R.J. and Garg, A. (2003) Zinc metalloproteinase, ZMPSTE24, is mutated in mandibuloacral dysplasia. *Hum. Mol. Genet.* **12:** 1995–2001.

Allsopp, R.C., Vaziri, H., Patterson, C., *et al.* (1992) Telomere length predicts replicative capacity of human fibroblasts. *Proc. Natl Acad. Sci. USA* **89:** 10114–10118.

Alvarado, C., Alvarez, P., Jimenez, L., *et al.* (2006) Dietary supplementation with antioxidants improves functions and decreases oxidative stress of leukocytes from prematurely aging mice. *Nutrition* **22:** 767–777.

Arimura, T., Helbling-Leclerc, A., Massart, C., *et al.* (2005) Mouse model carrying H222P-Lmna mutation develops muscular dystrophy and dilated cardiomyopathy similar to human striated muscle laminopathies. *Hum. Mol. Genet.* **14:** 155–169.

Arora, P., Muralikrishna, B. and Parnaik, V.K. (2004) Cell-type-specific interactions at regulatory motifs in the first intron of the lamin A gene. *FEBS Lett.* **568:** 122–128.

Bakay, M., Wang, Z., Melcon, G., *et al.* (2006) Nuclear envelope dystrophies show a transcriptional fingerprint suggesting disruption of Rb-MyoD pathways in muscle regeneration. *Brain* **129:** 996–1013.

Belmont, A.S., Zhai, Y., *et al.* (1993) Lamin B distribution and association with peripheral chromatin revealed by optical sectioning and electron microscopy tomography. *J. Cell Biol.* **123:** 1671–1685.

Benedetti, S., Bertini, E., Iannaccone, S., *et al.* (2005) Dominant LMNA mutations can cause combined muscular dystrophy and peripheral neuropathy. *J. Neurol. Neurosurg. Psychiat.* **76:** 1019–1021.

Bergo, M.O., Gavino, B., Ross, J., *et al.* (2002) Zmpste24 deficiency in mice causes spontaneous bone fractures, muscle weakness, and a prelamin A processing defect. *Proc. Natl Acad. Sci. USA* **99:** 13049–13054.

Bione, S., Maestrini, E., Rivella, S., *et al.* (1994) Identification of a novel X-linked gene responsible for Emery-Dreifuss muscular dystrophy. *Nat. Genet.* **8:** 323–327.

Bonne, G. and Levy, N. (2003) LMNA mutations in atypical Werner's syndrome. *Lancet* **362:** 1585–1586; author reply 1586.

Bonne, G., Di Barletta, M.R., Varnous, S., *et al.* (1999) Mutations in the gene encoding lamin A/C cause autosomal dominant Emery-Dreifuss muscular dystrophy. *Nat. Genet.* **21:** 285–288.

Bowley, E., O'Gorman, D.B. and Gan, B.S. (2007) Beta-catenin signaling in fibroproliferative disease. *J. Surg. Res.* **138:** 141–150.

Brack, A.S., Conboy, M.J., Roy, S., *et al.* (2007) Increased Wnt signaling during aging alters muscle stem cell fate and increases fibrosis. *Science* **317:** 807–810.

Bridger, J.M. and Kill, I.R. (2004) Aging of Hutchinson-Gilford progeria syndrome fibroblasts is characterised by hyperproliferation and increased apoptosis. *Exp. Gerontol.* **39:** 717–724.

Bridger, J.M., Kill, I.R. *et al.* (1993) Internal lamin structures within G1 nuclei of human dermal fibroblasts. *J. Cell Sci.* **104:** 297–306.

Broers, J.L., Machiels, B.M., Kuijpers, H.J., *et al.* (1997) A- and B-type lamins are differentially expressed in normal human tissues. *Histochem. Cell Biol.* **107:** 505–517.

Broers, J.L., Machiels, B.M., van Eys, G.J., *et al.* (1999) Dynamics of the nuclear lamina as monitored by GFP-tagged A-type lamins. *J. Cell Sci.* **112:** 3463–3475.

Broers, J.L., Peeters, E.A., Kuijpers, H.J., *et al.* (2004) Decreased mechanical stiffness in LMNA-/- cells is caused by defective nucleo-cytoskeletal integrity: implications for the development of laminopathies. *Hum. Mol. Genet.* **13:** 2567–2580.

Broers, J.L., Kuijpers, H.J., Ostlund, C., *et al.* (2005) Both lamin A and lamin C mutations cause lamina instability as well as loss of internal nuclear lamin organization. *Exp. Cell Res.* **304:** 582–592.

Broers, J.L., Ramaekers, F.C., Hutchison, C.J., *et al.* (2006) Nuclear lamins: laminopathies and their role in premature ageing. *Physiol. Rev.* **86:** 967–1008.

Buck, M. and Chojkier, M. (1996) Muscle wasting and dedifferentiation induced by oxidative stress in a murine model of cachexia is prevented by inhibitors of nitric oxide synthesis and antioxidants. *EMBO J.* **15:** 1753–1765.

Campisi, J. (2005) Senescent cells, tumor suppression, and organismal aging: good citizens, bad neighbors. *Cell* **120:** 513–522.

Candelario, J., Sudhakar, S., Navarro, S. *et al.* (2008) Perturbation of wild-type lamin A metabolism results in a progeroid phenotype. *Aging Cell* **7:** 355–367.

Cao, H. and Hegele, R.A. (2000) Nuclear lamin A/C R482Q mutation in Canadian kindreds with Dunnigan-type familial partial lipodystrophy. *Hum. Mol. Genet.* **9:** 109–112.

Cao, H. and Hegele, R.A. (2003) LMNA is mutated in Hutchinson-Gilford progeria (MIM 176670) but not in Wiedemann-Rautenstrauch progeroid syndrome (MIM 264090). *J. Hum. Genet.* **48:** 271–274.

Cao, K., Capell, B.C., Erdos, M.R. *et al.* (2007) A lamin A protein isoform overexpressed in Hutchinson-Gilford progeria syndrome interferes with mitosis in progeria and normal cells. *Proc. Natl Acad. Sci. USA* **104:** 4949–4954.

Capanni, C., Mattioli, E., Columbaro, M., *et al.* (2005) Altered pre-lamin A processing is a common mechanism leading to lipodystrophy. *Hum. Mol. Genet.* **14:** 1489–1502.

Caron, M., Auclair, M., Donadille, B., *et al.* (2007) Human lipodystrophies linked to mutations in A-type lamins and to HIV protease inhibitor therapy are both associated with prelamin A accumulation, oxidative stress and premature cellular senescence. *Cell Death Differ.* **14:** 1759–1767.

Caux, F., Dubosclard, E., Lascols, O., *et al.* (2003) A new clinical condition linked to a novel mutation in lamins A and C with generalized lipoatrophy, insulin-resistant diabetes, disseminated leukomelanodermic papules, liver steatosis, and cardiomyopathy. *J. Clin. Endocrinol. Metab.* **88:** 1006–1013.

Charniot, J.C., Pascal, C., Bouchier, C., *et al.* (2003) Functional consequences of an LMNA mutation associated with a new cardiac and non-cardiac phenotype. *Hum. Mutat.* **21:** 473–481.

Charniot, J.C., Bonnefont-Rousselot, D., Marchand, C., *et al.* (2007) Oxidative stress implication in a new phenotype of amyotrophic quadricipital syndrome with cardiac involvement due to lamin A/C mutation. *Free Radic. Res.* **41:** 424–431.

Chen, L., Lee, L., Kudlow, B.A., *et al.* (2003) LMNA mutations in atypical Werner's syndrome. *Lancet* **362:** 440–445.

Cohen, M., Lee, K.K., Wilson, K.L. and Gruenbaum, Y. (2001) Transcriptional repression, apoptosis, human disease and the functional evolution of the nuclear lamina. *Trends Biochem. Sci.* **26:** 41–47.

Csoka, A.B., English, S.B., Simkevich, C.P., *et al.* (2004a) Genome-scale expression profiling of Hutchinson-Gilford progeria syndrome reveals widespread transcriptional misregulation leading to mesodermal/mesenchymal defects and accelerated atherosclerosis. *Aging Cell* **3** 235–243.

Csoka, A.B., Cao, H., Sammak, P.J., *et al.* (2004b) Novel lamin A/C gene (LMNA) mutations in atypical progeroid syndromes. *J. Med. Genet.* **41** 304–308.

d'Adda di Fagagna, F., Reaper, P.M., *et al.* (2003) A DNA damage checkpoint response in telomere-initiated senescence. *Nature* **426:** 194–198.

D'Amico, A., Haliloglu, G., Richard, P., *et al.* (2005) Two patients with Dropped head syndrome'due to mutations in LMNA or SEPN1 genes. *Neuromusc. Disord.* **15:** 521–524.

Dahl, K.N., Kahn, S.M., *et al.* (2004) The nuclear envelope lamina network has elasticity and a compressibility limit suggestive of a molecular shock absorber. *J. Cell Sci.* **117:** 4779–4786.

Dahl, K.N., Scaffidi, P., Islam, M.F., Yodh, A.G., Wilson, K.L. and Misteli, T. (2006) Distinct structural and mechanical properties of the nuclear lamina in Hutchinson-Gilford progeria syndrome. *Proc. Natl Acad. Sci. USA* **103:** 10271–10276.

De Sandre-Giovannoli, A., Chaouch, M., Kozlov, S., *et al.* (2002) Homozygous defects in LMNA, encoding lamin A/C nuclear-envelope proteins, cause autosomal recessive axonal neuropathy in human (Charcot-Marie-Tooth disorder type 2) and mouse. *Am. J. Hum. Genet.* **70:** 726–736.

de Sandre-Giovannoli, A., Bernard, R., Cau, P., *et al.* (2003) Lamin a truncation in Hutchinson-Gilford progeria. *Science* **300:** 2055.

Dechat, T., Korbei, B., Vaughan, O.A., *et al.* (2000) Lamina-associated polypeptide 2 alpha binds intranuclear A-type lamins. *J. Cell Sci.* **113:** 3473–3484.

Dechat, T., Shimi, T., Adam, S.A., *et al.* (2007) Alterations in mitosis and cell cycle progression caused by a mutant lamin A known to accelerate human aging. *Proc. Natl Acad. Sci. USA* **104:** 4955–4960.

Delbarre, E., Tramier, M., Coppey-Moisan, M., *et al.* (2006) The truncated prelamin A in Hutchinson-Gilford progeria syndrome alters segregation of A-type and B-type lamin homopolymers. *Hum. Mol. Genet.* **15:** 1113–1122.

Dhe-Paganon, S., Werner, E.D., Chi, Y.I. and Shoelson, S.E. (2002) Structure of the globular tail of nuclear lamin. *J. Biol. Chem.* **277:** 17381–17384.

Dorner, D., Vlcek, S., Foeger, N., *et al.* (2006) Lamina-associated polypeptide 2 alpha regulates cell cycle progression and differentiation via the retinoblastoma-E2F pathway. *J. Cell Biol.* **173:** 83–93.

Duque, G. and Rivas, D. (2006) Age-related changes in lamin A/C expression in the osteoarticular system: laminopathies as a potential new aging mechanism. *Mech. Ageing Dev.* **127:** 378–383.

Eaton, P., Jones, M.E., *et al.* (2003) Reversible cysteine-targeted oxidation of proteins during renal oxidative stress. *J. Am. Soc. Nephrol.* **14(Suppl 3):** S290–S296.

Eriksson, M., Brown, W.T., Gordon, L.B., *et al.* (2003) Recurrent *de novo* point mutations in lamin A cause Hutchinson-Gilford progeria syndrome. *Nature* **423:** 293–298.

Espada, J., Varela, I., Flores, I., *et al.* (2008) Nuclear envelope defects cause stem cell dysfunction in premature-aging mice. *J. Cell Biol.* **181:** 27–35.

Faragher, R.G. and Kipling, D. (1998) How might replicative senescence contribute to human ageing? *Bioessays* **20:** 985–991.

Fatkin, D., MacRae, C., Sasaki, T., *et al.* (1999) Missense mutations in the rod domain of the lamin A/C gene as causes of dilated cardiomyopathy and conduction-system disease. *N. Engl. J. Med.* **341:** 1715–1724.

Favreau, C., Higuet, D., Courvalin, J.C. and Buendia, B. (2004) Expression of a mutant lamin A that causes Emery-Dreifuss muscular dystrophy inhibits *in vitro* differentiation of C2C12 myoblasts. *Mol. Cell Biol.* **24:** 1481–1492.

Forissier, J.F., Bonne, G., Bouchier, C., *et al.* (2003) Apical left ventricular aneurysm without atrio-ventricular block due to a lamin A/C gene mutation. *Eur. J. Heart Fail.* **5:** 821–825.

Frock, R.L., Kudlow, B.A., Evans, A.M., Jameson, S.A., Hauschka, S.D. and Kennedy, B.K. (2006) Lamin A/C and emerin are critical for skeletal muscle satellite cell differentiation. *Genes Dev.* **20:** 486–500.

Galderisi, U., Cipollaro, M. and Giordano, A. (2006) The retinoblastoma gene is involved in multiple aspects of stem cell biology. *Oncogene* **25:** 5250–5256.

Garg, A., Speckman, R.A. and Bowcock, A.M. (2002) Multisystem dystrophy syndrome due to novel missense mutations in the amino-terminal head and alpha-helical rod domains of the lamin A/C gene. *Am. J. Med.* **112:** 549–555.

Goizet, C., Yaou, R.B., Demay, L., *et al.* (2004) A new mutation of the lamin A/C gene leading to autosomal dominant axonal neuropathy, muscular dystrophy, cardiac disease, and leuconychia. *J. Med. Genet.* **41:** e29.

Goldberg, M.W. and Allen, T.D. (1995) Structural and functional organization of the nuclear envelope. *Curr. Opin. Cell Biol.* **7:** 301–309.

Goldman, R.D., Gruenbaum, Y., *et al.* (2002) Nuclear lamins: building blocks of nuclear architecture. *Genes Dev.* **16:** 533–547.

Goldman, R.D., Shumaker, D.K., Erdos, M.R., *et al.* (2004) Accumulation of mutant lamin A causes progressive changes in nuclear architecture in Hutchinson-Gilford progeria syndrome. *Proc. Natl Acad. Sci. USA* **101:** 8963–8968.

Gotzmann, J. and Foisner, R. (2006) A-type lamin complexes and regenerative potential: a step towards understanding laminopathic diseases? *Histochem. Cell Biol.* **125:** 33–41.

Gruber, J., Lampe, T., *et al.* (2005) RNAi of FACE1 protease results in growth inhibition of human cells expressing lamin A: implications for Hutchinson-Gilford progeria syndrome. *J. Cell Sci.* **118:** 689–696.

Haddad, J.J. (2004) Oxygen sensing and oxidant/redox-related pathways. *Biochem. Biophys. Res. Commun.* **316:** 969–977.

Haithcock, E., Dayani, Y., Neufeld, E., *et al.* (2005) Age-related changes of nuclear architecture in *Caenorhabditis elegans*. *Proc. Natl Acad. Sci. USA* **102:** 16690–16695.

Haque, F., Lloyd, D.J., Smallwood, D.T., *et al.* (2006) SUN1 interacts with nuclear lamin A and cytoplasmic nesprins to provide a physical connection between the nuclear lamina and the cytoskeleton. *Mol. Cell Biol.* **26:** 3738–3751.

Harborth, J., Elbashir, S.M., Bechert, K., *et al.* (2001) Identification of essential genes in cultured mammalian cells using small interfering RNAs. *J. Cell Sci.* **114:** 4557–4565.

Harman, D. (1956) Aging: a theory based on free radical and radiation chemistry. *J. Gerontol.* **11:** 298–300.

Hayflick, L. and Moorhead, P.S. (1961) The serial cultivation of human diploid cell strains. *Exp. Cell Res.* **25:** 585–621.

Hegele, R.A., Cao, H. and Liu, D.M., *et al.* (2006) Sequencing of the reannotated LMNB2 gene reveals novel mutations in patients with acquired partial lipodystrophy. *Am. J. Hum. Genet.* **79:** 383–389.

Hellemans, J., Preobrazhenska, O., Willaert, A., *et al.* (2004) Loss-of-function mutations in LEMD3 result in osteopoikilosis, Buschke-Ollendorff syndrome and melorheostosis. *Nat. Genet.* **36:** 1213–1218.

Herbig, U., Ferreira, M., Condel, L., *et al.* (2006) Cellular senescence in aging primates. *Science* **311:** 1257.

Holaska, J.M. and Wilson, K.L. (2007) An emerin þroteome: purification of distinct emerin-containing complexes from HeLa cells suggests molecular basis for diverse roles including gene regulation, mRNA splicing, signaling, mechanosensing, and nuclear architecture. *Biochemistry* **46:** 8897–8908.

Huang, S., Risques, R.A., Martin, G.M., Rabinovitch, P.S. and Oshima, J. (2008) Accelerated telomere shortening and replicative senescence in human fibroblasts overexpressing mutant and wild-type lamin A. *Exp. Cell Res.* **314:** 82–91.

Hutchison, C.J. (2002) Lamins: building blocks or regulators of gene expression? *Nat. Rev. Mol. Cell Biol.* **3:** 848–858.

Hutchison, C.J. and Worman, H.J. (2004) A-type lamins: guardians of the soma? *Nat. Cell Biol.* **6:** 1062–1067.

Hutchison, C.J., Alvarez-Reyes, M. and Vaughan, O.A. (2001) Lamins in disease: why do ubiquitously expressed nuclear envelope proteins give rise to tissue-specific disease phenotypes? *J. Cell Sci.* **114:** 9–19.

Hutter, E., Unterluggauer, H., Uberall, F., *et al.* (2002) Replicative senescence of human fibroblasts: the role of Ras-dependent signaling and oxidative stress. *Exp. Gerontol.* **37:** 1165–1174.

Imai, S., Nishibayashi, S., *et al.* (1997) Dissociation of Oct-1 from the nuclear peripheral structure induces the cellular aging-associated collagenase gene expression. *Mol. Biol. Cell* **8:** 2407–2419.

Ivorra, C., Kubicek, M., Gonzalez, J.M., *et al.* (2006) A mechanism of AP-1 suppression through interaction of c-Fos with lamin A/C. *Genes Dev.* **20:** 307–320.

Iwasa, H., Han, J. and Ishikawa, F. (2003) Mitogen-activated protein kinase p38 defines the common senescence-signalling pathway. *Genes Cells* **8:** 131–144.

Jagatheesan, G., Thanumalayan, S., *et al.* (1999) Colocalization of intranuclear lamin foci with RNA splicing factors. *J. Cell Sci.* **112:** 4651–4561.

Jeyapalan, J.C., Ferreira, M., Sedivy, J.M. and Herbig, U. (2007) Accumulation of senescent cells in mitotic tissue of aging primates. *Mech. Ageing Dev.* **128:** 36–44.

Johnson, B.R., Nitta, R.T., Frock, R.L., *et al.* (2004) A-type lamins regulate retinoblastoma protein function by promoting subnuclear localization and preventing proteasomal degradation. *Proc. Natl Acad. Sci. USA* **101:** 9677–9682.

Jones, W.K., Brown, M., Ren, X., *et al.* (2003) NF-kappaB as an integrator of diverse signaling pathways: the heart of myocardial signaling? *Cardiovasc. Toxicol.* **3:** 229–254.

Karlseder, J., Smogorzewska, A., *et al.* (2002) Senescence induced by altered telomere state, not telomere loss. *Science* **295:** 2446–2449.

Kennedy, B.K., Barbie, D.A., *et al.* (2000) Nuclear organization of DNA replication in primary mammalian cells. *Genes Dev.* **14:** 2855–2868.

Kim, E.S., Kim, M.S. and Moon, A. (2004) TGF-beta-induced upregulation of MMP-

2 and MMP-9 depends on p38 MAPK, but not ERK signaling in MCF10A human breast epithelial cells. *Int. J. Oncol.* **25:** 1375–1382.

Kipling, D., Davis, T., Ostler, E.L. and Faragher, R.G. (2004) What can progeroid syndromes tell us about human aging? *Science* **305:** 1426–1431.

Kirkland, J.L., Tchkonia, T., Pirtskhalava, T., *et al.* (2002) Adipogenesis and aging: does aging make fat go MAD? *Exp. Gerontol.* **37:** 757–767.

Kirkwood, T.B. (1977) Evolution of ageing. *Nature* **270:** 301–304.

Krimm, I., Ostlund, C., Gilquin, B., *et al.* (2002) The Ig-like structure of the C-terminal domain of lamin A/C, mutated in muscular dystrophies, cardiomyopathy, and partial lipodystrophy. *Structure* **10:** 811–823.

Lammerding, J., Schulze, P.C., Takahashi, T., *et al.* (2004) Lamin A/C deficiency causes defective nuclear mechanics and mechanotransduction. *J. Clin. Invest.* **113:** 370–378.

Lammerding, J., Hsiao, J., Schulze, P.C., *et al.* (2005) Abnormal nuclear shape and impaired mechanotransduction in emerin-deficient cells. *J. Cell Biol.* **170:** 781–791.

Lee, K.K., Haraguchi, T., Lee, R.S., *et al.* (2001) Distinct functional domains in emerin bind lamin A and DNA-bridging protein BAF. *J. Cell Sci.* **114:** 4567–4673.

Libotte, T., Zaim, H., Abraham, S., *et al.* (2005) Lamin A/C-dependent localization of Nesprin-2, a giant scaffolder at the nuclear envelope. *Mol. Biol. Cell* **16:** 3411–3424.

Lin, F., Morrison, J.M., Wu, W., and Worman, H.J. (2005) MAN1, an integral protein of the inner nuclear membrane, binds Smad2 and Smad3 and antagonizes transforming growth factor-beta signaling. *Hum. Mol. Genet.* **14:** 437–445.

Liu, J., Rolef Ben-Shahar, T., Riemer, D., *et al.* (2000) Essential roles for *Caenorhabditis elegans* lamin gene in nuclear organization, cell cycle progression, and spatial organization of nuclear pore complexes. *Mol. Biol. Cell* **11:** 3937–3947.

Liu, B., Wang, J., Chan, K.M., *et al.* (2005) Genomic instability in laminopathy-based premature aging. *Nat. Med.* **11:** 780–785.

Liu, Y., Rusinol, A., Sinensky, M., *et al.* (2006) DNA damage responses in progeroid syndromes arise from defective maturation of prelamin A. *J. Cell Sci.* **119:** 4644–4649.

Liu, H., Fergusson, M.M., Castilho, R.M., *et al.* (2007) Augmented Wnt signaling in a mammalian model of accelerated aging. *Science* **317:** 803–806.

Lloyd, D.J., Trembath, R.C. and Shackleton, S. (2002) A novel interaction between lamin A and SREBP1: implications for partial lipodystrophy and other laminopathies. *Hum. Mol. Genet.* **11:** 769–777.

Ly, D.H., Lockhart, D.H., *et al.* (2000) Mitotic misregulation and human aging. *Science* **287:** 2486–2492.

Mallampalli, M.P., Huyer, G., Bendale, P., *et al.* (2005) Inhibiting farnesylation reverses the nuclear morphology defect in a HeLa cell model for Hutchinson-Gilford progeria syndrome. *Proc. Natl Acad. Sci. USA* **102:** 14416–14421.

Mancini, M.A., Shan, B., Nickerson, J.A., *et al.* (1994) The retinoblastoma gene product is a cell cycle-dependent, nuclear matrix-associated protein. *Proc. Natl Acad. Sci. USA* **91:** 418–422.

Manju, K., Muralikrishna, B. and Parnaik, V.K. (2006) Expression of disease-causing lamin A mutants impairs the formation of DNA repair foci. *J. Cell Sci.* **119:** 2704–2714.

Mansharamani, M. and Wilson, K.L. (2005) Direct binding of nuclear membrane protein MAN1 to emerin *in vitro* and two modes of binding to barrier-to-autointegration factor. *J. Biol. Chem.* **280:** 13863–13870.
Mariappan, I. and Parnaik, V.K. (2005) Sequestration of pRb by cyclin D3 causes intranuclear reorganization of lamin A/C during muscle cell differentiation. *Mol. Biol. Cell* **16:** 1948–1960.
Markiewicz, E., Dechat, T., Foisner, R., *et al.* (2002) Lamin A/C binding protein LAP2alpha is required for nuclear anchorage of retinoblastoma protein. *Mol. Biol. Cell* **13:** 4401–4413.
Markiewicz, E., Ledran, M. and Hutchison, C.J. (2005) Remodelling of the nuclear lamina and nucleoskeleton is required for skeletal muscle differentiation *in vitro*. *J. Cell Sci.* **118:** 409–420.
Markiewicz, E., Tilgner, K., Barker, N., *et al.* (2006) The inner nuclear membrane protein emerin regulates beta-catenin activity by restricting its accumulation in the nucleus. *EMBO J.* **25:** 3275–3285.
Martin, G.M. (2007) The genetics and epigenetics of altered proliferative homeostasis in ageing and cancer. *Mech. Ageing Dev.* **128:** 9–12.
Martin, G.M. and Oshima, J. (2000) Lessons from human progeroid syndromes. *Nature* **408:** 263–266.
Mayans, O., Wuerges, J., Canela, S., *et al.* (2001) Structural evidence for a possible role of reversible disulphide bridge formation in the elasticity of the muscle protein titin. *Structure* **9:** 331–340.
McClintock, D., Ratner, D., *et al.* (2007) The mutant form of lamin A that causes Hutchinson-Gilford progeria is a biomarker of cellular ageing in human skin. *PLoS ONE* **2(12):** e1269.
Melcon, G., Kozlov, S., Cutler, D.A., *et al.* (2006) Loss of emerin at the nuclear envelope disrupts the Rb1/E2F and MyoD pathways during muscle regeneration. *Hum. Mol. Genet.* **15:** 637–651.
Mercuri, E. and Muntoni, F. (2001) What's new in neuromuscular disorders? Nuclear envelope and Emery-Dreifuss muscular dystrophy. *Eur. J. Paediatr. Neurol.* **5:** 3–5.
Mislow, J.M., Holaska, J.M., Kim, M.S., *et al.* (2002) Nesprin-1alpha self-associates and binds directly to emerin and lamin A *in vitro*. *FEBS Lett.* **525:** 135–140.
Mittnacht, S. and Weinberg, R.A. (1991) G1/S phosphorylation of the retinoblastoma protein is associated with an altered affinity for the nuclear compartment. *Cell* **65:** 381–393.
Moir, R.D., Spann, T.P., *et al.* (2000) Disruption of nuclear lamin organization blocks the elongation phase of DNA replication. *J. Cell Biol.* **149:** 1179–1192.
Mounkes, L.C. and Stewart, C.L. (2004) Aging and nuclear organization: lamins and progeria. *Curr. Opin. Cell Biol.* **16:** 322–327.
Mounkes, L.C., Kozlov, S., Hernandez, L., *et al.* (2003) A progeroid syndrome in mice is caused by defects in A-type lamins. *Nature* **423:** 298–301.
Mounkes, L.C., Kozlov, S.V., Rottman, J.N. and Stewart, C.L. (2005) Expression of an LMNA-N195K variant of A-type lamins results in cardiac conduction defects and death in mice. *Hum. Mol. Genet.* **14:** 2167–2180.
Muchir, A., Bonne, G., van der Kooi, A.J., *et al.* (2000) Identification of mutations in the gene encoding lamins A/C in autosomal dominant limb girdle muscular

dystrophy with atrioventricular conduction disturbances (LGMD1B). *Hum. Mol. Genet.* **9:** 1453–1459.

Muchir, A., van Engelen, B.G., Lammens, M., *et al.* (2003) Nuclear envelope alterations in fibroblasts from LGMD1B patients carrying nonsense Y259X heterozygous or homozygous mutation in lamin A/C gene. *Exp. Cell Res.* **291:** 352–362.

Muchir, A., Pavlidis, P., Decostre, V., *et al.* (2007) Activation of MAPK pathways links LMNA mutations to cardiomyopathy in Emery-Dreifuss muscular dystrophy. *J. Clin. Invest.* **117:** 1282–1293.

Muralikrishna, B. and Parnaik, V.K. (2001) SP3 and AP-1 mediate transcriptional activation of the lamin A proximal promoter. *Eur. J. Biochem.* **268:** 3736–3743.

Muralikrishna, B., Dhawan, J., Rangaraj, N. and Parnaik, V.K. (2001) Distinct changes in intranuclear lamin A/C organization during myoblast differentiation. *J. Cell Sci.* **114:** 4001–4011.

Narita, M., Nunez, S., *et al.* (2003) Rb-mediated heterochromatin formation and silencing of E2F target genes during cellular senescence. *Cell* **113:** 703–716.

Navarro, C.L., De Sandre-Giovannoli, A., Bernard, R., *et al.* (2004) Lamin A and ZMPSTE24 (FACE-1) defects cause nuclear disorganization and identify restrictive dermopathy as a lethal neonatal laminopathy. *Hum. Mol. Genet.* **13:** 2493–2503.

Niedernhofer, L.J., Garinis, G.A., Raams, A., *et al.* (2006) A new progeroid syndrome reveals that genotoxic stress suppresses the somatotroph axis. *Nature* **444:** 1038–1043.

Nikolova, V., Leimena, C., *et al.* (2004) Defects in nuclear structure and function promote dilated cardiomyopathy in lamin A/C-deficient mice. *J. Clin. Invest.* **113:** 357–369.

Nitta, R.T., Jameson, S.A., Kudlow, B.A., *et al.* (2006) Stabilization of the retinoblastoma protein by A-type nuclear lamins is required for INK4A-mediated cell cycle arrest. *Mol. Cell Biol.* **26:** 5360–5372.

Novelli, G. and D'Apice, M.R. (2003) The strange case of the lumper'lamin A/C gene and human premature ageing. *Trends Mol. Med.* **9:** 370–375.

Novelli, G., Muchir, A., Sangiuolo, F., *et al.* (2002) Mandibuloacral dysplasia is caused by a mutation in LMNA-encoding lamin A/C. *Am. J. Hum. Genet.* **71:** 426–431.

Ozaki, T., Saijo, M., Murakami, K., *et al.* (1994) Complex formation between lamin A and the retinoblastoma gene product: identification of the domain on lamin A required for its interaction. *Oncogene* **9:** 2649–2653.

Packer, L. and Fuehr, K. (1977) Low oxygen concentration extends the lifespan of cultured human diploid cells. *Nature* **267:** 423–425.

Padiath, Q.S., Saigoh, K., Schiffmann, R., *et al.* (2006) Lamin B1 duplications cause autosomal dominant leukodystrophy. *Nat. Genet.* **38:** 1114–1123.

Padmakumar, V.C., Libotte, T., Lu, W., *et al.* (2005) The inner nuclear membrane protein Sun1 mediates the anchorage of Nesprin-2 to the nuclear envelope. *J. Cell Sci.* **118:** 3419–3430.

Pan, D., Estevez-Salmeron, L.D., Stroschein, S.L., *et al.* (2005) The integral inner nuclear membrane protein MAN1 physically interacts with the R-Smad proteins to repress signaling by the transforming growth factor-β superfamily of cytokines. *J. Biol. Chem.* **280:** 1592–6001.

Pekovic, V. and Hutchison, C.J. (2008) Adult stem cell maintenance and tissue regeneration in the ageing context: the role for A-type lamins as intrinsic modulators of ageing in adult stem cells and their niches? *J. Anat.* **213:** 5-25.

Pekovic, V., Harborth, J., Broers, J.L., *et al.* (2007) Nucleoplasmic LAP2alpha-lamin A complexes are required to maintain a proliferative state in human fibroblasts. *J. Cell Biol.* **176:** 163–172.

Pendas, A.M., Zhou, Z., Cadinanos, J., *et al.* (2002) Defective prelamin A processing and muscular and adipocyte alterations in Zmpste24 metalloproteinase-deficient mice. *Nat. Genet.* **31:** 94–99.

Petrich, B.G., Molkentin, J.D. and Wang, Y. (2003) Temporal activation of c-Jun N-terminal kinase in adult transgenic heart via cre-loxP-mediated DNA recombination. *FASEB J.* **17:** 749–751.

Raffaele Di Barletta, M., Ricci, E., Galluzzi, G., *et al.* (2000) Different mutations in the LMNA gene cause autosomal dominant and autosomal recessive Emery-Dreifuss muscular dystrophy. *Am. J. Hum. Genet.* **66:** 1407–1412.

Rahman-Roblick, R., Roblick, U.J., Hellman, U., *et al.* (2007) p53 targets identified by protein expression profiling. *Proc. Natl Acad. Sci. USA* **104:** 5401–5406.

Rankin, J. and Ellard, S. (2006) The laminopathies: a clinical review. *Clin. Genet.* **70:** 261–274.

Reddy, K.L., Zullo, J.M., Bertolino, E. and Singh, H. (2008) Transcriptional repression mediated by repositioning of genes to the nuclear lamina. *Nature* **452:** 243–247.

Rober, R.A., Weber, K. and Osborn, M. (1989) Differential timing of nuclear lamin A/C expression in the various organs of the mouse embryo and the young animal: a developmental study. *Development* **105:** 365–378.

Rusinol, A.E. and Sinensky, M.S. (2006) Farnesylated lamins, progeroid syndromes and farnesyl transferase inhibitors. *J. Cell Sci.* **119:** 3265–3272.

Salpingidou, G., Smertenko, A., Hausmanowa-Petrucewicz, I., *et al.* (2007) A novel role for the nuclear membrane protein emerin in association of the centrosome to the outer nuclear membrane. *J. Cell Biol.* **178:** 897–904.

Sasseville, A.M. and Raymond, Y. (1995) Lamin A precursor is localized to intranuclear foci. *J. Cell Sci.* **108:** 273–285.

Sato, M. (2006) Upregulation of the Wnt/beta-catenin pathway induced by transforming growth factor-beta in hypertrophic scars and keloids. *Acta Derm. Venereol.* **86:** 300–307.

Scaffidi, P. and Misteli, T. (2005) Reversal of the cellular phenotype in the premature aging disease Hutchinson-Gilford progeria syndrome. *Nat. Med.* **11:** 440–445.

Scaffidi, P. and Misteli, T. (2006) Lamin A-dependent nuclear defects in human aging. *Science* **312:** 1059–1063.

Schirmer, E.C., Guan, T. and Gerace, L. (2001) Involvement of the lamin rod domain in heterotypic lamin interactions important for nuclear organization. *J. Cell Biol.* **153:** 479–489.

Sebillon, P., Bouchier, C., Bidot, L.D., *et al.* (2003) Expanding the phenotype of LMNA mutations in dilated cardiomyopathy and functional consequences of these mutations. *J. Med. Genet.* **40:** 560–567.

Semenza, G.L. (2001) Hypoxia-inducible factor 1: control of oxygen homeostasis in health and disease. *Pediatr. Res.* **49:** 614–617.

Serrano, M., Lin, A.W., McCurrach, M.E., Beach, D. and Lowe, S.W. (1997) Oncogenic ras provokes premature cell senescence associated with accumulation of p53 and p16INK4a. *Cell* **88:** 593–602.

Shackleton, S., Lloyd, D.J., Jackson, S.N., *et al.* (2000) LMNA, encoding lamin A/C, is mutated in partial lipodystrophy. *Nat. Genet.* **24:** 153–156.

Shalev, S.A., De Sandre-Giovannoli, A., Shani, A.A. and Levy, N. (2007) An association of Hutchinson-Gilford progeria and malignancy. *Am. J. Med. Genet. A.* **16:** 1821–1826.

Shay, J.W., Pereira-Smith, O.M. and Wright, W.E. (1991) A role for both RB and p53 in the regulation of human cellular senescence. *Exp. Cell Res.* **196:** 33–39.

Smith, E.D., Kudlow, B.A., Frock, R.L. and Kennedy, B.K. (2005) A-type nuclear lamins, progerias and other degenerative disorders. *Mech. Ageing Dev.* **126:** 447–460.

Smythe, C., Jenkins, H.E. and Hutchison, C.J. (2000) Incorporation of the nuclear pore basket protein nup153 into nuclear pore structures is dependent upon lamina assembly: evidence from cell-free extracts of *Xenopus* eggs. *EMBO J.* **19:** 3918–3931.

Spann, T.P., Moir, R.D., Goldman, A.E., *et al.* (1997) Disruption of nuclear lamin organization alters the distribution of replication factors and inhibits DNA synthesis. *J. Cell Biol.* **136:** 1201–1212.

Spann, T.P., Goldman, A.E., Wang, C., *et al.* (2002) Alteration of nuclear lamin organization inhibits RNA polymerase II-dependent transcription. *J. Cell Biol.* **156:** 603–608.

Stewart, C. and Burke, B. (1987) Teratocarcinoma stem cells and early mouse embryos contain only a single major lamin polypeptide closely resembling lamin B. *Cell* **51:** 383–392.

Stewart, C.L., Kozlov, S., Fong, L.G. and Young, S.G. (2007) Mouse models of the laminopathies. *Exp. Cell Res.* **313:** 2144–2156.

Stierle, V., Couprie, J., *et al.* (2003) The carboxyl-terminal region common to lamins A and C contains a DNA binding domain. *Biochemistry* **42:** 4819–4828.

Strelkov, S.V., Schumacher, J., *et al.* (2004) Crystal structure of the human lamin A coil 2B dimer: implications for the head-to-tail association of nuclear lamins. *J. Mol. Biol.* **343:** 1067–1080.

Stuurman, N., Heins, S. and Aebi, U. (1998) Nuclear lamins: their structure, assembly, and interactions. *J. Struct. Biol.* **122:** 42–66.

Sullivan, T., Escalante-Alcalde, D., Bhatt, H., *et al.* (1999) Loss of A-type lamin expression compromises nuclear envelope integrity leading to muscular dystrophy. *J. Cell Biol.* **147:** 913–920.

Taylor, M.R., Slavov, D., Gajewski, A., *et al.* (2005) Thymopoietin (lamina-associated polypeptide 2) gene mutation associated with dilated cardiomyopathy. *Hum. Mutat.* **26:** 566–574.

Toth, J.I., Yang, S.H., Qiao, X., *et al.* (2005) Blocking protein farnesyltransferase improves nuclear shape in fibroblasts from humans with progeroid syndromes. *Proc. Natl Acad. Sci. USA* **102:**, 12873–12878.

Toussaint, O., Medrano, E.E., *et al.* (2000) Cellular and molecular mechanisms of stress-induced premature senescence (SIPS) of human diploid fibroblasts and melanocytes. *Exp. Gerontol.* **35:** 927–945.

Tsukahara, T., Tsujino, S. and Arahata, K. (2002) CDNA microarray analysis of gene expression in fibroblasts of patients with X-linked Emery-Dreifuss muscular dystrophy. *Muscle Nerve* **25:** 898–901.

Ukekawa, R,, Miki, K,, Fujii, M., *et al.* (2007) Accumulation of multiple forms of lamin A with down-regulation of FACE-1 suppresses growth in senescent human cells. *Genes Cells* **12:** 397–406.

van Berlo, J.H., Voncken, J.W., Kubben, N., *et al.* (2005) A-type lamins are essential for TGF-beta1 induced PP2A to dephosphorylate transcription factors. *Hum. Mol. Genet.* **14:** 2839–2849.

van der Kooi, A,J,, Bonne, G,, Eymard, B., *et al.* (2002) Lamin A/C mutations with lipodystrophy, cardiac abnormalities, and muscular dystrophy. *Neurology* **59:** 620–623.

van der Pluijm, I., Garinis, G.A., Brandt, R.M., *et al.* (2006) Impaired genome maintenance suppresses the growth hormone – insulin-like growth factor 1 axis in mice with Cockayne syndrome. *PLoS Biol.* **5:** e2.

van Esch, H., Agarwal, A.K., Debeer, P., *et al.* (2005) A homozygous mutation in the lamin A/C gene associated with a novel syndrome of arthropathy, tendinous calcinosis, and progeroid features. *J. Clin. Endocrinol. Metab.* **91:** 517–521.

van Tintelen, J.P., Tio, R.A., Kerstjens-Frederikse, W.S., *et al.* (2007) Severe myocardial fibrosis caused by a deletion of the 5′ end of the lamin A/C gene. *J. Am. Coll. Cardiol.* **49:** 2430–2439.

Vantyghem, M.C., Pigny, P., Maurage, C.A., *et al.* (2004) Patients with familial partial lipodystrophy of the Dunnigan type due to a LMNA R482W mutation show muscular and cardiac abnormalities. *J. Clin. Endocrinol. Metab.* **89:** 5337–5346.

Varela, I., Cadinanos, J., Pendas, A.M., *et al.* (2005) Accelerated ageing in mice deficient in Zmpste24 protease is linked to p53 signalling activation. *Nature* **437:** 564–568.

Varga, R., Eriksson, M., Erdos, M.R., *et al.* (2006) Progressive vascular smooth muscle cell defects in a mouse model of Hutchinson-Gilford progeria syndrome. *Proc. Natl Acad. Sci. USA* **103:** 3250–3255.

Vaughan, A., Alvarez-Reyes, M., Bridger, J.M., *et al.* (2001) Both emerin and lamin C depend on lamin A for localization at the nuclear envelope. *J. Cell Sci.* **114:** 2577–2590.

Venables, R.S., McLean, S., *et al.* (2001) Expression of individual lamins in basal cell carcinomas of the skin. *Br. J. Cancer* **84:** 512–519.

Vergnes, L., Peterfy, M., Bergo, M.O., *et al.* (2004) Lamin B1 is required for mouse development and nuclear integrity. *Proc. Natl Acad. Sci. USA* **101:** 10428–10433.

Verrecchia, F., Mauviel, A. and Farge, D. (2006) Transforming growth factor-beta signaling through the Smad proteins: role in systemic sclerosis. *Autoimmun. Rev.* **5:** 563–569.

von Zglinicki, T. (2002) Oxidative stress shortens telomeres. *Trends Biochem, Sci,* **27:** 339–344.

von Zglinicki, T., Saretzki, G., *et al.* (1995) Mild hyperoxia shortens telomeres and inhibits proliferation of fibroblasts: a model for senescence? *Exp. Cell Res.* **220:** 186–193.

Wang, S.M., Nishigori, C., *et al.* (1991) Reduced DNA repair in progeria cells and effects of gamma-ray irradiation on UV-induced unscheduled DNA synthesis in normal and progeria cells. *Mutat. Res.* **256:** 59–66.

Wang, W., Chen, J.X., Liao, R., *et al.* (2002) Sequential activation of the MEK-extracellular signal-regulated kinase and MKK3/6-p38 mitogen-activated protein kinase pathways mediates oncogenic ras-induced premature senescence. *Mol. Cell Biol.* **22:** 3389–3403.

Wiesel, N., Mattout, A., Melcer, S., *et al.* (2008) Laminopathic mutations interfere with the assembly, localization, and dynamics of nuclear lamins. *Proc. Natl Acad. Sci. USA* **105:** 180–185.

Wilkinson, F.L., Holaska, J.M., *et al.* (2003) Emerin interacts *in vitro* with the splicing-associated factor, YT521-B. *Eur. J. Biochem.* **270:** 2459–2466.

Wilson, K.L. (2000) The nuclear envelope, muscular dystrophy and gene expression. *Trends Cell Biol.* **10:** 125–129.

Worman, H.J. and Bonne, G. (2007) Laminopathies: a wide spectrum of human diseases. *Exp. Cell Res.* **313:** 2121–2133.

Wright, W.E. and Shay, J.W. (2001) Cellular senescence as a tumor-protection mechanism: the essential role of counting. *Curr. Opin. Genet. Dev.* **11:** 98–103.

Wright, W.E. and Shay, J,W, (2002) Historical claims and current interpretations of replicative aging. *Nat. Biotechnol.* **20:** 682–688.

Yan, T., Li, S., *et al.* (1999) Altered levels of primary antioxidant enzymes in progeria skin fibroblasts. *Biochem. Biophys. Res. Commun.* **257:** 163–167.

Young, J., Morbois-Trabut, L., Couzinet, B., *et al.* (2005) Type A insulin resistance syndrome revealing a novel lamin A mutation. *Diabetes* **54:** 1873–1878.

Zastrow, M.S., Vlcek, S. and Wilson, K.L. (2004) Proteins that bind A-type lamins: integrating isolated clues. *J. Cell Sci.* **117:** 979–987.

Zhang, Q., Ragnauth, C.D., Skepper, J.N., *et al.* (2005) Nesprin-2 is a multi-isomeric protein that binds lamin and emerin at the nuclear envelope and forms a subcellular network in skeletal muscle. *J. Cell Sci.* **118:** 673–687.

Zhang, Q., Bethmann, C., Worth, N.F., *et al.* (2007) Nesprin-1 and -2 are involved in the pathogenesis of Emery Dreifuss muscular dystrophy and are critical for nuclear envelope integrity. *Hum. Mol. Genet.* **16:** 2816–2833.

Zhao, K., Harel, A., *et al.* (1996) Binding of matrix attachment regions to nuclear lamin is mediated by the rod domain and depends on the lamin polymerisation state. *FEBS Lett.* **380:** 161–164.

Zhen, Y.Y., Libotte, T., *et al.* (2002) NUANCE, a giant protein connecting the nucleus and actin cytoskeleton. *J. Cell Sci.* **115:** 3207–3222.

9

Role of the glyoxalase pathway in delaying plant senescence under stress conditions

Sneh L. Singla-Pareek, Sudesh K. Yadav, Ananda Mustafiz and S.K. Sopory

1 Introduction

During the exposure of plants to various environmental stress conditions, reactive oxygen species (ROS) and several other cytotoxic carbonyl compounds, including methylglyoxal (MG), are produced. The major source of MG is glycolysis but it is also produced by the catabolism of amino acids via aminoacetone or hydroxyacetone. MG is also synthesized by methylglyoxal synthase from dihydroxyacetone phosphate, an intermediate of glycolysis in some bacterial cells. However, the occurrence of methylglyoxal synthase in plants is still unreported. Over-accumulation of MG adversely affects cellular functions by reacting with guanyl residues of DNA, forming imidazopurinone adducts, and with lysine and arginine of proteins, forming imidazole derivatives (Thornalley, 1998). These modifications are sufficient to induce loss of cell viability, ultimately leading to the death of an organism (Booth *et al.*, 2003).

A diverse group of organisms including plants possess a ubiquitous enzymatic pathway, the glyoxalase system, for detoxification of MG (*Figure 1*). The glyoxalase system consists of glyoxalase I (Gly I) and glyoxalase II (Gly II). The former catalyses the conversion of MG to *S*-D-lactoylglutathione in the presence of glutathione. The latter hydrolyses the resultant glutathione thiolester to D-lactic acid and glutathione. Though both the enzymes of the glyoxalase system have been extensively studied in microbial and animal systems (Rhee *et al.*, 1987; Thornalley, 1990), very little work has been done in plants. Recently the presence of this pathway has been documented in plants and the two enzymes, Gly I and Gly II, of this pathway have been purified and characterized from different plant species. In this chapter, we will review the work on the characterization and functional significance of the glyoxalase system in plants, especially with respect to its role under abiotic stress conditions.

2 Glyoxalase I

After identifying the presence of Gly I in Douglas fir needles (Smits and Johnson, 1981), its enzyme activity has been detected in various plants like maize, jute, wheat,

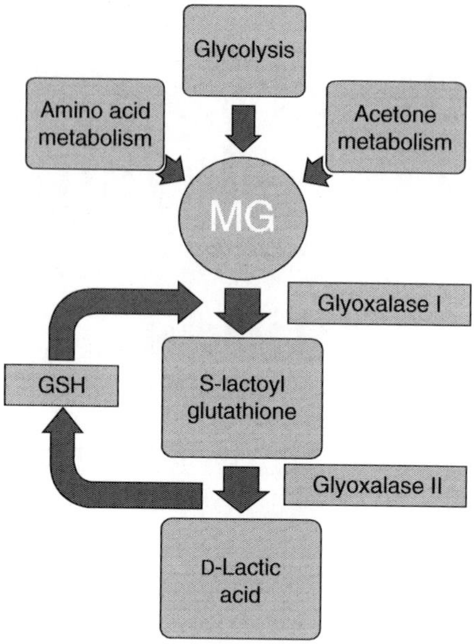

Figure 1. *Methylglyoxal synthesis and detoxification by the glyoxalase system (GSH, reduced glutathione).*

Sorghum, *Aloe vera*, soybean, *Cicer*, *Amaranthus*, *Nicotiana*, *Brassica*, *Datura* and pea (Chakravarty and Sopory, 1990; Deswal *et al.*, 1993; Norton *et al.*, 1990; Paulus *et al.*, 1993; Ramaswamy *et al.*, 1983, 1984; Seraj *et al.*, 1992). Gly I is a cytosolic enzyme (Deswal and Sopory, 1998) and a localization study of Gly I in tomato has shown its presence in all types of cell and tissue including hypocotyls, cotyledons, roots, floral organs and seeds. However, preferential accumulation has been observed in phloem and sieve cells (Deswal and Sopory, 1998; Espartero *et al.*, 1995).

The *Gly I* gene encoding Gly I enzyme has been isolated and characterized from a few plant species, such as tomato (Espartero *et al.*, 1995), *Brassica juncea* (Veena *et al.*, 1999) and soybean (Skipsey *et al.*, 2000). Contrary to the homodimeric and monomeric nature of Gly I in mammalian and microbial organisms (Allen *et al.*, 1993), this enzyme exists as a heterodimer in *Brassica*, with two subunits of 27–29 kDa, and in soybean, with two subunits of 26–29 kDa (Deswal and Sopory, 1991; Deswal *et al.*, 1993; Paulus *et al.*, 1993). However, in *Aloe vera*, it exists as a monomer with a molecular mass of 44 kDa (Norton *et al.*, 1990). In general, the pI values of plant Gly I are very similar to animal Gly I and range from 4.7 to 5.1 (Paulus *et al.*, 1993), while *Aloe vera* Gly I has a higher pI of 7.8 (Norton *et al.*, 1990). Similar to other organisms, Gly I from plants also follows Michaelis-Menton kinetics over a restricted range of hemithioacetal substrate (i.e. 0–0.67 mM). The Gly I showed a Km value of 460 μM in soybean (Paulus *et al.*, 1993) and 1400 μM in *Aloe vera* (Norton *et al.*, 1990) for hemithioacetal substrate. Gly I from *B. juncea* has a binding affinity for zinc ions in addition to hemithioacetal and histidine residues that plays a significant role in its catalysis (Deswal and Sopory, 1998).

Table 1. *Chromosomal location of putative* Gly I *genes in rice and Arabidopsis with indication of the existence of alternative spliced forms and total length of nucleotide and amino acids.*

Locus identifier	Chromosome	Alternative spliced forms	Total CDS length (bp)[a]	Polypeptide length (aa)[b]
LOC_Os01g07850	1	None	648	215
LOC_Os02g17920	2	2	1134/1038/873	377/345/290
LOC_Os03g45720	3	None	426	141
LOC_Os04g45590	4	None	528	175
LOC_Os05g14194	5	1	1053/825	350/274
LOC_Os05g07940	5	3	627/531/372/321	208/176/123/106
LOC_Os05g22970	5	None	714	237
LOC_Os07g46360	7	None	660	219
LOC_Os07g05800	7	None	576	191
LOC_Os08g09250	8	2	879/876/546	292/291/181
AT1G07645	1	None	414	137
AT1G08110	1	3	558/558/558/708	185/185/185/235
AT1G15380	1	1	525/525	174/174
AT1G64185	1	None	357	118
AT1G67280	1	1	1053/789	350/262
AT1G80160	1	1	504/426	167/141
AT2G28420	2	None	555	184
AT2G32090	2	1	408/342	135/113
AT5G41650	5	None	354	117
AT5G57040	5	None	594	197

LOC_OS, rice; AT, Arabidopsis; CDS, Coding Sequence
[a] *bp = base pair*
[b] *aa = amino acid*

Searching the rice and Arabidopsis genome database (www.arabidopsis.org; www.tigr.org) indicated that both genomes contain at least 10 different putative *Gly I* genes located on different chromosomes. Some of them go through alternative splicing resulting in the generation of CDS and corresponding polypeptide of varying size (*Table 1*). All the *Gly I* genes have two common signature sequences:

1. [HQ]-[IVT]-x-[LIVFY]-x-[IV]-x(4)-{E}-[STA]-x(2)-F-[YM]-x(2,3)-[LMF]-G-[LMF]
2. G-[NTKQ]-x(0,5)-[GA]-[LVFY]-[GH]-H-[IVF]-[CGA]-x-[STAGLE]-x(2)-[DNC]

A maximum of three genes were present on chromosome 5 in rice and six genes were present on chromosome 1 in Arabidopsis. Four genes in each of the plants showed alternate splice forms.

3 Glyoxalase II

Gly II has been purified partially from *Zea mays*, *Aloe vera* and spinach (Norton *et al.*, 1989, 1990; Talesa *et al.*, 1990) and characterized in detail from *Arabidopsis thaliana*

Table 2. *Chromosomal location of putative* Gly II *genes in rice and Arabidopsis with indication of the existence of alternative spliced forms and total length of nucleotide and amino acids.*

Locus identifier	Chromosome no.	Alternative spliced forms	Total CDS length (bp)[a]	Polypeptide length (aa)[b]
LOC_Os01g47690	1	2	921/846/738	306/281/245
LOC_Os03g21460	3	None	780	259
LOC_Os09g34100	9	None	1014	337
AT1G06130	1	1	996/993	331/330
AT1G53580	1	1	885	294

LOC_OS, rice; AT, Arabidopsis
[a] *bp = base pair*
[b] *aa = amino acid*

(Maiti *et al*., 1997; Ridderstrom and Mannervik, 1997), *Brassica* (Saxena *et al*., 2005) and rice (Yadav *et al*., 2007). The Gly II from *Z. mays*, with a relative molecular mass of about 26 kDa, is an acidic protein (pI 4.5; Norton *et al*., 1989) and that from *A. vera*, purified as three proteins, is also an acidic protein (pI 4.7, 4.8 and 5.0; Norton *et al*., 1990). However, the existence of multi-isomeric forms of Gly II have been revealed in spinach – cytosolic (pI 5.3, 5.8 and 6.2) and mitochondrial (pI 4.8) – and exist as a monomer of about 26 kDa (Talesa *et al*., 1990). These isoforms show similar kinetic properties when studied using several glutathione thiolesters as substrates (Talesa *et al*., 1990). Mutational study on Arabidopsis Gly II has demonstrated that its cytoplasmic form contains an iron-zinc binuclear metal centre that is essential for activity. These metals participate in substrate binding, transition state stabilization, and the hydrolysis reaction. Subtle alterations in the geometry and/or electrostatics of the binuclear centre have profound effects on the activity of the enzyme (Zang *et al*., 2001). Two genes encoding mitochondrial and cytoplasmic Gly II were found to be differentially expressed. The mitochondrial isozyme is most abundant in roots, while the cytoplasmic isozyme is highest in flower buds (Maiti *et al*., 1997). The differential expression of Gly II forms suggests that they may play different roles in the cell. Prior to this observation in plants, the presence of peculiar forms of Gly II in mitochondria had been revealed in mammalian systems. The structural and functional similarities between Gly II from *A. thaliana* and from human tissues suggest a common evolutionary origin for this enzyme (Ridderstrom and Mannervik, 1997).

There are three putative *Gly II* genes in the rice genome database, and two in the Arabidopsis genome database. Some of them also go through alternative splicing (*Table* 2). All putative *Gly II* genes fall into the metallo-beta-lactamase superfamily (pfam accession no. PF00753).

4 Regulation of the glyoxalase pathway

All metabolic pathways are generally under tight regulation for optimum performance of a system. The glyoxalase system in plants is under the regulation of developmental stage, hormonal changes, light and high salt exposure. The Gly I enzyme has shown modulation during cell division/proliferation (Deswal *et al.*, 1993; Sethi *et al.*, 1988). Similarly, a linear increase in its activity was seen with development in shoots and roots of pea seedlings (Ramaswamy *et al.*, 1983). However, Gly II is abundant in roots and flower buds compared to other tissues and expressed differentially in the mitochondria and in the cytosol of Arabidopsis. The differential regulation suggests a role for the Gly II isozymes in different compartments of a system (Maiti *et al.*, 1997).

Some studies have indicated there is hormonal regulation of glyoxalase enzymes. Benzylaminopurine (BAP), indole acetic acid (IAA) and kinetin application to pea callus resulted in a significant increase in Gly I activity, along with an increase in its cell number. A similar effect of hormones on Gly I activity was observed in *Datura* and *Nicotiana* calli where naphthaleneacetic acid (NAA) and BAP treatment enhanced callusing and, coordinately, Gly I activity (Ramaswamy *et al.*, 1984; Sethi *et al.*, 1988). In *Corchorus capsularis* and *Amaranthus* calli, an inverse relationship was seen between differentiation and the activity level of Gly I (Chakravarty *et al.*, 1991; Seraj *et al.*, 1992). However, to the contrary, a limited supply of auxin in the growth medium of soybean resulted in a decline of Gly I and Gly II enzyme activity as well as cell growth, which was regained upon auxin addition to the culture media (Paulus *et al.*, 1993).

Living organisms possess various strategies for perceiving and processing light signals to optimize their growth, development and behaviour. As plants are photosynthetic and sessile, the quality, intensity, duration and direction of light has a huge impact on their development and growth. Thus, plants possess a sophisticated light sensing and signalling system that ensures optimal photoperception and responses to their ever-changing light conditions. Red and blue light exposure stimulated Gly I activity in dark-grown *Amaranthus* calli suggesting a role of photoreceptors in the regulation of Gly I expression (Chakravarty and Sopory, 1990). Exogenous addition of calcium showed a similar effect to light, which was blocked by the inhibitors of calmodulin (Chakravarty and Sopory, 1990). Similarly, lithium chloride (an inhibitor of the phosphoinositide cycle) treatment inhibited cell proliferation and Gly I activity (Bagga *et al.*, 1987). These studies have documented that calcium mediates the response of light in activating Gly I enzymes in plants.

Modulation of free calcium in the cytosol is one of the general events that relay an external stimulus to the internal cellular machinery and allow eukaryotic organisms, including plants, to mount a specific biological response. Calcium acts as a second messenger that undergoes changes in response to adverse biotic and/or abiotic environmental factors. Decoding the information conveyed by calcium should allow the cell to generate an adaptive biological response. One of the well-known decoding molecules is calmodulin. Hence, calcium and calmodulin act in a coherent manner. Both calcium and calmodulin activate Gly I activity in *B. juncea* (Deswal and Sopory, 1999). During exposure of plants to conditions like hypoxia, temperature shock, water stress and phytohormones, calcium levels increase that transduce the signal by binding to specific target proteins, of which the most important are the protein kinases. Besides processing the signal via protein kinases, Ca^{2+} binds to calcium-binding proteins (CaBPs) like calmodulin (CaM, a ubiquitous Ca^{2+}-binding protein), which in turn may modulate the activity of Gly I along with various other proteins (Knight *et al.*, 1991; Snedden and Fromm, 1998). Recently, it has been reported that Gly I protein can be phosphorylated by GA3 in rice (Khan *et al.*, 2005).

4.1 Role of the glyoxalase pathway in senescence delay during abiotic stress

It was shown that *Gly I* transcript is increased in response to stress conditions (Espartero *et al.*, 1995). We also found up-regulation in transcript expression levels of *B. juncea Gly I* in response to salt, water and heavy metal (e.g. zinc) stresses, which indicated a possible role of the glyoxalase system in providing abiotic stress tolerance to plants. *Gly I* expression increased by as high as threefold after 72 h of salt exposure. Under these conditions, an increase in protein and Gly I activity were also observed. The role of Gly I under stress was confirmed by constitutive over-expression of *Gly I* from *B. juncea* in tobacco. The transgenic tobacco over-expressing *Gly I* showed relatively more tolerance to MG and high salt, as tested in detached-leaf disc senescence assay. The degree of tolerance to these stresses was found to correlate with *Gly I* expression levels (Veena *et al.*, 1999). This detailed study for the first time suggested a role for Gly I in conferring tolerance to plants under stress conditions.

Since the glyoxalase pathway involved Gly I and Gly II enzymes, and both enzymes are required for glutathione-based detoxification of MG, both the genes of this pathway were over-expressed in tobacco. The double-transgenic tobacco showed a better tolerance to that of individual over-expressors with either *Gly I* or *Gly II*. These transgenic plants were able to grow, flower and set normal viable seeds under continuous salinity-stress conditions (Singla-Pareek *et al.*, 2003). These transgenics were also evaluated for the accumulation of Na^+ and K^+ ions, where a detrimental Na^+ concentration was found to be accumulated in older leaves. Furthermore, seeds of transgenics showed negligible accumulation of Na^+, indicating their safer consumption. Comparison of various growth parameters and seed production demonstrated that there is hardly any yield penalty in the double transgenics under non-stress conditions, and that these plants suffered only 5% loss in total productivity when grown in 200 mM NaCl. These findings establish the potential of manipulation of the glyoxalase pathway for increased salinity tolerance without affecting yield in crop plants (*Figure 2* in colour plate section).

To test the suitability of the glyoxalase pathway for improved tolerance to other abiotic stresses, heavy metal tolerance in transgenic tobacco was investigated. Constitutively over-expressing glyoxalase transgenic tobacco plants were able to complete their life cycle in the presence of 5 mM $ZnCl_2$ without any yield penalty. The endogenous ion content measurements revealed roots to be the major sink for excess zinc accumulation, with negligible amounts in seeds of transgenic plants. Our preliminary observations also suggest that glyoxalase over-expression could confer tolerance to other heavy metals, such as cadmium or lead. However, this needs further experimental proof. Comparison of relative tolerance capacities of transgenic plants, over-expressing either *Gly I* or *Gly II* individually or together in double transgenics, evaluated in terms of various critical parameters such as survival, growth, and yield, reflected double transgenics to perform better than either of the single-gene transformants (Singla-Pareek *et al.*, 2006). Recent work has also shown that these transgenic plants could also tolerate dehydration stress (unpublished work). Together, these findings suggest an important role of glyoxalases in delaying senescence and allowing the plants to complete their life cycle under stress conditions. Glyoxalase-pathway engineering thus presents a novel strategy to develop multiple stress tolerance, implicating its potential use in engineering agriculturally important crop plants to grow on rapidly deteriorating lands with multiple unfavourable edaphic factors.

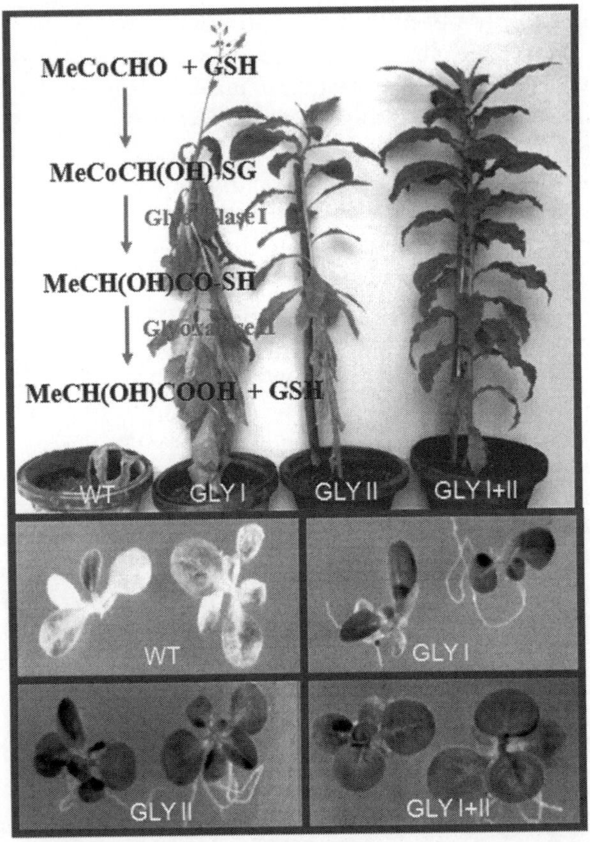

Figure 2. *Representative picture of non-transgenic and glyoxalase transgenic tobacco plants showing their relative tolerance to salt stress. Upper panel shows the relative growth of wild type (WT), Gly I, Gly II and Gly I + II transgenic plants in the continuous presence of 200 mM NaCl. Lower panel shows the growth of seedlings in the media containing 200 mM NaCl. Note the better growth and survival of the double transgenic plants.*

The feasibility of glyoxalase over-expression in conferring salinity tolerance has also been studied in one of the most agronomically important crops, rice, through over-production of Gly II. Transgenic rice showed higher constitutive expression and activity of Gly II, that increased further upon salt stress reflecting the up-regulation of endogenous Gly II. The transgenic plants showed higher tolerance to toxic concentrations of MG and NaCl. Similar to transgenic tobacco, transgenic rice at the T1 generation exhibited sustained growth and a more favourable ion balance compared with non-transgenics under salt-stress conditions (Singla-Pareek *et al.*, 2008).

4.2 Mechanism of delay in plant senescence by over-expression of glyoxalases

A great degree of success has been achieved in the production of herbicide-, virus- and fungal-resistant plants and plants with fortified nutritional values using transgenic tools. However, there has been little breakthrough in production of abiotic-stress-tolerant crops. This is largely because of the complex molecular and biochemical mechanisms that govern abiotic-stress tolerance in plants. The genes that have been proven to be effective in providing stress tolerance using a transgenic approach belong to both structural and regulatory gene categories. Abiotic-stress-tolerant transgenics have been produced over-expressing several genes, like ornithine decarboxylase, S-adenosylmethionine decarboxylase, mannitol-1-phosphate dehydrogenase, codA, hsp100, pyruvate decarboxylase1, osmotin, connexin, inositol synthase, *hal1*, catalase and superoxide dismutase, S-adenosyl-L-methionine synthetase, chloroplastic fructose-1,6-bisphosphate, trehalose, Na^+/H^+ antiporter, and so on, individually or in combination. Therefore, several mechanisms have been proposed for abiotic-stress tolerance in plants. The selective websites that contain information on different aspects of abiotic-stress-related studies on plants are http://www.stress-genomics.org/, http://www.uoguelph.ca and http://www.plantstress.com.

As mentioned in the previous section, the over-production of two enzymes of the glyoxalase pathway that act coordinately to convert MG to D-lactic acid has been useful in providing abiotic-stress tolerance in plants. The detailed mechanism of abiotic-stress tolerance in transgenic tobacco over-expressing glyoxalase pathway genes has been studied. It was found that salinity stress induced a higher level of accumulation of MG (Yadav *et al.*, 2005a). In plants like *Oryza sativa*, *Pennisetum glaucum*, *Nicotiana tabacum*, and *Brassica juncea*, MG levels ranged from 40 to 75 μM under normal conditions and from 75 to 200 μM under salinity stress. This indicates that the plants have a higher level of MG compared to yeast and other animal systems. This study has also documented that the enhanced accumulation of MG under salinity is a universal response in plants (Yadav *et al.*, 2005a). Further studies on MG accumulation in rice under drought, salinity and cold conditions have suggested that an increase in MG levels seems to be a general response of plants towards all abiotic stresses. The MG level increased (300 μM per gFW) to a greater extent in shoots of two rice varieties (PB1 and IR64) during drought stress compared to salinity and cold stress. The increase in MG level was more pronounced in shoots compared to roots. However, roots showed a considerably higher level of MG under drought stress compared to salinity and cold stress. This emphasizes the existence of an organ-specific variable response to abiotic stresses seen in terms of degree of MG accumulation (Yadav *et al.*, 2005a).

The transgenic tobacco plants over-expressing glyoxalase-pathway enzymes, resist an increase in the level of MG that increased to over 70% in wild-type (WT) plants under salinity stress. Since MG is used as a substrate by the glyoxalase pathway as well as by aldose reductases, the contribution of glyoxalase enzymes in regulating the levels of endogenous MG was monitored. For this, transgenic tobacco plants that either over-produced (NtSgly I) or under-produced (NtASgly I) the Gly I enzyme were analysed. Both types of transgenic plants showed a growth pattern similar to the WT during the T0 generation. However, when T1 seeds were germinated, the NtSgly I and WT tobacco seeds showed normal growth patterns whereas the NtASgly I seeds failed to germinate. When MG content was measured, it was found that NtASgly I seeds contained 132% more MG compared to WT and NtSgly I seeds. This suggests that NtASgly I plants accumulated more MG due to the low levels of Gly I enzyme (Yadav *et al.*, 2005b). It is possible that some other detoxification processes may be involved during the later stages of development. Also, the level to which MG accumulates during normal and unstressed conditions may not be inhibitory to growth and development beyond seed germination. The role of aldose/aldehyde reductase (ALR) enzyme, which uses NADPH (unlike Gly I, which uses GSH) for the reduction of reactive aldehydes such as 4-hydroxynon-2-enal and MG, has been uncovered recently in tobacco over-expressing ALR (Hideg *et al.*, 2003; Oberschall *et al.*, 2000).

When plants are exposed to various environmental stresses, it leads to generation of reactive oxygen species (ROS). These ROS are detoxified either directly by non-enzymatic antioxidants, such as reduced glutathione (GSH), ascorbate (ASH), tocopherols and carotenoids, or by antioxidative enzymes such as superoxide dismutase, ascorbate peroxidase (APX), glutathione peroxidase (GPX) and glutathione-S-transferase (GST; Kumar *et al.*, 2003; Wang *et al.*, 2003; Xiong and Zhu, 2002). Transgenic plants over-expressing glyoxalase genes showed enhanced basal activity of various glutathione-related antioxidative enzymes, such as GR, GST, GPX and APX, that increased further upon salinity stress indicating a higher antioxidative capacity of the transgenics. These plants suffered minimal salt- and zinc-stress-induced oxidative damage measured in terms of lipid peroxidation. It has been reported that GSH is important for several metabolic functions, including protection of cells from peroxides under oxidative stress. GSH is the most abundant low molecular thiol compound in plants. It is a powerful cellular reductant and protects against a range of peroxides, xenobiotics and heavy metals. GSH has also been shown to regulate the expression of genes whose products are involved in redox regulation and/or in enhancement of stress tolerance. Almost all cells exhibit two forms of glutathione, oxidized (GSSG) and reduced (GSH), interchanging with each other to maintain the redox potential under different conditions (Kumar *et al.*, 2003).

During detoxification of MG by the glyoxalase pathway, glutathione is recycled. To decipher its involvement in the mechanism of stress tolerance, the concentration of GSH as well as the GSH:GSSG ratio was measured in WT and glyoxalase transgenic plants. High sodium chloride treatment was not found to perturb the GSH content in WT plants, but induced a decline in the GSH:GSSG ratio, suggesting an increase in GSSG levels in these plants. In contrast, salt induced significant accumulation of GSH in transgenics, as high as 40% more in double-glyoxalase transgenic plants, while only a 10% decrease in the GSH:GSSG ratio was observed (Yadav *et al.*, 2005b). Further proof that glyoxalase transgenics maintain the glutathione ratio was obtained by exogenous alteration of GSSG. Exogenous application of GSSG retarded the growth of non-transgenic plants whereas transgenic plants sustained their growth. These results suggest that over-expression of glyoxalase-pathway enzymes resist an increase in MG and maintain higher reduced glutathione levels, which could be the mechanism of stress tolerance in transgenic plants (Yadav *et al.*, 2005b).

5 Role of MG in signal transduction

Generally, one wonders why such a toxic compound as MG is formed via the ubiquitous energy-generating pathway. As mentioned previously, the MG level in plants varies from 30 to 75 μM in various plant species and it increases two- to sixfold in response to salinity, drought and cold stress (Yadav *et al.*, 2005a). However, despite having an efficient detoxification mechanism, the level of MG does not fall below its

30 μM basal level even in glyoxalase over-expressing transgenic plants. It is indicative of the fact that MG might have some other important role to play. As the level of MG increases in response to stress, is there a possibility that MG is one of the stress signals and can up-regulate the stress-responsive genes?

In *Saccharomyces cerevisiae*, a possible signalling pathway for the induction of the *Gly I* gene under high osmotic stress conditions has been suggested (Inoue *et al.*, 1998). The *Gly I* gene has been shown to be expressed via the HOG-MAPK pathway to scavenge MG that may be increased by the osmotic stress (Inoue *et al.*, 1998). Furthermore, it has been shown that MG can act as a signal initiator as it can activate the HOG-MAP kinase cascade (Maeta *et al.*, 2005). Two osmosensors, Sln1 and Sho1, have been identified to function upstream of the HOG-MAP kinase cascade, and it was revealed that MG initiates signal transduction to this MAP kinase cascade through the Sln1 branch. It has also been demonstrated that MG activates the Msn2 transcription factor. Moreover, MG was found to activate the uptake of Ca^{2+} in yeast cells, thereby stimulating the calcineurin/Crz1-mediated Ca^{2+}-signalling pathway. Therefore, it has been proposed that MG functions as a signal initiator in yeast (Maeta *et al.*, 2005). It was also shown that MG acts as a signal initiator for activation of the stress-activated protein kinase in *Schizosaccharomyces pombe*. Here an alternative module directs the MG signal to the SAPK pathway via Mcs4 (Takatsume *et al.*, 2006).

6 Does MG act as a signal molecule in plants?

Our recent experiments have indicated that exogenous application of MG regulates the transcript level of *Gly I* and *Gly II* (unpublished work). However, the mechanism by which MG regulates expression has not been studied yet. It was recently shown that over-expression of NAC resulted in expression of many genes, one of which was found to be *Gly I* (Fujita *et al.*, 2004). NACs are transcription factors containing a plant-specific highly conserved N-terminal domain (NAC, for NAM, ATAF1, 2 and CUC2) reported by Aida *et al.* (1997). Whether MG affects NAC-like transcription factors to regulate the expression of *Gly I* and *Gly II* needs to be studied further. Our preliminary results show that exogenous application of MG does increase the transcript level of one of the NAC transcription factors in rice.

Plants accumulate higher levels of MG in response to abiotic stresses (Yadav *et al.*, 2005a). In addition, there is an increase in ROS under such adverse conditions. However, ROS are a key mediator in activating the members of the mitogen-activated protein kinase (MAPK) superfamily pathway, such as ASK1 and JNK (Seger and Krebs, 1995), and seem to be responsible for apoptosis. Additionally, MAPK proteins play a major role in integration and transmission of a wide range of signals for cell growth and differentiation (Mishra *et al.*, 2006). Therefore, the presence of MAPK family proteins in plants suggests the possibility of MG-mediated signal transduction through the same pathway, at least up to MAPK (*Figure 3*). However, this needs to be experimentally verified. Furthermore, MG-mediated redox-sensitive signalling may also interact with more classical signalling pathways and influence many other cellular events, such as activation of transcription factors, regulation of gene expression and cell proliferation. As described above, abiotic-stress-induced MG disturbs the GSH homeostasis; this could be due to the generation of oxidative stress in the form of H_2O_2, which has been

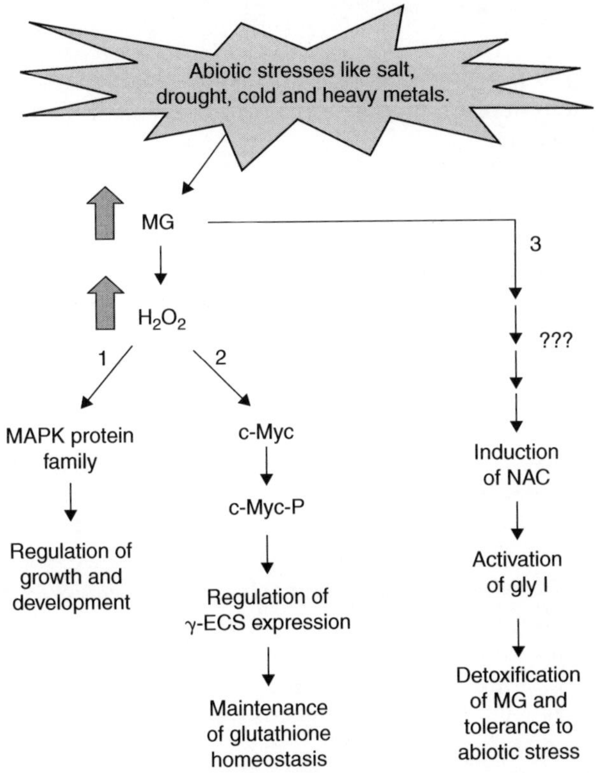

Figure 3. *MG as a signalling molecule: exposure of plants to abiotic stresses enhances MG levels in the cell. High levels of MG transduce the signal through three probable routes. High levels of MG might induce H_2O_2, that may initiate the MAP kinase cascade (1) or induce c-Myc phosphorylation and, by regulating γ-glutamylcysteine synthetase (γ-ECS) expression, maintain glutathione homeostasis (2). MG might also directly induce some NAC-like transcription factor and thus induce Gly I, which leads to tolerance to abiotic stress (3).*

shown to induce phosphorylation of c-Myc protein. The c-Myc protein binds to the promoter of the γ-glutamyl cysteine synthetase (γ-ECS) gene, encoding a rate-limiting enzyme of GSH biosynthesis (Benassi *et al.*, 2006; *Figure 3*).

The glyoxalase pathway is well studied in animal systems and yeast. Recent studies in plant systems have revealed that it plays a crucial role in response to abiotic stress and in the developmental process. MG, the substrate of the glyoxalase pathway, might have roles in abiotic stress signalling. In the Arabidopsis and rice genomes, multiple putative *Gly I* and *Gly II* genes can be found. Some of them also go through alternative splicing. The role of each of these needs to be studied with respect to tissue and cell specificity, and regulation in response to environmental cues. It appears that the glyoxalase pathway in plants might not be linear; rather, it is an intricate network with complex regulation. Further in-depth study is required to resolve the regulation of this pathway, and also the role of MG as a signalling molecule.

References

Aida, M., Ishida, T., Fukaki, H., Fujisawa, H. and Tasaka, M. (1997) Genes involved in organ separation in Arabidopsis: analysis of the cup-shaped cotyledon mutant. *Plant Cell* **9:** 841–857.

Allen, R.E., Lo, T.W. and Thornalley, P.J. (1993) A simplified method for the purification of human red blood cell glyoxalase I: characterization, immunoblotting and inhibitor studies. *J. Protein Chem.* **12:** 111–119.

Bagga, S., Das, R. and Sopory, S.K. (1987) Inhibition of cell proliferation and glyoxalase I activity by calmodulin inhibitors and lithium in *Brassica oleracea. J. Plant Physiol.* **129:** 149–153.

Benassi, B. Fanciulli, M., Fiorentino, F., Porrello, A., Chiorino, G., Loda, M., Zupi, G. and Biroccio, A. (2006) c-Myc phosphorylation is required for cellular response to oxidative stress. *Mol. Cell* **21:** 509–519.

Booth, I.R., Ferguson, G.P., Miller, S., Li, C. and Gunasekera, B. (2003) Bacterial production of methylglyoxal: a survival strategy or death by misadventure? *Biochem. Soc. Trans.* **31:** 1406–1408.

Chakravarty, T.N. and Sopory, S.K. (1990) Light stimulated cell proliferation and glyoxalase I activity in callus cultures of *Amaranthus paniculatus*. In: *Progress in Plant Cellular and Molecular Biology* (eds H.J.J. Nijkamp, L.H.W. Van-Derplas and V.J. Asrtrijk). Kluwer Academic, Dordrecht, pp. 379–384.

Chakravarty, P.K., Ghosh, S.K. and Sinha, D.K. (1991) Pregnancy-induced cytotoxicity of splenocytes against mammary tumor cells in rats. *Oncology* **48:** 425–430.

Deswal, R. and Sopory, S.K. (1991) Purification and partial characterization of glyoxalase I from a higher plant *B. juncea. FEBS Lett.* **282:** 277–280.

Deswal, R. and Sopory, S.K. (1998) Biochemical and immunochemical characterization of *Brassica juncea* glyoxalase I. *Phytochemistry* **49:** 2245–2253.

Deswal, R. and Sopory, S.K. (1999) Glyoxalase I from *Brassica juncea* is a calmodulin-stimulated protein. *Biochim. Biophys. Acta* **1450:** 460–467.

Deswal, R., Chakaravarty, T.N. and Sopory, S.K. (1993) The glyoxalase system in higher plants: regulation in growth and differentiation. *Biochem. Soc. Trans.* **21:** 527–530.

Espartero, J., Sanchez-Aguayo, I. and Pardo, J.M. (1995) Molecular characterization of glyoxalase I from a higher plant: upregulation by stress. *Plant Mol. Biol.* **29:** 1223–1233.

Fujita, M., Fujita, Y., Maruyama, K., Seki, M., Hiratsu, K., Ohme-Takagi, M., Tran, L.S.P., Yamaguchi-Shinozaki, K. and Shinozaki, K. (2004) A dehydration-induced NAC protein, RD26, is involved in a novel ABA-dependent stress-signaling pathway. *Plant J.* **39:** 863–876.

Hideg, E., Nagy, T., Oberschall, A., Dudits, D. and Vass, I. (2003). Detoxification function of aldose/aldehyde reductase during drought and ultraviolet-B (280–320) stresses. *Plant Cell Environ.* **26:** 513–522.

Inoue, Y., Tsujimoto, Y. and Kimura, A. (1998) Expression of the glyoxalase I gene of *Saccharomyces cerevisiae* is regulated by high osmolarity glycerol mitogen-activated protein kinase pathway in osmotic stress response. *J. Biol. Chem.* **273:** 2977–2983.

Khan, M.M., Jan, A., Karibe, H. and Komatsu, S. (2005) Identification of phosphoproteins regulated by gibberellin in rice leaf sheath. *Plant Mol. Biol.* **58:** 27–40.

Knight, M.R., Campbell, A.K., Smith, S.M. and Trewavas, A.J. (1991) Transgenic plant aequorin reports the effects of touch and cold-shock and elicitors on cytoplasmic calcium. *Nature* **352:** 524–526.

Kumar, S., Singla-Pareek, S.L., Reddy, M.K. and Sopory, S.K. (2003) Glutathione: biosynthesis, homeostasis and its role in abiotic stresses. *J. Plant Biol.* **30:** 179–187.

Maeta, K., Izawa, S. and Inoue, Y. (2005) Methylglyoxal, a metabolite derived from glycolysis, functions as a signal initiator of the high osmolarity glycerol-mitogen-activated protein kinase cascade and calcineurin/Crz1-mediated pathway in *Saccharomyces cerevisiae. J. Biol. Chem.* **280:** 253–260.

Maiti, M.K., Krishnasamy, S., Owen, H.A. and Makaroff, C.A. (1997) Molecular characterization of glyoxalase II from *Arabidopsis thaliana. Plant Mol. Biol.* **35:** 471–481.

Mishra, N.S., Tuteja, R. and Tuteja, N. (2006) Signaling through MAP kinase networks in plants. *Arch. Biochem. Biophys.* **452:** 55–68.

Norton, S.J., Principato, G.B., Talesa, V., Lupattelli. M. and Rosi, G. (1989) Glyoxalase II from *Zea mays*: properties and inhibition study of the enzyme purified by use of a new affinity ligand. *Enzyme* **42:**189–196.

Norton, S.J., Talesa, V., Yuan, W.J. and Principato, G.B. (1990) Glyoxalase I and glyoxalase II from *Aloe vera*: purification, characterization and comparision with animal glyoxalases. *Biochem. Int.* **22:** 411–418.

Oberschall, A., Deák, M., Török, K., Sass, L., Vass, I., Kovács, I., Fehér, A., Dudits, D. and Horváth, G.V. (2000) A novel aldose/aldehyde reductase protects transgenic plants against lipid peroxidation under chemical and drought stresses. *Plant J.* **24:** 437–446.

Paulus, C., Köllner, B. and Jacobsen, H.J. (1993) Physiological and biochemical characterization of glyoxalase I, a general marker for cell proliferation, from a soybean cell suspension. *Planta* **189:** 561–566.

Ramaswamy, O., Guha-Mukherjee, S. and Sopory, S.K. (1983) Presence of glyoxalase I in pea. *Biochem. Int.* **7:** 307–318.

Ramaswamy, O., Pal, S., Guha-Mukherjee, S. and Sopory, S.K. (1984) Correlation of glyoxalase I activity with cell proliferation in *Datura* callus culture. *Plant Cell Rep.* **3:** 121–124.

Rhee, H., Murata, K. and Kimura, A. (1987) Molecular cloning of the *Pseudomonas putida* glyoxalase I gene in *Escherichia coli. Biochem. Biophys. Res. Commun.* **147:** 831–838.

Ridderstrom, M. and Mannervik, B. (1997) Molecular cloning and characterization of the thiolesterase glyoxalase II from *Arabidopsis thaliana. Biochem. J.* **322:** 449–454.

Saxena, M., Bisht, R., Roy, S.D., Sopory, S.K. and Bhalla-Sarin, N. (2005) Cloning and characterization of a mitochondrial glyoxalase II from *Brassica juncea* that is upregulated by NaCl, Zn, and ABA. *Biochem. Biophys. Res. Commun.* **336:** 813–819.

Seger, R. and Krebs, E.G. (1995) The MAPK signaling cascade. *FASEB J.* **9:** 726–735.

Seraj, Z.I., Sarker, A.B. and Islam, A.S. (1992) Plant regeneration in a jute species (*C. capsularis*) and its possible relationship with glyoxalase-I. *Plant Cell Rep.* **12:** 29–33.

Sethi, U., Basu, A. and Guha-Mukherjee, S. (1988) Control of cell proliferation and differentiation by regulating polyamine biosynthesis in cultures of *Brassica* and its correlation with glyoxalase I activity. *Plant Sci.* **56:** 167–175.

Singla-Pareek, S.L., Reddy, M.K. and Sopory, S.K. (2003) Genetic engineering of the glyoxalase pathway in tobacco leads to enhanced salinity tolerance. *Proc. Natl Acad. Sci.* USA **100:** 14672–14677.

Singla-Pareek, S.L., Yadav, S.K., Pareek, A., Reddy, M.K. and Sopory, S.K. (2006) Transgenic tobacco overexpressing glyoxalase pathway enzymes grow and set viable seeds in zinc spiked soils. *Plant Physiol.* **140:** 613–623.

Singla-Pareek, S.L., Yadav, S.K., Pareek, A., Reddy, M.K. and Sopory, S.K. (2008) Enhancing salt tolerance in a crop plant by overexpression of glyoxalase II. *Trans. Res.* **17:** 171–180.

Skipsey, M., Andrews, C.J., Townson, J.K., Jepson, I. and Edwards, R. (2000) Cloning and characterization of glyoxalase I from soybean. *Arch. Biochem. Biophys.* **374:** 261–268.

Smits, M.M. and Johnson, M.A. (1981) Methylglyoxal: enzyme distributions relative to its presence in Douglas-fir needles and absence in Douglas-fir needle callus. *Arch. Biochem. Biophy.* **208:** 431–439.

Snedden, W.A. and Fromm, H. (1998) Calmodulin, calmodulin-related proteins and plant responses to the environment. *Trends Plant Sci.* **3:** 299–304.

Takatsume, Y., Izawa, S. and Inoue, Y. (2006) Methylglyoxal as a signal initiator for activation of the stress-activated protein kinase cascade in the fission yeast *Schizosaccharomyces pombe*. *J. Biol. Chem.* **281:** 9086–9092.

Talesa, V., Rosi, G., Contenti, S., Mangiabene, C., Lupattelli, M., Norton, S.J., Giovannini, E. and Principato, G.B. (1990). Presence of glyoxalase II in mitochondria from spinach leaves: comparison with the enzyme from cytosol. *Biochem. Int.* **22:** 1115–1120.

Thornalley, P.J. (1990) The glyoxalase system: new developments towards functional characterization of a metabolic pathway fundamental to biological life. *Biochem. J.* **269:** 1–11.

Thornalley, P.J. (1998) Glutathione-dependent detoxification of α-oxoaldehyde by the glyoxalase system: involvement in disease mechanisms and antiproliferative activity of glyoxalase I inhibitors. *Chem. Biol. Interact.* **111–112:** 137–151.

Veena, Reddy, V.S. and Sopory, S.K. (1999) Glyoxalase I from *Brassica juncea*: molecular cloning, regulation and its over-expression confer tolerance in transgenic tobacco under stress. *Plant J.* **17:** 385–395.

Wang, W., Vinocur, B. and Altman, A. (2003) Plant responses to drought, salinity and extreme temperatures: towards genetic engineering for stress tolerance. *Planta* **218:** 1–14.

Xiong, L. and Zhu, J.K. (2002) Molecular and genetic aspects of plant responses to osmotic stress. *Plant Cell Environ.* **25:** 131–139.

Yadav, S.K., Singla-Pareek, S.L., Ray, M., Reddy, M.K. and Sopory, S.K. (2005a) Methylglyoxal levels in plants under salinity stress are dependent on glyoxalase I and glutathione. *Biochem. Biophys. Res. Commun.* **337:** 61–67.

Yadav, S.K., Singla-Pareek, S.L., Reddy, M.K. and Sopory, S.K. (2005b) Transgenic tobacco plants overexpressing glyoxalase enzymes resist an increase in methylglyoxal and maintain higher reduced glutathione levels under salinity stress. *FEBS Lett.* **579:** 6265–6271.

Yadav, S.K., Singla-Pareek, S.L., Kumar, M., Pareek, A., Saxena, M., Sarin, N.B. and Sopory, S.K. (2007) Characterization and functional validation of glyoxalase II from rice. *Protein Expr. Purif.* **51:** 126–132.

Zang, T.M., Hollman, D.A., Crawford, P.A., Crowder, M.W. and Makaroff, C.A. (2001) Arabidopsis glyoxalase II contains a zinc/iron binuclear metal center that is essential for substrate binding and catalysis. *J. Biol. Chem.* **276:** 4788–4795.

10

Catalase regulation during leaf senescence of Arabidopsis

Ulrike Zentgraf

1 Introduction

One of our future challenges will be to cope with global climatic changes and to develop crop plants that are adapted to these new environmental conditions. These plants will have to be able to react over a wide range to sudden and extreme changes. Therefore, plants with abiotic stress resistance are attracting more and more attention. However, it is well known that abiotic stress can induce premature senescence, which can lead to massive quantitative and qualitative crop losses. Many different agriculturally important traits, like number and quality of seeds, timing of seed set, fruit ripening, and so on, are affected by senescence. Therefore, an understanding of senescence-regulation processes and their connection to abiotic stress management might be a helpful tool to reach this goal.

Despite the importance of senescence processes, our knowledge of the regulatory mechanisms of senescence is still poor. It is well established that senescence is not a chaotic breakdown, but an orderly loss of normal cell function that is under the control of the nucleus. In contrast to ageing processes, which have a passive and non-regulated degenerative character (for review see Krupinska *et al.*, 2003), senescence is an active and highly regulated process. Senescence is triggered by exogenous and endogenous parameters of which the most important endogenous factors are the age of the leaves and the age and developmental stage of the plant. However, how these two parameters are sensed and translated into molecular signals is still unclear. The photosynthetic activity of leaves of annual plants decreases continuously after full expansion (Batt and Woolhouse, 1975; Hensel *et al.*, 1993). In fast-ageing plants like Arabidopsis, the photosynthesis rate of the leaves decreases by 50% within four to six days of full leaf expansion under continuous light conditions (Hensel *et al.*, 1993). A decline in photosynthetic activity under a certain threshold may act as a senescence-inducing signal (Matile *et al.*, 1992; Smart, 1994). However, autumn senescence in free-growing aspen (*Populus tremula*) is exclusively initiated by the photoperiod (Keskitalo *et al.*, 2005). Sugar accumulation, as well as sugar starvation, are discussed as signals to induce senescence; however, changes in the sugar content relative to nitrogen content of leaves during the sink/source transition could also play a role in the induction of leaf senescence (Masclaux *et al.*, 2000; Pourtau *et al.*, 2004; van Doorn, 2004; Wingler *et al.*,

2004). Besides photoperiod and sugar, senescence is most likely triggered by the interplay of all plant hormones in specific concentrations acting in synergistic or antagonistic ways, and in concert with other signals like sugar, calcium and especially reactive oxygen species (ROS) (Zentgraf 2007).

2 Antioxidative capacity and lifespan

It has been known for many years that there is a striking correlation between tolerance to oxidative stress and lifespan ranging from *Caenorhabditis* to mammals (Harman 1956, 1998; Martin *et al.*, 1996; Orr and Sohal, 1994; Sohal and Weindruch, 1996). Similarly, a loss of antioxidative capacity during the progression of senescence has been reported for many different plants (Chia *et al.*, 1981; Dhinsda *et al.*, 1981; Jiménez *et al.*, 1998; McRae and Thompson, 1983; Orendi *et al.*, 2001; Panavas and Rubinstein, 1998; Pastori and del Rio, 1994; Pauls and Thompson, 1984), implying that this might be a more general phenomenon for many aerobic organisms. Oxidative stress arises when the critical balance between generation and elimination of ROS is disturbed, either by depletion of antioxidants or excess accumulation of ROS, which can often lead to the death of cells or even whole organisms. Due to their toxic nature, it is suggested that the accumulation of ROS might be involved in degradation processes associated with senescence. Obviously, oxidative damage in plant tissues is not always unwanted (Halliwell and Gutteridge, 1989; Kar and Feierabend, 1984; Thompson *et al.*, 1987). In addition, ROS can also function as signalling molecules in certain concentrations (Alvarez *et al.*, 1998; Dat *et al.*, 2000; Foyer *et al.*, 1998; Grant and Loake, 2000; Harding *et al.*, 1997; Kovtun *et al.*, 2000; Kuźniak and Urbanek, 2000; Mullineaux and Karpinski, 2002; Orozco-Cardenas and Ryan, 1999; Vandenabeele *et al.*, 2003).

An investigation of different late-flowering mutants of *Arabidopsis thaliana* revealed that flowering time and longevity are tightly correlated with the resistance to oxidative stress (Kurepa *et al.*, 1998). However, the relationship between lifespan and oxidative stress tolerance in plants is poorly understood. The most extended longevity mutant was *gigantea3*, which also showed the highest resistance to oxidative stress. The GIGANTEA protein appears to be involved in red-light signalling, central clock function and flowering-time regulation. Recent results indicate that GIGANTEA also acts in blue-light signalling and has biochemically separable roles in circadian-clock and flowering-time regulation (Martin-Tryon *et al.*, 2007) but the link between this nuclear-localized protein and resistance to oxidative stress is still unclear. Clock-controlled genes like CATALASE2 (CAT2) and CATALASE3 (CAT3) might be candidates for this link. Catalases are peroxisomal enzymes that are able to detoxify hydrogen peroxide even at high concentrations. Elevated enzyme activity of CAT3 in the *gigantea3* mutant plants is probably responsible for an elevated antioxidative capacity (Zentgraf and Hemleben, 2008). The delayed-leaf-senescence mutants of *Arabidopsis thaliana*, *ore1*, *ore3*, and *ore9*, are also more tolerant to various types of oxidative stress, but the activities of antioxidant enzymes are similar or lower in the mutants. It is still unclear where the stress tolerance originates from, but oxidative stress tolerance appears to be genetically linked to control of leaf longevity (Woo *et al.*, 2004).

3 The Arabidopsis catalases

Catalases are tetrameric heme-containing enzymes and are present in all aerobic organisms. Due to a very high apparent Michaelis constant, catalases are not easily saturated with substrate and can act over a wide range of H_2O_2 levels, maintaining a controlled intracellular H_2O_2 concentration. Catalases provide the cell with a very energy-efficient mechanism for the removal of hydrogen peroxide because they decompose H_2O_2 without consuming cellular reducing equivalents. Catalases detoxify hydrogen peroxide extremely rapidly due to a catalytic centre activity of about 10^7 min^{-1}. However, depending upon the concentration of H_2O_2, catalases exert a dual function. At high concentrations they function as 'catalases', using H_2O_2 as acceptor and donor of hydrogen molecules ($2H_2O_2 \rightarrow 2H_2O + O_2$). At low concentrations (<1 μM) they act as 'peroxidases', using a variety of molecules (e.g. ethanol, ascorbic acid) as hydrogen donors ($RH_2 + H_2O_2 \rightarrow R + 2H_2O$; Scandalios, 2005). In plants, catalase is the most important enzyme scavenging H_2O_2 produced by photorespiration, but the function of catalase is not limited to peroxisomal H_2O_2 production (Dat *et al.*, 2000; Foyer and Noctor, 2000; Willekens *et al.*, 1997). The enzyme appears to be critical for maintaining the whole cell redox balance during oxidative stress and is indispensable for stress defence in C_3 plants (Willekens *et al.*, 1997). Whereas animals have only one form of catalase, plants have evolved small gene families encoding catalases. The plant catalases can be grouped into three classes depending on their expression patterns and physiological effects (Willekens *et al.*, 1995). Class I catalases are characterized by their strong, light-dependent expression in leaves. They are very abundant in photosynthetically-active cells, and, therefore, presumably involved in the scavenging of H_2O_2 generated during photo- respiration. Class II catalases are also found in mature plants, with a strong expression in the vascular tissue, which is light independent. In addition, they are considered to fulfil roles in lignification and in stress responses (Dat *et al.*, 2001; Orendi *et al.*, 2001). However, the precise function of this class is still unknown. Class III catalases comprise enzymes mainly expressed in seeds and young seedlings; they are in some cases also present in the late stages of leaf senescence. The function of this class of catalases is the removal of H_2O_2 produced during fatty acid degradation in the glyoxysomes (Dat *et al.*, 2000; Willekens *et al.*, 1995). In Arabidopsis, a small catalase gene family has been characterized: class III catalase *CAT1*, class I catalase *CAT2* and class II catalase *CAT3* (Frugoli *et al.*, 1996; Zentgraf and Zinkernagel, 1996). *CAT2* and *CAT3* mRNA expression patterns display circadian rhythms with distinct phases. Consistent with its photorespiratory role, *CAT2* mRNA has its maximum abundance in the early morning, whereas *CAT3* mRNA shows its maximum in the evening (Zhong and McClung, 1996). The enzyme activities of the different isoforms can be distinguished after activity staining on native PAGE and have been identified by inhibition experiments with 3-amino-1,2,4-triazole (Chandlee and Scandalios, 1983; Orendi *et al.*, 2001). Therefore, the activities of the different catalases can easily be monitored and quantified. Oxidative stress leads to an increase in *CAT2* and *CAT3* mRNA levels; however, only CAT3 activity increased in parallel so CAT3 appears to be the stress-responsive form. Since CAT3 is mainly expressed in vascular tissue (Zimmermann *et al.*, 2006), there seems to be a higher demand for oxidative stress protection in this tissue. Most likely CAT2 activity, which can be detected in high mounts in photosynthetically active tissue, is sufficient to detoxify even highly increasing amounts of hydrogen peroxide.

4 Senescence-specific regulation of catalases in Arabidopsis

4.1 CAT2 down regulation and APX1 inhibition

Without exogenous stress factors, leaf senescence is governed by the age of the leaf and the age and developmental stage of the plant (Zentgraf *et al.*, 2004). Since the protein content of the leaves decreases continuously during senescence and the protein pattern also changes the basis on which enzyme activities are assessed is by reference to two different parameters: per leaf, indicating the absolute status of the cells, and per microgramme soluble protein, indicating regulatory effects. Referring to both parameters, all three Arabidopsis catalases show senescence-specific changes in expression and activity. *CAT1* mRNA and activity can only be detected in very late stages of leaf and plant senescence. This has also been characterized for pumpkin (Esaka *et al.*, 1997), maize (Scandalios *et al.*, 1997) and rape (Buchanan-Wollaston and Ainsworth, 1997). On the basis of a leaf, the activities of both leaf isoforms, CAT2 and CAT3, declined at the same time point, when chlorophyll loss became detectable. Referring to the same amount of soluble proteins, two different regulatory effects can be observed: CAT2 activity declined when plants started to bolt, whereas CAT3 activity first increased and reached its maximum in 9-week-old plants. No further increase could be seen in older plants. Most likely, the vascular tissue needs very efficient protection in senescing regions because it is a prerequisite for mobilization processes. The activity profile of the catalases is directly mirrored by the hydrogen peroxide content measured in leaf discs. When CAT2 activity is lowered in 6-week-old plants, hydrogen peroxide content increases, and when CAT3 activity increases substantially in 8-week-old plants, hydrogen peroxide content decreases again.

However, catalase is not the only enzyme that can detoxify hydrogen peroxide. Another key enzyme scavenging H_2O_2 is ascorbate peroxidase (APX). Because the properties and requirements of CAT and APX are different, they function in parallel. Catalases are predominantly located in the peroxisomes, with the exception of maize CAT3, which is located in the mitochondria (Scandalios *et al.*, 1997). APX is the most important enzyme scavenging H_2O_2 produced in the chloroplast. Eight types of APX located in the cytosol, the chloroplast and the peroxisome have been described for Arabidopsis (Jespersen *et al.*, 1997; Panchuk, *et al.*, 2002, 2005). These genes are also differentially regulated in response to stress and development (Morita *et al.*, 1999; Panchuk *et al.*, 2002; Ye *et al.*, 2000). In contrast to CAT, APX requires ascorbate as a reductant. APX has a high affinity for H_2O_2 and is able to detoxify low concentrations of H_2O_2. Since APXs are extremely unstable enzymes, only two isoforms, the cytosolic APX1 and probably cytosolic APX2, can be visualized in crude extracts by activity staining following native PAGE (Mittler and Zilinskas, 1993; Panchuk *et al.*, 2002). Other APX isoenzymes tend to loose their activities during isolation procedures and subsequent electrophoresis. With regard to development-dependent activity profiles, the activity of the cytosolic isoform APX1 declined dramatically during bolting, in parallel with the decrease in CAT2 activity, a finding that supports the hypothesis of an increase of the hydrogen peroxide content during bolting and senescence (Ye *et al.*, 2000; Zimmermann *et al.*, 2006). In contrast to *CAT2*, whose expression was down-regulated (Zimmermann *et al.*, 2006), no decrease of *APX1* mRNA was detected during bolting and senescence (Panchuk *et al.*, 2005), indicating a post-transcriptional regulation of APX1 activity. However, there is some evidence that APX activity can be inhibited by the substrate, hydrogen peroxide (Miyake and Asada, 1996; Orendi *et al.*, 2001). Moreover, the sensitivity of APX1 to inhibition by hydrogen peroxide appears to be dependent on the developmental stage of the plants, being very pronounced in leaves at the time of bolting. The sensitivity of APX1 to hydrogen peroxide decreased markedly when the plants started to flower (Zimmermann *et al.*, 2006). This indicates that there might be additional factors that influence the age-dependent sensitivity of APX1 to hydrogen peroxide. Taken together, this suggests a feedback amplification loop in which CAT2 down-regulation would be the initial step to produce an elevated level of hydrogen peroxide. This increased hydrogen peroxide level might lead to the inactivation of APX1 activity, which is rendered sensitive for hydrogen peroxide inhibition by a so-far-unknown mechanism. This would lead in a feedback amplification loop to a further increase in accumulation of hydrogen peroxide, which in turn could then lead to the induction of *CAT3* expression. As plant development progresses, APX1 will become insensitive to hydrogen peroxide and APX1 activity will be restored. This coordinated regulation of the hydrogen peroxide-scavenging enzymes at the transcriptional and posttranscriptional levels could create a distinct increase in tissue hydrogen peroxide precisely at that time when the plants start to bolt, leading to a coordination of the senescence process of all rosette leaves (see *Figure 1*).

This hydrogen peroxide peak could indeed be measured in leaves of Arabidopsis plants at bolting (Miao *et al.*, 2004; Zimmermann *et al.*, 2006). Continuously removing

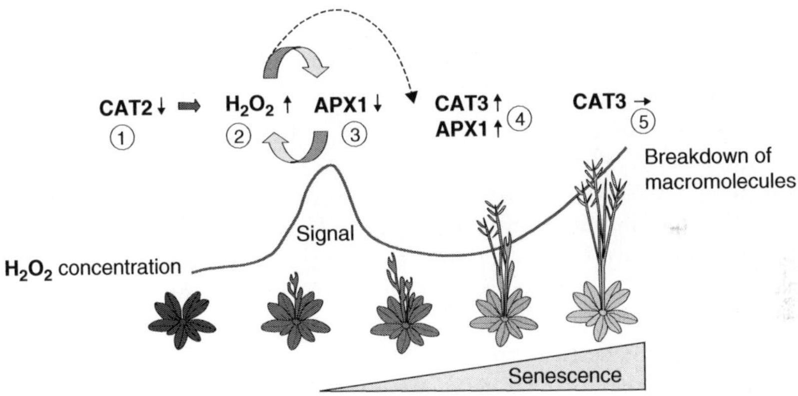

Figure 1. *Model of the coordinate regulation of antioxidative enzymes during senescence, leading to a hydrogen peroxide peak during bolting. (1) CAT2 down-regulation on the transcriptional level appears to be the initial step to produce an elevated level of hydrogen peroxide (2). (3) This then might lead to the inactivation of APX1 activity, which in turn increases the hydrogen peroxide level creating a feedback amplification regulation by the posttranscriptional inhibition of APX1 activity by hydrogen peroxide. (4) This increased hydrogen peroxide level then leads to the induction of CAT3 expression and activity, which then lowers the hydrogen peroxide level again and leads together with developmental factors to a restoration of APX1 activity. (5) During later stages of development, CAT3 activity can no longer be induced by oxidative stress leading again to an increase of hydrogen peroxide.*

the bolt could delay chlorophyll loss in the leaves (Ye *et al.*, 2000), indicating that bolting time is correlated with onset of leaf senescence. Furthermore, timing of senescence is also altered in the ascorbate-deficient Arabidopsis mutant *vtc1* (Barth *et al.*, 2004) indicating the participation of ascorbate and ascorbate peroxidase in regulation of leaf senescence. In addition, antisense expression of either *CAT* or *APX*, or both genes, also revealed that there appears to be a sensitive balance between the antioxidant enzymes, since double-antisense plants are more tolerant to oxidative stress than single-antisense plants (Rizhsky *et al.* 2002). Compensation for the lack of enzymatic protection by CAT or APX appears more efficient if both enzyme activities are affected. The extensive antioxidant system is composed of many enzymatic and non-enzymatic components. Many different peroxidases are modulated by internal and external stimuli and are probably regulated by a fine-tuning that meets the demands of plants during stress conditions and senescence, and has yet to be elucidated (Passardi *et al.*, 2005). A range of other components, such as ascorbate, glutathione, tocopherols, metallothioneins, peroxiredoxins and selenium, are involved in redox balancing. However, ascorbate and glutathione are multifunctional compounds with functions that extend beyond the antioxidative system. The flexibility of coupling between the antioxidant pools of the ascorbate and glutathione redox couples is crucial to differential signalling by ascorbate and glutathione (Foyer *et al.*, 2005; Noctor *et al.*, 2002). Furthermore, even a radical like nitric oxide (NO) is able to scavenge ROS and can strongly counteract many ROS-mediated cytotoxic processes in plants (Beligni and Lamattina, 1999). In Arabidopsis, NO can also act as a negative regulator of leaf senescence (Mishina *et al.*, 2007).

4.2 Loss of CAT3 induction

CAT3 is up-regulated during leaf and plant senescence. In addition, a decrease in the oxidative stress responsiveness of CAT3 is observed during leaf senescence. This is not part of a general loss of stress responsiveness occurring in senescing tissue but rather it appears to be restricted to the oxidative stress response (Orendi *et al.*, 2001). Remarkably, this is not a unique phenomenon for Arabidopsis catalases. The induction of superoxide dismutase (SOD) by severe paraquat treatment in Arabidopsis or by light in barley leaves was significantly lower in mature-senescent than in young leaves (Casano *et al.*, 1994; Pan and Yau, 1992). In senescing rye leaves, transfer from low to high light conditions failed to induce glutathione reductase (Kar *et al.*, 1993). Variability in photooxidative stress tolerance has been reported during plant development (Amsellem *et al.*, 1991) and with age of spinach and barley leaves (Boese and Huner, 1992; Casano *et al.*, 1999). Therefore, changes in the responsiveness of the ROS scavenging enzymes to different environmental conditions might be one of the specific regulatory mechanisms promoting leaf senescence in various plants.

4.3 CAT2 regulators

The catalase protein is light sensitive and it undergoes a process of continuous turnover in the light (Hertwig *et al.*, 1992). Under conditions that inhibit translation, such as heat shock (Feierabend *et al.*, 1992), salt (Streb *et al.*, 1993) or senescence (Kar *et al.*, 1993), the catalase protein level can be rapidly decreased. Therefore, a decline in catalase activity is one of the early symptoms of stress damage. Furthermore, this rapid turnover of catalase protein necessitates continuous transcription and translation of the mRNA. P_{CAT2}:*GUS* fusions revealed that the decline in CAT2 activity during bolting is accompanied by a decrease in gene expression, indicating that the down-regulation might be achieved on the transcriptional level. Since down-regulation of *CAT2* appears to be the initial event in creating the hydrogen peroxide peak during onset of leaf senescence, analyses of the factors that regulate catalase transcription might also give some new clues on regulation of leaf senescence.

Computer analyses of the *CAT2* promoter sequence and deletion analyses of the P_{CAT2}:*GUS* plants revealed that an approximately 200-bp fragment is mainly responsible for transcription of the *CAT2* gene. Using the yeast one-hybrid system, a G-box binding factor 1 (GBF1) was identified as binding to this fragment, which contains a G-box (P. Zimmerman, A. Smykowsky and U. Zentgraf, in preparation). Expression analyses by Northern blot and P_{GBF1}:*GUS* plants revealed that *GBF1* is up-regulated in senescing tissue, indicating that GBF1 could be a negative regulator of *CAT2* expression since *CAT2* is down-regulated. Transient expression assays using a *GBF1* over-expressing construct in protoplasts of P_{CAT2}:*GUS* plants confirmed this function. If GBF1 is a negative regulator of *CAT2* and *CAT2* down-regulation is involved in regulating the onset of senescence, then *GBF1* knockout (GBF1-ko) plants should show a delayed onset of leaf senescence. That is exactly what is observed when GBF1-ko plants are analysed phenotypically. Biochemical analyses confirm that *CAT2* down-regulation is abolished in these plants and the hydrogen peroxide peak disappears (P. Zimmerman, A. Smykowsky and U. Zentgraf, in preparation).

4.4 H_2O_2 as signal in leaf senescence

If the hydrogen peroxide peak is an important signal to turn on leaf senescence, what are the effectors of this signal? Senescence is accompanied by a massive change in the transcriptome, implying an important role for transcriptional regulators (Buchanan-Wollaston *et al.*, 2003; Guo *et al.*, 2004; Zentgraf *et al.*, 2004). Many NAC and WRKY transcription factors are up-regulated during leaf senescence in Arabidopsis (Guo *et al.*, 2004). *WRKY53* appears to have an important role during early phases of leaf senescence and there is strong evidence that *WRKY53* expression during leaf senescence is triggered by hydrogen peroxide (Miao *et al.*, 2004). The hydrogen peroxide peak during bolting coincides with a switch of *WRKY53* expression from a leaf-age-dependent pattern to a high systemic expression in all rosette leaves. Since *WRKY53* expression can be induced by hydrogen peroxide, this strongly suggests that the hydrogen peroxide peak is the signal that induces this switch in *WRKY53* expression (Miao *et al.*, 2004). *WRKY53* was shown to be involved in regulation of leaf senescence as *WRKY53* knockout or over-expressing plants show a senescence phenotype. In addition, target gene analyses revealed that many senescence-associated genes (SAGs) – but also many other transcription factors, including other WRKY factors – are direct target genes of WRKY53, indicating an upstream position of WRKY53 in signal transduction. However, the binding motifs of WRKY factors (W-boxes) are present in the promoters of almost all WRKY-factors (Dong *et al.*, 2003) and microarray analyses of different WRKY knockout and over-expressing plants suggest that WRKY factors act in a regulatory network, influencing the transcription of each other, rather than in a linear signal-transduction pathway (Dong *et al.*, 2003; Li *et al.*, 2004; Miao *et al.*, 2004; Robatzek and Somssich, 2002). If hydrogen peroxide is involved in the transcriptional regulation of *WRKY53* it should act on an upstream regulatory factor that acts on the promoter of *WRKY53*. Several proteins have already been shown to bind to the promoter of *WRKY53* and influence its expression. First, WRKY53 itself can bind to its own promoter and influence its own expression in a negative feedback loop. Since a P_{WRKY53}:GUS construct can no longer be induced by hydrogen peroxide in the *WRKY53*-knockout background it can be concluded that the WRKY53 protein itself is involved in the response of the *WRKY53* promoter to hydrogen peroxide (Miao *et al.*, 2007). The redox-sensitive zinc-finger DNA-binding domain of the WRKY53 protein, in which two cysteines together with two histidines interact electrostatically with a zinc atom to form a 'zinc finger', makes it an excellent candidate for direct redox regulation (Arrigo, 1999). If the protein structure and thereby the function of the WRKY53 protein is disturbed by hydrogen peroxide, the expression of *WRKY53* or a reporter gene driven by the *WRKY53* promoter would increase, since WRKY53 negatively regulates its own expression (Miao *et al.*, 2004). This is exactly what is observed. However, this is not the only connection between hydrogen peroxide and *WRKY53* expression. Surprisingly, a mitogen-activated protein kinase kinase kinase (MEKK1) was identified as a DNA-binding protein in a yeast one-hybrid screen with the *WRKY53* promoter. Remarkably, MEKK1 was also able to interact with WRKY53 on the protein level. The complex formation of MEKK1 and WRKY53 could be confirmed *in vivo* and was predominately localized in the nucleus of Arabidopsis cells. Furthermore, MEKK1 could also phosphorylate WRKY53 and this phosphorylation increased the DNA-binding activity of WRKY53 *in vitro* and transcription of a *WRKY53*-promoter-driven reporter gene *in vivo*. These results suggest that MEKK1 has a dual function in WRKY53 regulation: it binds to the promoter of the *WRKY53* gene modulating its expression and it phosphorylates the WRKY53 protein thereby increasing its DNA-binding activity. Thus, MEKK1 might be able to take a very direct short cut in MAPK signalling by directly phosphorylating a transcription factor (Miao *et al.*, 2007). Interestingly, expression of *MEKK1* is also triggered by hydrogen peroxide suggesting that there might be additional upstream factors involved in the signal transduction from hydrogen peroxide to *WRKY53* expression (Miao *et al.*, 2007). Since MEKK1 is also involved in the signal transduction pathways induced by pathogens and cold or salt stress, MEKK1 might be one of the nodes of convergence between senescence and stress response. Enhanced hydrogen peroxide levels have been observed in all these stress responses.

However, there is another link between hydrogen peroxide and WRKY53 regulation. The degradation of WRKY53 also appears to be regulated, since an E3-ubiquitin ligase (UPL5) is also able to interact with WRKY53. WRKY53 protein levels decrease in the presence of high amounts of UPL5, and UPL5 can ubiquitinate WRKY53 *in vitro*. *UPL5*-knockout plants show a senescence phenotype (Y. Miao, D. Kolb and U. Zentgraf, submitted). *UPL5* expression is negatively regulated by hydrogen peroxide and P_{UPL5}:*GUS* plants reveal that UPL5 expression is down-regulated during bolting when hydrogen peroxide levels are high. This clearly illustrates that WRKY53 is tightly regulated on different levels – expression, activity and degradation – and that hydrogen peroxide is involved in the regulation of the whole network (see *Figure 2*).

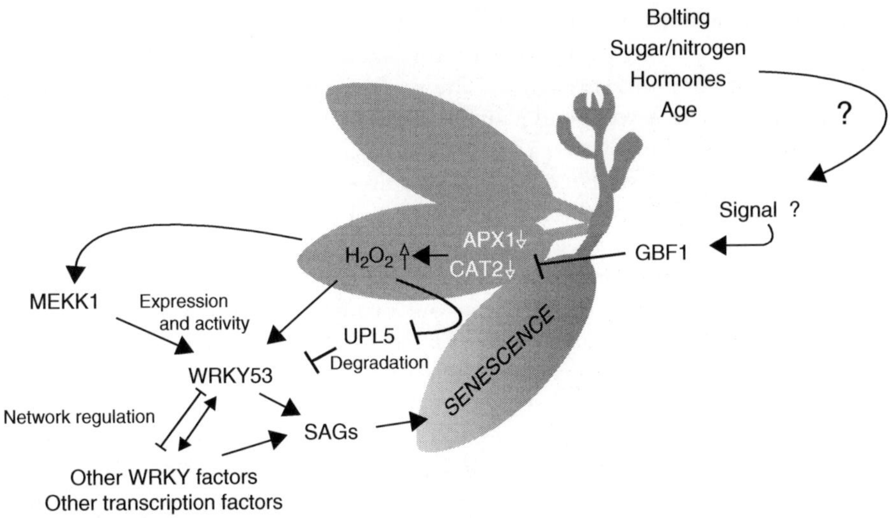

Figure 2. *Model of how hydrogen peroxide might be used as a signal to induce senencence (SAGs, senescence-associated genes; MEKK1, mitogen-activated kinase kinase kinase1; UPL5, ubiquitin protein ligase 5; GBF1, G-box binding factor1; WRKY53, transcription factor of the WRKY family).*

4.5 CAT knockout plants do not have an early senescence phenotype

Since the CAT2 regulator *GBF1* knockout shows a senescence phenotype, one may ask what would happen if CAT2 is absent? CAT2 is expressed in the photosynthetically active leaves, but also in roots, stems and flowers. In leaves, CAT2 is the most active isoform, contributing substantially to the protection of the cell against hydrogen peroxide. One would predict therefore, that the knock out of this gene would have severe effects on plants. If CAT2 knockout plants are viable then we can assume that there must be other systems that can compensate for the loss of catalase. However, one would predict that any CAT2 knockout plants should be more sensitive to stress and there is likely to be an early senescence phenotype. The lack of peroxisomal catalase CTL-2 in *Caenorhabditis elegans* caused a progeric phenotype, whereas the lack of the cytosolic catalase CTL-1 had no effect on nematode ageing (Petriv and Rachubinski, 2004). In yeast, under repressing conditions on glucose media, catalase T activity, but not catalase A activity, was necessary to assure longevity. However, under derepressing conditions, on ethanol media, both catalases were required for longevity assurance (Van Zandycke *et al.*, 2002) indicating a correlation between CAT activity and longevity in animal systems. Surprisingly, the CAT2 knockout plants have little phenotype. They are slightly impaired in germination but the plants develop normally. While leaf and plant senescence does not seem to be altered, the population does not senesce as homogenously as wild-type plants. The leaf hydrogen peroxide content was slightly increased at the beginning of development but thereafter no differences in the mean values of hydrogen peroxide accumulation were detected compared to wild-type plants. This result indicates that a very efficient compensation system for the loss of

CAT2 activity had been activated. However, there is considerable variation in hydrogen peroxide contents between different leaves and plants and thus the standard deviation is high in such measurements. The peak in leaf hydrogen peroxide contents occurring in the 6- to 7-week-old plants could no longer be detected. The system that compensates for the loss of *CAT2* remains to be characterized, but CAT3 and APX1 activity were not increased. Even *CAT2/3* double-knockout plants appeared to be similar to the wild type, although a slight activation of *CAT1* was observed. More importantly, tissue hydrogen peroxide contents were not significantly different to those measured in wild-type plants. The only clear difference that was observed was in stress tolerance. The single- and double-knockout plants were more sensitive to different stress treatments (P. Zimmermann and U. Zentgraf, in preparation). As mentioned above, antisense expression of *CAT*, *APX* or both genes simultaneously revealed the presence of a sensitive balance in the different antioxidant enzymes, since double-antisense plants are more tolerant to oxidative stress than single-antisense plants (Rizhsky *et al.*, 2002). Taken together, these studies show that ROS levels are tightly regulated on many levels with extensive possibilities for compensation if one or more detoxifying systems fail.

5 Concluding remarks

The timing of plant senescence affects many different agriculturally important traits, such as the timing of seed set, the number and quality of seeds and also fruit ripening. In addition, increased plant longevity can have an important impact on human health, as fungi and other microorganisms thrive on senescing tissue. Therefore, the understanding of senescence mechanisms provides new tools for the further improvement of crop plants. ROS play an important role in at least two different aspects of leaf senescence, through signalling and through effects on degradation. Although many proteins, particularly those containing thiol groups, are known to react directly with H_2O_2, to date no complete ROS signal transduction pathways have been described (Foyer and Nocter, 2005). Plants have developed a very fine-tuned network of enzymatic and low-molecular-weight antioxidative components in different cell compartments, and different plants have different strategies to balance their redox potential and regulate their ROS status. In Arabidopsis, a network of at least 152 genes is involved in managing the level of ROS. This network is highly dynamic and redundant, and encodes ROS-scavenging and ROS-producing proteins (Mittler *et al.*, 2004). Catalase regulation in concert with APX regulation is part of this dynamic network. However, many questions related to the integration of single components into this network, its mode of action and cross-talk, and its compensatory mechanisms remain unanswered. New approaches, such as metabolic modelling, are being developed as tools with which to analyse network regulation involving, for example, redox reactions in the SOD-ascorbate-glutathione pathway (Polle, 2001). These kinds of modelling approaches are not yet sufficiently robust to make exact predictions but they can contribute to the theoretical understanding of the functioning of antioxidant systems by pointing out questions that need to be validated (Polle, 2001). Moreover, the development of *in vivo* imaging systems for different ROS in order to visualize local changes in ROS levels in different cellular compartments will also help in the integration of current knowledge in a holistic view of the cells.

References

Alvarez, M.E., Pennell, R.I., Meijer, P.J., Ishikawa, A., Dixon, R.A. and Lamb, C. (1998) Reactive oxygen intermediates mediate a systemic signal network in the establishment of plant immunity. *Cell* **92:** 773–784.

Amsellem, Z., Jansen, M.A.K., Driesenaar, A.R.J. and Gressel, J. (1991) Developmental variability of photooxidative stress tolerance in paraquat resistant *Conyza. Plant Physiol.* **103:** 1097–1106.

Arrigo, A.P. (1999) Gene expression and the thiol redox state. *Free Radic. Biol. Med.* **27:** 936–944.

Barth, C., Moeder, W., Klessig, D.F. and Conklin, P.L. (2004) The timing of senescence and response to pathogens is altered in the ascorbate-deficient Arabidopsis mutant vitamin c-1. *Plant Physiol.* **134:** 1784–1792.

Batt, T. and Woolhouse, H.W. (1975) Changing activities during senescence and sites of synthesis of photosynthetic enzymes in leaves of labiate, *Perilla frutenscens* (L.). *Br. J. Exp. Bot.* **26:** 569–579.

Beligni, M.V. and Lamattina, L. (1999) Nitric oxide counteracts cytotoxic processes mediated by reactive oxygen species in plant tissues. *Planta* **208:** 337–344.

Boese, S.R. and Huner, N.P.A. (1992) Developmental history affect susceptibility of spinach leaves to in vivo low temperature photoinhibition. *Plant Physiol.* **99:** 1141–1145.

Buchanan-Wollaston, V. and Ainsworth, C. (1997) Leaf senescence in *Brassica napus*: cloning of senescence related genes by subtractive hybridization. *Plant Mol. Biol.* **33:** 821–834.

Buchanan-Wollaston, V., Earl, S., Harrison, E., Mathas, E., Navabpour, S., Page, T. and Pink, D. (2003) The molecular analysis of leaf senescence – a genomics approach. *Plant Biotechnol. J.* **1:** 3–22.

Casano, L.M., Martin, M. and Sabater, B. (1994) Sensitivity of superoxide dismutase transcript levels and activities to oxidative stress is lower in mature-senescent than in young barley leaves. *Plant Physiol.* **106:** 1033–1039.

Casano, L.M., Martin, M., Zapata, J.M. and Sabater, B.J. (1999) Leaf age- and paraquat concentration-dependent effects on the level of enzymes protecting against photooxidative stress. *Plant Sci.* **149:** 13–22.

Chandlee, J.M. and Scandalios, J.G. (1983) Gene expression during early kernel development in *Zea mays. Dev. Genet.* **4:** 99–115.

Chia, L.S., Thompson, J.E. and Dumbroff, E.B. (1981) Simulation of the effects of leaf senescence on membranes by treatment with paraquat. *Plant Physiol.* **67:** 415–420.

Dat, J.F., Vandenabeele, S., Vranová, E., Van Montagu, M., Inzé, D. and Van Breusegem, F. (2000) Dual action of the active oxygen species during plant stress responses. *Cell. Mol. Life Sci.* **57:** 779–795.

Dat, J.F., Inzé, D. and Van Breusegem, F. (2001) Catalase-deficient tobacco plants: tools for *in planta* studies on the role of hydrogen peroxide. *Redox Rep.* **6**: 37–42.

Dhinsda, R.J., Dhinsda, P.P. and Thorpe, T.A. (1981) Leaf senescence: correlation with increased levels of membrane permeability and lipid peroxidation, and decreased levels of superoxide dismutase and catalase. *J. Exp. Bot.* **32:** 93–101.

Dong, J., Chen, C. and Chen, Z. (2003) Expression profile of the WRKY gene superfamily during plant defence. *Plant Mol. Biol.* **51:** 21–37.

Esaka, M., Yamada, N., Kitabayashi, M., Setoguchi, Y., Tsugeki, R., Kondo, M. and Nishimura, M. (1997) cDNA cloning and differential gene expression of three catalases in pumpkin. *Plant Mol. Biol.* **33:** 141–155.

Feierabend, J., Schaan, C. and Hertwig, B. (1992) Photoinactivation of catalase occurs under both high and low temperature stress conditions and accompanies photoinhibition of photosystem II. *Plant Physiol.* **100:** 1554–1561.

Foyer, C.H. and Noctor, G. (2000) Oxygen processing in photosynthesis: regulation and signalling. *New Phytol.* **146:** 359–388.

Foyer, C,H. and Noctor, G. (2005) Oxidant and antioxidant signalling in plants: a re-evaluation of the concept of oxidative stress in a physiological context. *Plant Cell Environ.* **28:** 1056–1071.

Foyer, C.H., Lopez-Delgado, H., Dat, J. and Scott, I. (1998) Parallel changes in H_2O_2 and catalase during thermotolerance induced by salicylic acid or heat acclimation in mustard seedlings. *Plant Physiol.* **116:** 1351–1357

Foyer, C.H., Trebstm, A. and Noctor, G. (2005) Protective and signalling functions of ascorbate, glutathione and tocopherol in chloroplasts. In: *Advances in Photosynthesis and Respiration: Photoprotection, Photoinhibition, Gene Regulation, and Environment*, Vol 19 (eds B. Demmig-Adams and W.W. Adams). Kluwer Academic, Dortrecht, pp. 241–268.

Frugoli, J.A., Zhong, H.H., Nuccio, M.L., McCourt, P., McPeek, M.A., Thomas, T.L. and McClung, C.R. (1996) Catalase is encoded by a multigene family in *Arabidopsis thaliana* (L.) Heynh. *Plant Physiol.* **112:** 327–336.

Grant, J. and Loake, G. (2000) Role of reactive oxygen intermediates and cognate redox signaling in disease resistance. *Plant Physiol.* **124**: 21–29.

Guo, Y., Cai, Z. and Gan, S. (2004) Transcriptome of Arabidopsis leaf senescence. *Plant Cell Environ.* **27:** 521–549.

Halliwell, B. and Gutteridge, J.M.C. (1989) *Free Radicals in Biology and Medicine*, 2nd Edn. Oxford University Press (Clarendon), Oxford.

Harding, S.A., Oh, S.H. and Roberts, D.M. (1997) Transgenic tobacco expressing a foreign calmodulin gene shows an enhanced production of active oxygen species. *EMBO J.* **16:** 1137–1144.

Harman, D. (1956) Aging: a theory based on free radical and radiation chemistry. *J. Gerontol.* **11:** 298–300.

Harman, D. (1998) Extending functional life span. *Exp. Gerontol.* **33:** 95–112.

Hensel, L.L., Grbić, V., Baumgarten, D. and Bleecker, A.B. (1993) Developmental and age-related processes that influence the longevity and senescence of photosynthetic tissues in Arabidopsis. *Plant Cell* **5:** 553–564.

Hertwig, B., Streb, P. and Feierabend, J. (1992) Light dependence of catalase synthesis and degradation in leaves and the influence of interfering stress conditions. *Plant Physiol.* **100:** 1547–1553.

Jespersen, H.M., Kjærsgård, V.H., Østergaard, L. and Welinder, K.G. (1997) From sequence analysis of three novel ascorbate peroxidases from *Arabidopsis thaliana* to structure, function and evolution of seven types of ascorbate peroxidase. *Biochem. J.* **326:** 305–310.

Jiménez, A., Hernandez, J.A., Pastori, G., del Rio, L.A. and Sevilla, F. (1998) Role of the ascorbate-glutathione cycle of mitochondria and peroxisomes in the senescence of pea leaves. *Plant Physiol.* **118:** 1327–1335.

Kar, M. and Feierabend, J. (1984) Metabolism of activated oxygen in detached wheat and rye leaves and its relevance to the initiation of senescence. *Planta* **160:** 385–391.

Kar, M., Streb, P., Hertwig, B. and Feierabend, L. (1993) Sensitivity to photodamage increases during senescence in excised leaves. *J. Plant Physiol.* **141:** 538–544.

Keskitalo, J., Bergquist, G., Gardestrom, P. and Jansson, S. (2005) A cellular timetable of autumn senescence. *Plant Physiol.* **139:** 1635–1648.

Kovtun, Y., Chiu, W.L., Tena, G. and Sheen, J. (2000) Functional analysis of oxidative stress-activated mitogen-activated protein kinase cascade in plants. *Proc. Natl Acad. Sci. USA* **97:** 2940–2945.

Krupinska, K., Falk, J. and Humbeck, K. (2003) Genetic, metabolic and environmental factors associated with ageing in plants. In: *Aging of Organisms* (ed. H.D. Osiewacz). Kluwer Academic, Dordrecht, pp. 55–78.

Kurepa, J., Smalle, J., Van Montagu, M. and Inzé, D. (1998) Oxidative stress tolerance and longevity in Arabidopsis: the late flowering mutant *gigantea* is tolerant to paraquat. *Plant J.* **14:** 759–764.

Kuźniak, E. and Urbanek, H. (2000) The involvement of hydrogen peroxide in plant responses to stresses. *Acta Physiol. Plant.* **22:** 195–203.

Li, J., Brader, G. and Palva, E.T. (2004) The WRKY70 transcription factor: a node of convergence of jasmonate-mediated and salicylate-mediated signals in plant defense. *Plant Cell* **16:** 319–331.

Martin, G.M., Austad, S.N. and Johnson, T.E. (1996) Genetic analysis of aging: role of oxidative damage and environmental stresses. *Nat. Genet.* **13:** 25–34.

Martin-Tryon, E.L., Kreps, J.A. and Harmer, S.L. (2007) GIGANTEA acts in blue light signaling and has biochemically separable roles in circadian clock and flowering time regulation. *Plant Physiol.* **144:** 495–502.

Masclaux ,C., Valadie, M.H., Brugière, N., Morot-Gaudry, J.F. and Hirel, B. (2000) Characterization of the sink/source transition in tobacco (*Nicotiana tabacum* L) shoots in relation to nitrogen management and leaf senescence. *Planta* **211:** 510–518.

Matile, P., Schellenberg, M. and Peisker, C. (1992) Production and release of a chlorophyll catabolite in isolated senescent chloroplasts. *Planta* **187:** 230–235.

McRae, D.G. and Thompson, J.E. (1983) Senescence dependent changes in superoxide anion production by illuminated chloroplast from bean leaves. *Planta* **158:** 185–193.

Miao, Y., Laun, T., Zimmermann, P. and Zentgraf, U. (2004) Targets of the WRKY53 transcription factor and its role during leaf senescence in *Arabidopsis*. *Plant Mol. Biol.* **55:** 853–867.

Miao, Y., Laun, T., Smykowski, A. and Zentgraf, U. (2007) Arabidopsis MEKK1 can take a short cut: it can directly interact with senescence-related WRKY53 transcription factor on the protein level and can bind to its promoter. *Plant Mol. Biol.* **65:** 63–76.

Mishina, T.E., Lamb, C. and Zeier, J. (2007) Expression of a nitric oxide degrading enzyme induces a senescence programme in Arabidopsis. *Plant Cell Environ.* **30:** 39–52.

Mittler, R. and Zilinskas, B.A. (1993) Detection of ascorbate peroxidase activity in native gels by inhibition of the ascorbate-dependent reduction of nitroblue tetrazolium. *Analyt. Biochem.* **212:** 540–546.

Mittler, R., Vanderauwera, S., Gollery, M. and Van Breusegem, F. (2004) Reactive oxygen gene network of plants. *Trends Plant Sci.* **9:** 490–498.

Miyake, C. and Asada, K. (1996) Inactivation mechanism of ascorbate peroxidase at low concentrations of ascorbate: hydrogen peroxide decomposes compound I of ascorbate peroxidase. *Plant Cell Physiol.* **37:** 423–430.

Morita, S., Kaminaka, H., Masumura, T. and Tanaka, K. (1999) Induction of rice cytosolic ascorbate peroxidase mRNA by oxidative stress; the involvement of hydrogen peroxide in oxidative stress signalling. *Plant Cell Physiol.* **40:** 417–422.

Mullineaux, P. and Karpinski, S. (2002) Signal transduction in response to excess light: getting out of the chloroplast. *Curr. Opin. Plant Biol.* **5:** 43–48.

Noctor, G., Veljovic-Jovanovic, S., Driscol, S., Novitskaya, L. and Foyer, C.H. (2002) Drought and oxidative load in wheat leaves: a predominant role for photorespiration? *Ann. Bot.* **89:** 841–850.

Orendi, G., Zimmermann, P., Baar, C. and Zentgraf, U. (2001) Loss of stress-induced expression of catalase3 during leaf senescence in *Arabidopsis thaliana* is restricted to oxidative stress. *Plant Sci.* **161:** 301–314.

Orozco-Cardenas, M. and Ryan, C.A. (1999) Hydrogen peroxide is generated systemically in plant leaves by wounding and systemin via the octadecanoid pathway. *Proc. Natl Acad. Sci. USA* **96:** 6553–6557.

Orr, W.C. and Sohal, R.S. (1994) Extension of life-span by overexpression of superoxide dismutase and catalase in *Drosophila melanogaster*. *Science* **263:** 1128–1130.

Pan, S.M., and Yau, Y.Y. (1992) Characterisation of superoxide dismutase in *Arabidopsis*. *Taiwania* **37:** 58–66.

Panavas, T. and Rubinstein, B. (1998) Oxidative events during programmed cell death of daylily (*Hemerocallis* hybrid) petals. *Plant Sci.* **133:** 125–138.

Panchuk, I.I., Volkov R.A. and Schöffl F. (2002) Heat stress- and heat shock transcription factor-dependent expression and activity of ascorbate peroxidase in Arabidopsis. *Plant Physiol.* **129:** 838–853.

Panchuk, I.I., Zentgraf, U. and Volkov, R.A. (2005) Expression of the *Apx* gene family during leaf senescence of *Arabidopsis thaliana. Planta* **222:** 926–932.

Passardi, F., Cosio, C., Penel, C. and Dunand, C. (2005) Peroxidases have more functions than a Swiss army knife. *Plant Cell Rep.* **24:** 255–265.

Pastori, G.M. and del Rio, L.A. (1994) An activated-oxygen-mediated role for peroxisomes in the mechanism of senescence of *Pisum sativum* L. leaves. *Planta* **193:** 385–391.

Pauls, K.P. and Thompson, J.E. (1984) Evidence for the accumulation of peroxidized lipids in membranes of senescing cotyledons. *Plant Physiol.* **75:** 1152–1157.

Petriv, O.I. and Rachubinski, R.A. (2004) Lack of peroxisomal catalase causes a progeric phenotype in *Caenorhabditis elegans*. *J. Biol. Chem.* **279:** 19996–20001.

Polle, A. (2001) Dissecting the superoxide dismutase-ascorbate-glutathione-pathway in chloroplasts by metabolic modeling. Computer simulations as a step towards flux analysis. *Plant Physiol.* **126:** 445–462.

Pourtau, N., Marès, M., Purdy, S., Quentin, N., Ruël, A. and Wingler, A. (2004) Interactions of abscisic acid and sugar signalling in the regulation of leaf senescence. *Planta* **219:** 765–772.

Rizhsky, L., Hallak-Herr, E., Van Breusegem, F., Rachmilevitch, S., Barr, J.E., Rodermel, S., Inzé, D. and Mittler, R. (2002) Double antisense plants lacking ascorbate peroxidase and catalase are less sensitive to oxidative stress than single antisense plants lacking ascorbate peroxidase or catalase. *Plant J.* **32:** 329–342.

Robatzek, S., and Somssich, I.E. (2002) Targets of *AtWRKY6* regulation during plant senescence and pathogen defense. *Genes Dev.* **16:** 1139–1149.

Scandalios, J.G. (2005) Oxidative stress: molecular perception and transduction of signals triggering antioxidant gene defenses. *Braz. J. Med. Biol. Res.* **38:** 995–1014.

Scandalios, J.G., Guan, L. and Polidoros, A.N. (1997) Catalases in plants: gene structure, properties, regulation and expression. In: *Oxidative Stress and the Molecular Biology of Antioxidant Defences* (ed. J.G. Scandalios). Cold Spring Harbor Press, New York, NY, pp. 343–406.

Smart, C.M. (1994) Gene expression during leaf senescence. *New Phytol.* **126:** 419–448.

Sohal, R.S. and Weindruch, R. (1996) Oxidative stress, caloric restriction, and aging. *Science* **273:** 59–63.

Streb, P., Michael-Knauf, A. and Feierabend, J. (1993) Preferential photoinactivation of catalase and photoinhibition of photosystem II are common early symptoms under various osmotic and chemical stress conditions. *Physiol. Plant.* **88:** 590–598.

Thompson, J.E., Ledg, R.L. and Barber, R.F. (1987) The role of free radicals in senescence and wounding. *New Phytol.* **105:** 317–344.

van Doorn, W.G. (2004) Is petal senescence due to sugar starvation? *Plant Physiol.* **134:** 35–42.

van Zandycke, S.M., Sohier, P.J. and Smart, K.A. (2002) The impact of catalase expression on the replicative lifespan of *Saccharomyces cerevisiae*. *Mech. Ageing Dev.* **123:** 365–373.

Vandenabeele, S., Van Der Kelen, K., Dat, J., *et al.* (2003) A comprehensive analysis of hydrogen peroxide-induced gene expression in tobacco. *Proc. Natl Acad. Sci. USA* **100:** 16113–16118.

Willekens, H., Inzé, D., Van Montagu, M. and van Camp, W. (1995) Catalases in plants. *Mol. Breed.* **1:** 207–228.

Willekens, H., Chamnongpol, S., Davey, M., Schraudner, M., Langebartels, C., Van Montagu, M., Inzé, D. and Van Camp, W. (1997) Catalase is a sink for H_2O_2 and is indispensable for stress defence in C_3 plants. *EMBO J.* **16:** 4806–4816.

Wingler, A., Marès, M. and Pourtau, N. (2004) Spatial patterns and metabolic regulation of photosynthetic parameters during leaf senescence. *New Phytol.* **161:** 781–789.

Woo, H.R., Kim, J.H., Nam, H.G. and Lim, P.O. (2004) The delayed leaf senescence mutants of Arabidopsis, ore1, ore3, and ore9 are tolerant to oxidative stress. *Plant Cell Physiol.* **45:** 923–932.

Ye, Z.Z., Rodriguez, R., Tran, A., Hoang, H., de los Santos, D., Brown, S. and Vellanoweth, R.L. (2000) The developmental transition to flowering represses ascorbate peroxidase activity and induces enzymatic lipid peroxidation in leaf tissue in *Arabidopsis thaliana*. *Plant Sci.* **158:** 115–127.

Zentgraf, U. (2007) Oxidative stress and leaf senescence. In: *Senescence Processes in Plants*. Annual Plant Reviews, Vol. 26 (ed. S. Gan). Blackwell Publishing, Oxford, pp. 69–86.

Zentgraf, U. and Hemleben, V. (2008) Molecular cell biology: are reactive oxygen species regulators of leaf senescence? In: *Progress in Botany*, Vol. 68. Springer Verlag, Berlin pp. 117–138.

Zentgraf, U. and Zinkernagel, I. (1996) A gene encoding a catalase isoform from *Arabidopsis thaliana* (PGR 96-005). *Plant Physiol.* **110:** 1047.

Zentgraf, U., Jobst, J., Kolb, D. and Rentsch, D. (2004) Senescence-related gene expression profiles of rosette leaves of *Arabidopsis thaliana*: leaf versus plant age. *Plant Biol.* **6:** 178–183.

Zhong, H.H. and McClung, C.R. (1996) The circadian clock gates expression of two catalase genes to distinct and opposite circadian phase. *Mol. Gen. Genet.* **251:** 196–203.

Zimmermann, P., Heinlein, C., Orendi, G. and Zentgraf, U. (2006) Senescence specific regulation of catalases in *Arabidopsis thaliana* (L.) Heynh. *Plant Cell Environ.* **29:** 1049–1060.

11

Atmospheric CO_2 signalling, cellular redox state and plant growth and development

Anneke Prins, Karl Kunert and Christine H. Foyer

1 Introduction

The mechanisms that control plant growth and development, from seed germination to senescence, have been extensively studied, but the concepts of lifespan and ageing remain poorly defined compared to those in animals (Thomas *et al.*, 2003). Lifespan, like growth, is controlled by the complex interplay of genetic, developmental and metabolic triggers that respond to environmental inputs and cues. Plant species with a low relative growth rates (RGR) are often found in nutrient-poor environments, whereas inherently fast-growing species inhabit nutrient-rich, less stressful environments (Grime, 2001). High RGRs may be advantageous in some circumstances as they enable plants to achieve high biomass, reach sexual maturity and complete the life cycle, and so there may have been positive selection for high RGRs during evolution. However, plants in natural environments often cannot achieve RGRs close to their potential maxima because of environmental restrictions and impacts. Moreover, in stressful conditions, conservation of existing biomass may be an important constraint.

Studies on the nature and integration of information from metabolism and the environment have revealed a complex network of interacting pathways and controls brokered by the interplay of phytohormone and nutritional/metabolic signals (Aarts and Fiers, 2003; Chinnusamy *et al.*, 2004; Potters *et al.*, 2007). Information from metabolism and the environment also primes the plant 'innate immune defence' response and the 'stress-induced morphogenic' responses (Potters *et al.*, 2007). Similarly, exposure to single stresses such as chilling and drought can induce common signalling pathways that increase tolerance to a wide range of biotic and abiotic stresses, a phenomenon known as 'cross-tolerance' (Boudsocq and Lauriere, 2005; Chinnusamy *et al.*, 2004; Ho *et al.*, 2001; Pastori and Foyer, 2002).

Of the many signals perceived by plants, the atmospheric level of CO_2 is particularly important because of concerns related to climate change and its impact on crop production and on the environment (Nemani *et al.*, 2003). The average global atmospheric CO_2 concentration in 2007 was 382 $\mu l\ l^{-1}$, a figure that is approximately 30% higher than the pre-industrial level. Atmospheric CO_2 concentrations are projected to rise even further to values between 600–800 μl l^{-1} by the end of the 21st century (Cox *et al.*, 2000). Carbon dioxide is the substrate for photosynthetic carbon assimilation (*Figure 1*). Hence, future increases in atmospheric CO_2 levels will have immediate effects on photosynthesis and associated processes (Moore *et al.*, 1999). Through effects on photosynthesis, the abundance of CO_2 in the atmosphere will have a profound effect on plant growth, morphology and biomass production (Evans, 2004). In addition, plants may also respond to changing atmospheric CO_2 levels by as yet uncharacterized signalling mechanisms that are independent of photosynthesis. Here, we consider the mechanisms that contribute to the control of plant growth and development in response to changes in atmospheric CO_2, focusing on the involvement of sugars, cellular redox state and plant growth regulators/hormones.

2 The perception of atmospheric CO_2 and associated signalling systems

There is an extensive literature on plant responses to CO_2 enrichment, which has been shown to exert a wide range of effects, from altered cell-cycle progression (Jitla *et al.*, 1997; Kinsman *et al.*, 1997), stomatal densities (Woodward *et al.*, 2002), leaf structure and composition (Martin and Glover, 2007; Woodward, 2002) to modifications in whole-plant morphology (Lewis *et al.*, 1999, 2000). Plant acclimation responses to the abundance of atmospheric CO_2 occur in a systemic manner (Sims *et al.*, 1998) with signals transported between different plant organs. Sugar signalling is an important component of long distance CO_2 signalling pathways operating from source leaves to developing leaves (Lake *et al.*, 2002). However, the mechanisms involved in the perception and signalling of atmospheric CO_2 abundance remain poorly characterized (Gray *et al.*, 2000; Long *et al.*, 2004).

Long-term growth at elevated CO_2 can result in a down-regulation of photosynthesis (Jacob *et al.*, 1995; Nie *et al.*, 1995) and an up-regulation of processes that use assimilate, such as carbohydrate synthesis and respiration (Lewis *et al.*, 2000; Long *et al.*, 2004; Robertson and Leech, 1995; Robertson *et al.*, 1995; Stitt, 1991). The CO_2-dependent regulation of gene expression is considered to be central to effects on photosynthesis. As discussed below, sucrose and hexoses are strong repressors of photosynthetic gene expression. Hence, increases in leaf sugars and starch in high-CO_2-grown plants have been linked to repression of photosynthesis (Jang and Sheen, 1997; Sheen *et al.*, 1999; Stitt, 1991). The decrease in photosynthesis that can occur in plants grown with CO_2 enrichment has also been linked to redistribution of the N from photosynthetic proteins, especially ribulose-1,5-bisphosphate carboxylase/oxygenase (RuBisCO), from the mature sucrose-exporting source leaves to developing sucrose-importing sink tissues (Long *et al.*, 2004; Stitt, 1991;). Developing leaves are less susceptible than mature source leaves to the high CO_2-dependent loss of photosynthetic proteins (Nie *et al.*, 1995). One reason for the differential response is that developing leaves already have repressed photosynthetic gene expression because of their active sucrose import. The altered signalling in developing leaves may thus prevent redistribution of N away from RuBisCO (Moore *et al.*, 1999). However, RuBisCO subunit transcripts are highest in developing leaves (Moore *et al.*, 1999). The observed effects of high CO_2 may also be linked to altered hormone signalling (Jang and Sheen, 1994, 1997). Sugar, nitrogen and hormone signalling pathways are closely interlinked in the orchestration of the plant C/N interactions that are central to the responses to high CO_2 (Finkelstein and Gibson, 2002; León and Sheen, 2003).

The immediate effect of CO_2 enrichment on C_3 plants is an enhancement of photosynthetic CO_2 assimilation and a decrease in photorespiration (*Figure 1*). This is because the activity of the enzyme responsible for the incorporation of atmospheric CO_2 into sugar phosphate (RuBisCO) is limited by CO_2 availability under current atmospheric conditions in C_3 plants. RuBisCO catalyses the incorporation of CO_2 into ribulose-1,5-bisphosphate (RuBP) producing two molecules of 3-phosphoglycerate (3PGA). However, RuBisCO can also catalyse the incorporation of molecular oxygen into RuBP, a reaction that produces one molecule of phosphoglycolate (PGlyco) and one molecule of 3PGA. Phosphoglycolate is the substrate for photorespiration, which provides an alternate pathway for the production of 3PGA but it uses energy and leads to the release of CO_2 and ammonia (*Figure 1*). Crucially, the photorespiratory pathway leads to high rates of hydrogen peroxide production in the peroxisomes (*Figure 1*; Foyer and Noctor, 2000, 2003). Although RuBisCO has a higher affinity for CO_2 than O_2, the rate of photorespiration is very high in air in C_3 plants such as wheat and Arabidopsis. Moreover, the rate of oxygenation increases in C_3 plants at high temperatures and in stress situations that lead to stomatal closure. Plants with the C_4 pathway of photosynthesis, such as maize and sorghum, are able to achieve higher rates of photosynthetic CO_2 assimilation because they have a metabolic ATP-dependent CO_2-pumping mechanism that facilitates saturation of RuBisCO with CO_2 in the bundle sheath cells. Unlike C_3 plants, in which photosynthesis can be limited by photorespiration by up to 50% in warm climates, C_4 plants have very low levels of photorespiration (2–5%; Sage, 2001). C_4 plants are not only able to exploit the higher RuBisCO activities that occur at warmer temperatures but they invest less nitrogen in the RuBisCO protein (Sage, 2001). While RuBisCO constitutes up to 50% of the total soluble protein in the leaves of C_3 plants (Spreitzer and Salvucci, 2002), the amount of RuBisCO protein is much lower in C_4 leaves, where it constitutes up to 30% of total soluble protein (Sugiyama *et al.*, 1984). Since C_4 plants already have a metabolic CO_2-enrichment mechanism, the effects of enhanced atmospheric CO_2 abundance are predicted to be much less pronounced in C_4 plants than C_3 plants (Long *et al.*, 2004).

The down-regulation of photosynthesis that is observed in C_3 plants in response to long-term growth with CO_2 enrichment is generally not observed in C_4 leaves, where photosynthetic rates are similar in plants grown in air and high CO_2 (Leakey *et al.*, 2006; Soares *et al.*, 2008). Despite the absence of any marked effects on photosynthesis and hence the capacity to drive sugar metabolism, the leaves of C_4 plants such as maize still show CO_2-dependent changes in epidermal structure, stomatal distribution and carbohydrate metabolism (Driscoll *et al.*, 2006). While the stomatal density of maize leaves was not affected by growth with high CO_2, stomatal area, epidermal cell size and stomatal index were significantly increased when plants were grown at high CO_2 (Driscoll *et al.*, 2006). Similarly, maize plants grown in air and in high CO_2 had comparable growth, development and morphologies (*Figure 2* in colour plate section), but their leaves also show a significant decrease in leaf soluble protein and chlorophyll, with the highest degree of acclimation observed in leaves on the middle of the stem (Driscoll *et al.*, 2006). The thickness of the bundle sheath was changed in sorghum leaves grown at high CO_2 and the CO_2-saturated rates of photosynthesis were decreased (Watling *et al.*, 2000). However, with the exception of sorghum, the

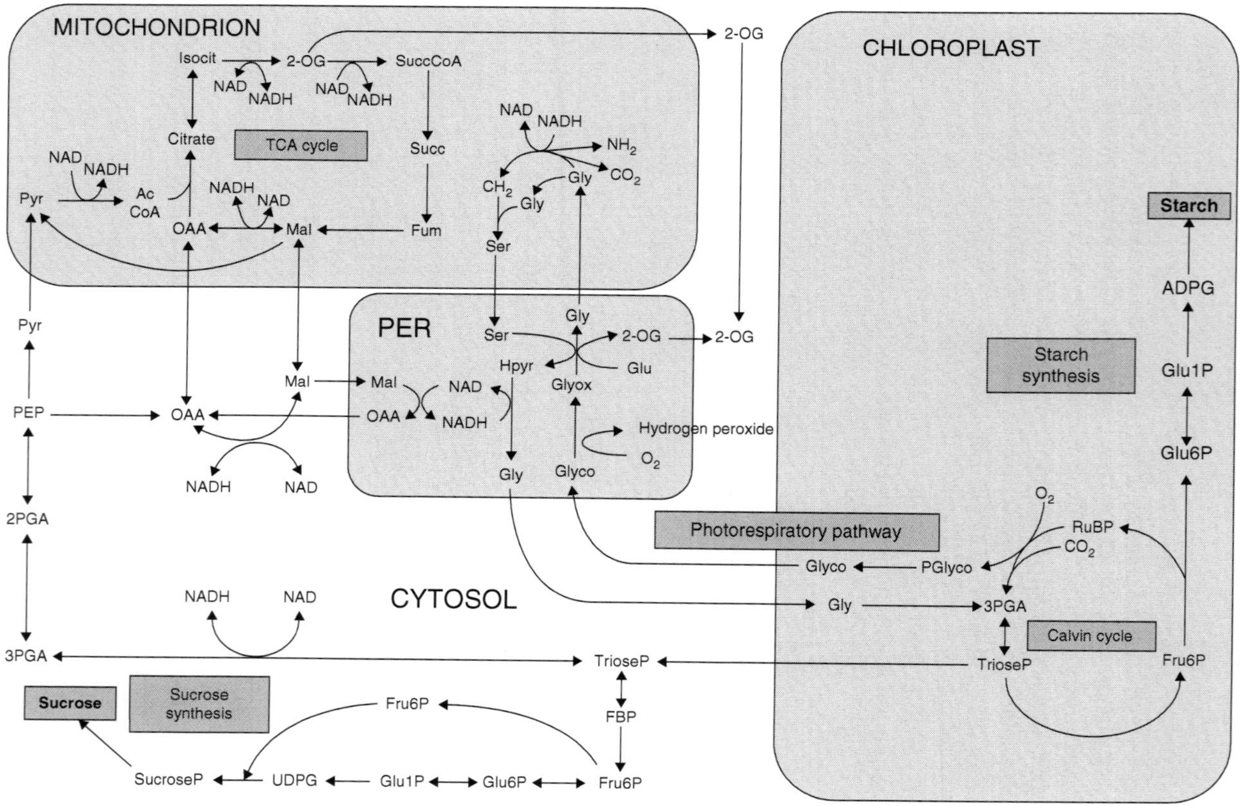

Figure 1. *Diagrammatic representation of the interacting pathways of photosynthetic CO_2 assimilation, photorespiration, sucrose and starch synthesis and respiratory carbon oxidation in the tricarboxylic acid (TCA) cycle (Ribulose-1,5-bisphosphate, RuBP; fructose 6-phosphate, Fru6P; triose phosphate, TrioseP; 3-phosphoglycerate, 3PGA; 2-phosphoglycerate, 2PGA; glucose 6-phosphate, Glu6P; glucose 1-phosphate, Glu1P; ADPglucose, ADPG; fructose-1,6-bisphosphate, FBP; UDP glucose, UDPG; sucrose phosphate, sucroseP; phosphoglycolate, PGlyco; glycolate, Glyco; glyoxalate, Glyox; glycine, Gly; serine, Ser; hydroxypyruvate, Hpyr; phosphoenolpyruvate, PEP; pyruvate, Pyr; acetyl coenzyme A, AcCoA; isocitrate, Isocit; 2-oxoglutarate, 2-OG; succinyl coenzyme A, SuccCoA; succinate, Succ; fumarate, Fum; malate, Mal; oxaloacetate, OAA;*

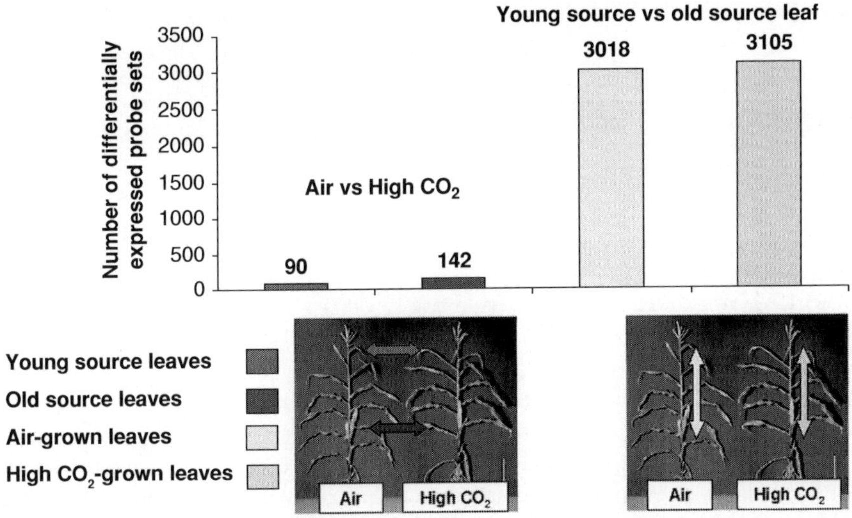

Figure 2. *A comparison of the effect of age (young verses old source leaves; as illustrated in the lower figure on the right) and growth CO_2 availability (air: 350 μL L^{-1} CO_2 vs high: 700 μL L^{-1} CO_2; as illustrated in the lower figure on the left). Young source leaves, rank 12; mature source leaves, rank 3.*

CO_2-signal leading to acclimatory changes in leaf structure and composition in many C_4 leaves is not directly related to changes in the rate of photosynthesis. The effects of CO_2 enrichment on the maize leaf transcriptome are much less pronounced than the effect of leaf rank and developmental stage (*Figure* 2). While the age of the maize leaf had a marked effect on the abundance of large numbers of leaf mRNAs, only a hundred or so transcripts were altered in leaves of equivalent positions on the stems of maize plants grown in high CO_2 compared to those grown in air (*Figure* 2).

Since the rates of photosynthetic CO_2 assimilation are similar in maize plants grown in high CO_2 to those grown in ambient air (Leakey *et al.*, 2006), key questions remain unanswered regarding how CO_2 availability is perceived and transmitted in C_4 leaves. Even though the high CO_2 signal cannot arise from photosynthetic CO_2 assimilation, it is probable that the cellular oxidative load associated with photosynthesis is decreased in photosynthetic tissues as CO_2 availability increases. While photorespiration, the major metabolic source of leaf hydrogen peroxide, is already low in C_4 leaves compared to the level observed in C_3 counterparts, it may be repressed even further in plants grown at high CO_2. Similarly, superoxide and hydrogen peroxide production by the photosynthetic and respiratory electron transport chains may also be repressed when C_4 plants are grown with CO_2 enrichment. Growth with CO_2 enrichment may also alter the abundance of hormones such as abscisic acid (ABA) which is known to produce superoxide and hydrogen peroxide as second messengers (Kwak *et al.*, 2003). Hence, it is probable that redox signals are important in the orchestration of responses of C_4 plants to high CO_2. The activities of a large number of carbon assimilation enzymes and enzymes involved in carbohydrate metabolism/partitioning are regulated by redox changes mediated by thioredoxin. It is likely therefore that a CO_2-induced change in cellular redox state will also lead to changes in sugar metabolism and

Table 1. *The relative abundance of different hexokinase transcripts in young source leaves (rank 12) and mature source leaves (rank 3) in maize plants grown either in air (350 μL L^{-1} CO_2) or at high (700 μL L^{-1}) CO_2 and expressed as a $\log_2$ ratio.*

Probe ID	Description	$\log_2$ (rank 12/rank3)	
		350 μL L^{-1}	700 μL L^{-1}
Zm.5206.1.S1_x_at	Hexokinase	–0.571	–
Zm.5206.2.S1_a_at	Hexokinase	–0.710	–0.512
Zm.5206.3.A1_a_at	Hexokinase	–0.799	–0.525
Zm.8365.1.A1_at	Putative hexokinase 1	–	–0.608

signalling (Hendriks *et al.*, 2003; Kolbe *et al.*, 2005). The CO_2-induced effect on redox and sugar signalling might explain, at least in part, why atmospheric CO_2 enrichment can increase plant stress tolerance. Growth with elevated CO_2 enhanced the resistance of tobacco plants to viral infection (Matros *et al.*, 2006) and increased the concentration of anti-viral compounds even in sink tissues such as bulbs (Idso *et al.*, 2000).

3 Sugar sensing and signalling

Sucrose and hexoses (particularly glucose) are important local and systemic signalling molecules in plants (Gibson, 2005; León and Sheen, 2003; Pego *et al.*, 2000; Rolland *et al.*, 2002, 2006). Elevated sugar levels repress photosynthetic gene expression and confer feedback inhibition of photosynthesis, and induce storage processes (Jang and Sheen, 1994; Jang *et al.*, 1997; Moore *et al.*, 2003; Rolland *et al.*, 2006; Rook *et al.*, 2001). Hexokinases are evolutionarily conserved glucose sensors in eukaryotes (Rolland *et al.*, 2006). Hexokinase (HXK1) in Arabidopsis remains the only well-identified sugar sensor (Moore *et al.*, 2003) although there is evidence for HXK1-independent signalling pathways (Chen and Jones, 2004; Johnston *et al.*, 2007; Xiao *et al.*, 2000). Hexokinase-dependent glucose signalling involves ABA signal transduction as the two signalling pathways interact in a positive manner (Rolland *et al.*, 2006; Smeekens, 2000). The effects of developmental stage and growth CO_2 level on the abundance of hexokinase transcripts in maize leaves as illustrated in *Figure 2*, are shown in *Table 1*. Hexokinase transcripts are more abundant in the older source leaves. However, growth at high CO_2 decreased the age-dependent repression of hexokinase transcripts in the older source leaves (*Table 1*).

The glucose sensing HXK1 (Claeyssen and Rivoal, 2007) is particularly important in the sugar-mediated repression of photosynthetic genes such as the RuBisCO small subunit (Rbcs; Ehness *et al.*, 1997). CO_2 enrichment leads to general hexokinase-mediated decrease in *Rbcs* expression (Cheng *et al.*, 1998; Sun *et al.*, 2002). In addition to ABA (Huijser *et al.*, 2000), hexokinase-dependent sugar signalling pathways are also involved in the sensing of plant hormones such as auxin and cytokinin (Moore *et al.*, 2003), ethylene (Alonso *et al.*, 1999; Yanagisawa *et al.*, 2003), jasmonic acid (JA; Alonso *et al.*, 1999) and salicylate (Sheen *et al.*, 1999). Growth at high CO_2 had little effect on the abundance of transcripts involved in hormone metabolism (*Table 2*). However, transcripts encoding proteins involved in ABA and gibberellic acid metabolism tended to be higher in young source leaves compared to old source leaves (*Table 2*).

Table 2. *The relative abundance of phytohormone-related transcripts in young source leaves (rank 12) and mature source leaves (rank 3) in maize plants grown either in air (350 μL L^{-1} CO_2) or at high (700 μL L^{-1}) CO_2 and expressed as a log_2 ratio.*

Phytohormone/ Probe/ID	Gene	log_2 (rank 12/rank3)	
		350 μL L^{-1}	700 μL L^{-1}
ABA			
This hormone is important in plant stress and development responses. It also plays a key role in the regulation of stomatal closure			
Zm.6532.7.A1_x_at	ABA- and ripening-inducible-like protein	1.263	1.103
Zm.6532.1.A1_at	ABA- and ripening-inducible-like protein	0.988	1.053
Zm.6532.1.A1_a_at	ABA- and ripening-inducible-like protein	0.980	1.045
Zm.8843.1.A1_at	Putative ABA-responsive protein	0.615	–
Zm.17704.1.A1_at	Putative ABA-responsive protein	–1.127	–1.388
Zm.2757.1.A1_at	B73 terpene synthase 9 (TPS9)	2.326	2.681
Zm.18108.1.S1_at	Delprim terpene synthase 4 (TPS4)	1.321	1.533
Auxin (indoleacetic acid)			
This hormone is important in organ initiation, cell division and cell elongation. It also delays leaf senescence			
Zm.283.1.S1_at	Bx1: Indole synthase	2.294	1.439
Zm.6178.1.A1_at	Auxin efflux carrier protein-like	–0.649	–0.690
Zm.8331.1.A1_at	Auxin-induced protein-related-like protein	0.600	0.612
Zm.5530.2.A1_at	Auxin response transcription factor 3-like protein	1.292	1.347
Zm.2574.1.A1_at	Auxin response factor 2	–0.666	–0.810
Zm.6483.1.A1_at	Putative auxin response factor 7a	–1.880	–2.242
Ethylene			
This hormone is important in plant defence responses and in senescence			
Zm.11590.1.A1_at	Ethylene-forming-enzyme-like dioxygenase	1.568	1.325
Zm.8181.1.A1_at	Putative ethylene-forming enzyme	–1.603	–1.645
Zm.17428.1.A1_at	Putative ethylene-inducible protein	1.390	1.373
Zm.5042.1.A1_a_at	W64A acc oxidase (ACO20)	0.769	0.966
Zm.18900.1.S1_at	acc oxidase (ACO20)		–1.119
Zm.8714.1.A1_at	acc oxidase (ACO31)	–0.519	–
Zm.7909.1.A1_at	B73 acc oxidase (ACO31)	–0.736	–
Zm.12317.1.S1_at	Putative ethylene-responsive element binding factor	–	–0.921
Gibberellin			
This hormone is important in plant growth and defence responses			
Zm.6316.1.A1_at	Putative ent-kaurene synthase	1.118	0.892
Zm.228.1.S1_at	Kaurene synthase A	0.748	0.787

The HXK-independent pathways involve hexose transporters that function as glucose sensors (Ashe *et al.*, 2000; Rolland *et al.*, 2002). Plants also have glucose-sensing systems similar to the yeast Gpr1-Gpa2 system (Chen and Jones, 2004; Perfus-Barbeoch *et al.*, 2004).

Sugar signalling is also an important component of plant stress responses (Rolland *et al.*, 2002). Sugar availability influences the expression of genes involved in secondary metabolism and defence as well as primary metabolism (Price *et al.*, 2004; Rolland *et al.*, 2006; Thibaud *et al.*, 2004). Glucose and sucrose enhance the expression of pathogen- and stress-response genes (Roitsch *et al.*, 2003; Tajima and Koizumi, 2006) such as potato inhibitor II (*pinII*), chalcone synthase, superoxide dismutase and PR proteins (Roitsch *et al.*, 2003; Tsukaya *et al.*, 1991). Similarly, the local and systemic expression of genes associated with carbohydrate metabolism is modified upon exposure to stresses such as wounding (Quilliam *et al.*, 2006).

4 Sugar and hormone signalling in the control of growth and defence responses

Cross-talk between sugar and hormone signalling pathways has been demonstrated for gibberellins (GA; Gibson, 2004; Perata *et al.*, 1997), ethylene (Gazzarrini and McCourt, 2001), auxin and cytokinin (Hartig and Beck, 2006). Moreover, the different types of hormones found in plants frequently interact with each other in a complex manner to facilitate effective growth and defence responses. Many opposing hormone activities are regulated via modulation of Skp, Cullin, F-box containing complex (SCF) Enztme (E3) ubiquitin ligase complexes and protein degradation (Dreher and Callis, 2007; Rogg and Bartel, 2001; Thomann *et al.*, 2005). The DELLA family of proteins are transcriptional regulators that restrain plant growth and they also have important roles in protection against a range of different stresses, including nutrient (phosphate and N) deficiencies (Feng *et al.*, 2008). They are thus an important central hub for the integration of growth and defence responses in plants. The lifetime of DELLA proteins is controlled by hormones such as GA and by environmental factors such as light through interactions with phytochrome-interacting transcription factor 3 (PIF3; Feng *et al.*, 2008).

GAs are essential endogenous regulators of plant growth and development. They stimulate elongation or expansion of organs via enhancement of cell elongation and, in some cases, also cell division. Furthermore, GAs may induce developmental switches, such as between the juvenile and adult phases or between vegetative and reproductive development. They also promote differentiation in meristems where they suppress the maintenance of stem cells. This mode of action is antagonistic to that of cytokinins, whose action they inhibit. GAs promote the degradation of DELLA proteins by favouring binding of DELLAs to a soluble, nuclear protein gibberellin-insensitive dwarf 1 (GID1), thereby facilitating the association of DELLAs with the F-box component of an SCF E3 ubiquitin ligase. The resulting polyubiquitination targets DELLAs for degradation by the 26S proteasome. Protein degradation via the ubiquitin/26S proteosome system is important in regulating responses to other plant hormones such as auxin, ethylene and brassinosteroids (Binder *et al.*, 2007; Zhang *et al.*, 2007).

Gibberellin signalling responds to environmental cues, including changes in light conditions, temperature or stress, allowing GAs to translate these extrinsic signals into developmental changes. Stress resistance is increased when tissue GA contents are decreased by treatment with biosynthesis inhibitors or by mutation (Achard *et al.*, 2006). Arabidopsis mutants with either reduced GA contents or containing a GA-insensitive form of DELLA proteins are more tolerant of salt stress than wild-type plants, while plants with loss-of-function DELLA mutations have increased salt susceptibility (Achard *et al.*, 2006). Furthermore, over-expression of C-repeat binding protein (CBF)/apetala (AP) transcription factors, which are involved in many stress responses, results in GA-reversible growth inhibition and a reduction in bioactive GA concentration. Nutrient stress also impacts on the GA-DELLA signalling system. Phosphate limitation-dependent changes in root architecture, particularly increased root hair growth, are at least partly the result of a reduction in the levels of bioactive GAs, leading to DELLA accumulation (Jiang *et al.*, 2007). Similarly, we have recently obtained evidence that GA metabolism is also influenced by N availability, particularly when the rate of primary N assimilation is enhanced as a result of alterations in cellular redox state caused by the loss of mitochondrial Complex I (Pellny *et al.*, 2008). Complex I is the major NADH dehydrogenase of the respiratory electron transport chain. However, when it is absent through mutation, plants can sustain electron flow via the engagement of non-phosphorylating alternative dehydrogenases that have a low affinity for NADH. The resultant change in mitochondrial NADH use causes a large increase in tissue pyridine nucleotide contents and the resultant enhanced availability of NAD(P)H drives N assimilation and amino acid production. Hence, plants lacking Complex I have a N-rich phenotype, particularly when grown under N limitation, as well as enhanced stress tolerance. Moreover, Complex I deficiency results in a conditional slow-growth extended-lifespan phenotype (Pellny *et al.*, 2008). The effect of CO_2 enrichment on the regulated stabilization of DELLAs is unknown but it is likely that sugar, redox and hormone changes arising from growth with high CO_2 have an impact on this central system controlling plant growth/defence relationships.

5 ABA and stress sugar signalling

Accumulating evidence of the extensive cross-talk between the sugar and ABA signalling pathways has suggested that they have a pivotal relationship in plant growth and defence responses with multiple sites of reciprocal control. ABA has important functions in plant development, metabolism and physiology, particularly through the control of stomatal closure, and also in biotic and abiotic stress responses (Kaliff *et al.*, 2007; Leung and Giraudat, 1998; Rock, 2000). The germination and seedling establishment phases of plant growth are sensitive to both ABA and sugars, which together control developmental events (Rolland *et al.*, 2006; Sheen *et al.*, 1999; Xiao *et al.*, 2000). Moreover, ABA synthesis and signalling are important in the control of lateral root growth in response to sugar and nitrogen signals (De Smet *et al.*, 2003; Signora *et al.*, 2001). The reciprocal action of ABA and GA on plant growth and defence is well documented, the ABA/GA balance interfacing with cellular sugar metabolism (Arenas-Huertero *et al.*, 2000; Rook *et al.*, 2001). Sugars and ABA activate 'sink-related' genes (Morikami *et al.*, 2005; Ohto *et al.*, 1992), while GA has the opposite effect (Ehness and Roitsch, 1997). ABA tends to inhibit phloem loading while GA promotes assimilate export. ABA also down-regulates other hormones such as JA and salicylic acid (SA), an interaction that is very important in plant pathogen responses (Beckers and Spoel,

2006; Robert-Seilaniantz *et al.*, 2007). Many sugar-induced genes are also responsive to jasmonate, which is a major signalling molecule activated upon stress perception (Berger *et al.*, 1995; Mason *et al.*, 1992).

Much of our current understanding of the pathway of glucose signalling has come from the analysis of Arabidopsis mutants that have impaired responses to sugar (Cheng *et al.*, 2002; León and Sheen, 2003). Similarly, these mutants provided evidence for cross-talk between ABA and metabolic signalling pathways (Arenas-Huertero *et al.*, 2000; Smeekens, 2000). ABA signal transduction pathways operate in HXK-dependent glucose signalling (Dijkwel *et al.*, 1997; Huijser *et al.*, 2000; Kim *et al.*, 2004). For example, the *sun6/abi4* mutant is insensitive to sucrose, glucose and mannose such that photosynthesis genes are not repressed in the presence of high sugars (Finkelstein *et al.*, 1998; Huijser *et al.*, 2000).

The pathways of ABA- and HXK-dependent glucose signal transduction are extensively interlinked with multiple sites of reciprocal control (Rook *et al.*, 2001). The sucrose non-fermenting-1 (SNF1)-related protein kinases (SnRKs) are involved in both the ABA and stress signalling. ABA and GA differentially regulate the transcription of SnRK genes (Bradford *et al.*, 2003). SnRK1 is a metabolic regulator that is activated in response to high cellular sucrose/low cellular glucose and regulates carbon metabolism through the modulation of enzyme activity and gene expression. It is structurally and functionally related to the yeast protein kinase, SNF1, which gives the family its name, and to mammalian AMP-activated protein kinase. SnRK1 phosphorylates and inactivates 3-hydroxy-3-methylglutaryl-Coenzyme A reductase (HMG-CoA reductase), sucrose phosphate synthase and nitrate reductase (NR). In addition, SnRK1 is required for redox modulation of ADP-glucose pyrophosphorylase activity in response to sucrose and, like its fungal and animal counterparts, is involved in regulating the expression of genes involved in carbon metabolism. A further link between plant stress responses and sugar signalling involves trehalose metabolism and SnRKs (Rolland *et al.*, 2006). Trehalose is widely believed to be a major determinant of stress resistance in many organisms including plants, and trehalose 6-phosphate is a regulator of sugar utilization (Paul and Pellny, 2003; Schluepmann *et al.*, 2003; Van Dijken *et al.*, 2004). The trehalose biosynthetic enzymes serve a regulatory function in the control of HXK activity and signalling in yeast (Thevelein and Hohmann, 1995). While a number of studies have shown that trehalose metabolism is integrated with sugar sensing in plants (Eastmond *et al.*, 2002; Penna, 2003; Wingler *et al.*, 2000), the sites of trehalose 6-phosphate action remain poorly characterized. For example, HXKs are insensitive to trehalose 6-phosphate (Paul *et al.*, 2001; Schluepmann *et al.*, 2003). While trehalose induces the expression of sucrose synthase and alkaline invertase activities (Müller *et al.*, 1998), trehalose 6-phosphate modifies AGPase activity by promoting the redox activation of the enzyme in response to sucrose, thereby controlling starch synthesis (Hendriks *et al.*, 2003; Kolbe *et al.*, 2005).

Transcriptome analyses have identified more than a thousand genes that are differentially regulated by ABA. Studies with mutants have shown that these are regulated by a large number of ABA signalling components (Hoth *et al.*, 2002; Sánchez, *et al.* 2004). These include two ABA-signalling protein phosphatase 2C (PP2C) isoforms (*abi1* and *abi2*) that have overlapping functions. However, certain transcription-factor-gene families are differentially induced in the two genotypes (Hoth *et al.*, 2002). In particular, the ABA-insensitive (ABI4) transcription factor is differentially modulated in the *abi1* and *abi2* mutants (Söderman *et al.*, 2000). ABI4 is considered to be a 'master-switch' in the regulation of nuclear genes in response to developmental cues as well as environmental signals (Koussevitzky *et al.*, 2007).

A recent study has shown that human granulocytes can also produce and respond to ABA (Bruzzone *et al.*, 2007). The similar cellular and molecular responses to ABA observed in human and plant cells suggest common perception and signalling mechanisms (Bruzzone *et al.*, 2007). ABA signalling events linked to stomatal movement and to plant defence responses involve the generation of reactive oxygen species (ROS), changes in cellular redox state and ROS-induced oscillations of cytosolic Ca^{2+} (Klüsener *et al.*, 2002). The ABA-dependent formation of the secondary messenger hydrogen peroxide is dependent on the production of superoxide via the NADPH oxidases respiratory burst oxidase homologue (Rboh) D and RbohF (Kwak *et al.*, 2003; Torres and Dangl, 2005; Torres *et al.*, 2002).

6 Cellular redox state as a convergence regulator of information controlling growth and stress responses

Reactive oxygen species produced as a result of energy metabolism are important components of survival signalling in plants, as in animals (Gong *et al.*, 2007; Pastori and Foyer, 2002; Urbano *et al.*, 2005). Cellular redox state plays a crucial role as a convergence regulator of biotic and abiotic stress responses (Fujita *et al.*, 2006). Plants maintain a robust antioxidant defence system that buffers the cell in most instances against the extreme oxidation that triggers programmed cell death (Noctor and Foyer, 1998). ROS generated in response to hormone action often influence processes that are also regulated or affected by sugars (Couée *et al.*, 2006; Loreti *et al.*, 2005; Price *et al.*, 2004). Biotic and abiotic stresses favour ROS accumulation (Polle, 2001) and lead to specific effects on gene expression (Mittler, 2002; Mittler *et al.*, 2004). ROS accumulation has been observed following exposure to drought, desiccation, salt (Tsugane *et al.*, 1999), chilling (Prasad *et al.*, 1994), pathogen attack (Grant and Loake, 2000; Hammond-Kosack and Jones, 1996) and high light stress (Polle, 2001). While supplying sugars such as glucose to plant cells often favours over-reduction (Ryu *et al.*, 2004; Zhang *et al.*, 2000), sugar starvation can enhance ROS accumulation and increases the expression of antioxidant genes such as catalase (Contento *et al.*, 2004). However, sugars also favour repression of certain metabolic processes that generate ROS, such as fatty acid mobilization and peroxisomal β-oxidation (Brouquisse *et al.*, 1991).

Reactive oxygen species can also activate plant defence pathways (Apel and Hirt, 2004; Mittler 2002; Orozco-Cardenas and Ryan, 1999). While ROS accumulation is often viewed as an indicator of stress, it can also play a central role in survival signalling (Torres and Dangl, 2005; Torres *et al.*, 2002). This view is supported by their generation as secondary messengers during growth and movement responses initiated by plant hormones such as auxin and ABA. Such signalling cascades often employ protein kinases, with calcium ions and ROS as secondary messengers. Sugars play a central role in the metabolic processes that generate ROS and in plant responses to environmental stress (Roitsch, 1999). The interfacing relationships between sugars, ROS and antioxidant processes implicate sugar signalling and sugar-modulated gene expression in the control of oxidative stress (Loreti *et al.*, 2005; Nishikawa *et al.*, 2005; Sulmon *et al.*, 2004).

The innate immune response of plants involves common elements of signal transduction and/or amplification, particularly the activation of enzymes such as NADPH oxidases that leads to an 'oxidative burst' of ROS generation, enzymes involved in protein turnover and enzymes involved in hormone metabolism and signalling (Apel and Hirt, 2004; Bolwell, 1999; Glazebrook, 1999; Hayashi *et al.*, 2001). Sucrose and stress cause specific interactions and patterns of gene expression different from sugar-only responses (Loreti *et al.*, 2005; Price *et al.*, 2004; Thibaud *et al.*, 2004). For example, there is a clear interaction between sugars and anoxia, where the expression of heat-shock proteins and oxidative-stress-response genes were shown to be induced by an interaction between the two treatments (Loreti *et al.*, 2005). ROS and glutathione participate in plant pathogen responses that lead to the expression of pathogenesis-related (PR) proteins (Fellbrich *et al.*, 2002). The expression of PR proteins is enhanced by exogenous sugar treatments and by an increase in endogenous sugar levels in a signal-dependent manner (Badur *et al.*, 1994; Herbers and Sonnewald, 1998; Thibaud *et al.*, 2004). Many stress situations, such as pathogen challenge, drought, salt stress, ABA treatment, low temperatures, or excess light, which favour ROS accumulation also lead to sugar accumulation, which is generally considered to be an adaptive response to stress conditions (Roitsch, 1999). Like the sugar/hormone interface, sugar and redox signalling pathways appear to have many interconnections in plant growth and defence (Couée *et al.*, 2006).

7 Regulation of protein turnover

The selective degradation of proteins is an important aspect of the acclimatory responses that allow plants to optimize metabolism and growth with respect to changing environmental conditions, such as atmospheric CO_2 availability. One important mechanism involved in the regulation of the rate of protein turnover is selective inhibition by proteases (Solomon *et al.*, 1999). The expression of endogenous protease inhibitors, such as cysteine protease inhibitors (phytocystatins) and serine protease inhibitors (serpins), is increased in response to stresses such as wounding and pathogens and is orchestrated by the relative abundance of JA, ethylene and salicylic acid (Bent, 1996; Delaney *et al.*, 1994).

Phytocystatins fulfil multiple roles in plant growth and defence responses in plants, where they regulate the activities of cysteine proteases (CPs). Cysteine proteases are involved in the control of plant development, and participate in a wide variety of processes such as the degradation of seed storage proteins, fruit ripening and legume nodule development. Some CPs have specific characteristics such as a C-terminal KDEL motif. This motif, which is an endoplasmic-reticulum retention signal for soluble proteins, allows cysteine proteinase propeptides to be stored in KDEL vesicles before transport to vacuoles through a Golgi complex-independent route. The papain-type CPs are particularly important in the senescence of leaves and flowers and in germinating seeds where the breakdown of proteins is required for nutrient remobilization to other developing parts of the plant (Callis, 1995). The expression of certain CP genes, such as the senescence-associated gene, *SAG12*, which is expressed exclusively in senescent leaves, can be used as markers for senescence. The activities of CPs are enhanced in response to stresses such as wounding, cold and drought and also in programmed cell death (Solomon *et al.*, 1999). Phytocystatin expression is also induced by wounding and by oxylipins such as JA (Bolter, 1993; Botella *et al.*, 1996). Phytocystatins also exert effects on exogenous proteases such as those in the guts of phytophagous organisms (Liang *et al.*, 1991; Zhao *et al.*, 1996). Ectopic expression of phytocystatins in transgenic plants enhances resistance to insects and nematodes (Leple *et al.*, 1995; Urwin *et al.*, 1995).

The expression of photosynthesis genes such as those encoding the chlorophyll *a*/*b* binding proteins and RuBisCO decreases during leaf senescence. There is an important redistribution of the N stored in proteins such as RuBisCO to drive protein accumulation in storage tissues. Although RuBisCO is the major protein in leaves, much uncertainty still remains concerning the mechanisms regulating RuBisCO turnover, with different pathways described in the chloroplasts and the vacuole. The constitutive expression of the rice cystatin, OC-1, in transgenic tobacco plants has shown that CPs are important in RuBisCO degradation particularly during senescence and in stress conditions (Prins *et al.*, 2008b). Such studies have shown that the RuBisCO protein is transported from the chloroplasts in vesicles containing RuBisCO-degrading proteases (Prins *et al.*, 2008b). Since CPs have now been shown to be central to the turnover of photosynthetic proteins, it is probable that they are important in the acclimation of photosynthesis to CO_2 enrichment (Prins *et al.*, 2008b).

Serpins regulate serine protease cascades by forming suicide substrates with their cognate substrates (Laskowski and Kato, 1980). They are also important in plant defence responses, as inhibition of serpin expression modifies SA- and JA-dependent signalling as well as ROS accumulation (Bowler and Fluhr, 2000; Budai-Hadrian *et al.*, 2006). It has long been recognized that serpins inhibit serine proteases in the guts of phytophagous insects and thus they have an important anti-nutritional action (Applebaum, 1985; Brovosky, 1986). The potato inhibitor II, which inhibits both trypsin and chymotrypsin (Broadway and Duffey, 1986) accumulates during tuber development and increases in response to wounding (Graham and Ryan, 1981; Hollander-Czytko *et al.*, 1985). Jasmonic acid and sugars, which have been shown to induce serpin expression, may be involved in this response (Koiwa *et al.*, 1997). Serine proteases are also important in the acclimation of plants to high CO_2. The expression of a Bowman-Birk-type inhibitor and a serpin was modified in maize leaves from plants grown with CO_2 enrichment compared to those grown in air (Prins *et al.*, 2008a). The expression of the genes encoding the maize Bowman-Birk-type inhibitor and the maize serpin was modified by sugars and by oxidative stress (Prins *et al.*, 2008a) suggesting that sugar and redox signals might participate in serpin regulation in response to high CO_2.

8 Conclusions

Photosynthesis not only supports plant biomass production but it is also an important driver of ecosystem sustainability in the face of climate change. Through photosynthesis and photorespiration, the amount of CO_2 in the atmosphere has a major impact on plant growth and biomass accumulation, as it not only alters sugar availability and carbohydrate metabolism but also cellular redox status and related hormone signalling (*Figure 3*). The signal transduction network that conveys information concerning CO_2 availability in order to modulate gene expression and protein turnover is thus highly complex, involving interactive pathways of sugar, hormone and redox signalling with multiple sites of reciprocal control. CO_2 signalling regulates plant growth, morphology

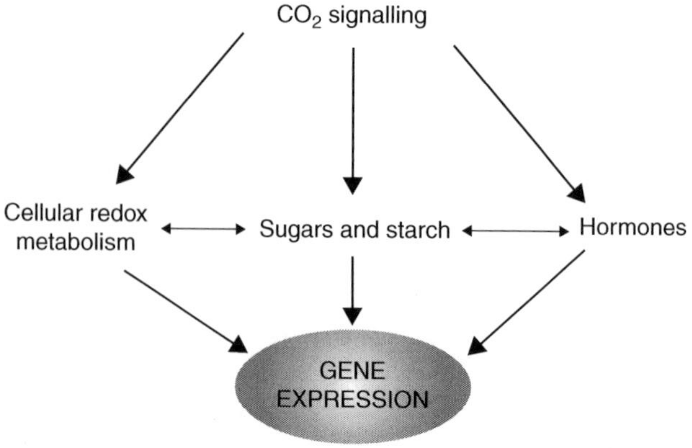

Figure 3. *Diagrammatic representation of the metabolic and signalling pathways contributing to the CO_2 acclimation response in plants.*

and development. While growth with CO_2 enrichment can lead to premature leaf senescence if sugars accumulate to levels that inhibit photosynthetic gene expression (Dai *et al.*, 1999; Diaz *et al.*, 2005; Moore *et al.*, 2003; Xiao *et al.*, 2000) this response is not observed in the leaves of C_4 plants where photosynthesis rates are similar in air and high-CO_2-grown plants. Redox-dependent mechanisms linked to photorespiration and the photosynthetic and respiratory electron transport chains are also implicated in the CO_2 acclimation responses that control gene expression and protein turnover rates. Cellular redox regulation forms an integrating hub of information from metabolism and the environment, and it is central to the signalling network that controls gene expression and related growth responses in C_3 and C_4 plants during acclimation to high CO_2.

References

Aarts, M.G.M. and Fiers, M.W.E.J. (2003) What drives plant stress genes? *Trends Plant Sci.* **8:** 99–102.

Achard, P., Cheng, H., De Grauwe, L., Decat, J., Schoutteten, H., Moritz, T., Van Der Straeten, D., Peng, J. and Harberd, N.P. (2006) Integration of plant responses to environmentally activated phytohormonal signals. *Science* **301:** 91–94.

Alonso, J.M., Hirayama, T., Roman, G., Nourizadeh, S. and Ecker, J.R. (1999) EIN2, a bifunctional transducer of ethylene and stress responses in *Arabidopsis. Science* **284:** 2148–2152.

Apel, K. and Hirt, H. (2004) Reactive oxygen species: metabolism, oxidative stress, and signal transduction. *Annu. Rev. Plant Biol.* **55:** 373–399.

Applebaum, S.W. (1985) Biochemistry of digestion. In: *Comprehensive Insect Physiology, Biochemistry and Pharmacology*, Vol. 4 (eds G.A. Kerkot and L.I. Gilbert). Pergamon Press, New York, NY, pp. 279–311.

Arenas-Huertero, F., Arroyo, A., Zhou, L., Sheen, J. and Leon, P. (2000) Analysis of Arabidopsis glucose insensitive mutants, *gin5* and *gin6*, reveals a central role of the plant hormone ABA in regulation of plant vegetative development by sugar. *Gene Dev.* **14:** 2085–2096.

Ashe, M.P., De Long, S.K. and Sachs, A.B. (2000) Glucose depletion rapidly inhibits translation initiation in yeast. *Mol. Biol. Cell* **11:** 833–848.

Badur, R., Herbers, K., Monke, G., Ludewig, F. and Sonnewald, U. (1994) Induction of pathogenesis-related proteins in sugar accumulating tobacco leaves. *Photosynthetica* **30:** 575–582.

Beckers, G.J.M. and Spoel, S.H. (2006) Fine-tuning plant defence signalling: salicylate versus jasmonate. *Plant Biol.* **8:** 1–10.

Bent, A.F. (1996) Plant disease resistance genes: function meets structure. *Plant Cell* **8:** 1757–1771.

Berger, S., Bell, E., Sadka, A. and Mullet, J.E. (1995) *Arabidopsis thaliana Atvsp* is homologous to soybean *VspA* and *VspB*, genes encoding vegetative storage protein acid phosphatases, and is regulated similarly by methyl jasmonate, wounding, sugars, light and phosphate. *Plant Mol. Biol.* **27:** 933–942.

Binder, B.M., Walker, J.M., Gagne, J.M., Emborg, T.J., Hemmann, G., Blecker, A.B. and Viestra, R.D. (2007) The Arabidopsis EIN3 binding F-Box proteins EBF1 and EBF2 have distinct but overlapping roles in ethylene signalling. *Plant Cell* **19:** 509–523.

Bolter, C.J. (1993) Methyl jasmonate induces papain inhibitor(s) in tomato leaves. *Plant Physiol.* **103:** 1347–1353.

Bolwell, G.P. (1999) Role of active oxygen species and NO in plant defence responses. *Curr. Opin. Plant Biol.* **2:** 287–294.

Botella, M.A., Xu, Y., Prabha, T.N., Zhao, Y., Narasimhan, M.L., Wilson, K.A., Nielsen, S.S., Bressan, R.A. and Hasegawa, P.M. (1996) Differential expression of soybean cysteine proteinase inhibitor genes during development and in response to wounding and methyl jasmonate. *Plant Physiol.* **112:** 1201–1210.

Boudsocq, M. and Lauriere, C. (2005) Osmotic signaling in plants. Multiple pathways mediated by emerging kinase families. *Plant Physiol.* **138:** 1185–1194.

Bowler, C. and Fluhr, R. (2000) The role of calcium and activated oxygens as signals for controlling cross-tolerance. *Trends Plant Sci.* **5:** 241–246.

Bradford, K.J., Downie, A.B., Gee, O.H., Alvardo, V., Yang, H. and Dahal, P. (2003) Abscisic acid and gibberellin differentially regulate expression of genes of the SNF1-related kinase complex in tomato seeds. *Plant Physiol.* **132:** 1–17.

Broadway, R.M. and Duffey, S.S. (1986) The effect of dietary protein on the growth and digestive physiology of larval *Heliothis zea* and *Spodoptera exigua*. *J. Insect Physiol.* **32:** 673–680.

Brouquisse, R., James, F., Raymond, P. and Pradet, A. (1991) Study of glucose starvation in excised maize root tips. *Plant Physiol.* **96:** 619–626.

Brovosky, D. (1986) Proteolytic enzymes and blood digestion in the mosquito *Culex nigripalpus*. *Arch. Insect Biochem.* **3:** 147–160.

Bruzzone, S., Moreschi, I., Usai, C., *et al.* (2007) Absiscic acid is an endogenous cytokine in human granulocytes with cyclic ADP-ribose as a second messenger. *Proc. Natl Acad. Sci. USA* **104:** 5759–5764.

Budai-Hadrian, O., Davydov, O., Wasternack, C. and Fluhr, R. (2006) SAS1 Serpin is at a node of cross-talk between jasmonic and salicylic acid pathways in Arabidopsis. *Israel J. Plant Sci.* **54:** 55–86.

Callis, J. (1995) Regulation of protein degradation. *Plant Cell* **7:** 845–857.

Chen, J.G. and Jones, A.M. (2004) AtRGS1 function in *Arabidopsis thaliana*. *Method Enzymol.* **389:** 338–350.

Cheng, S.H., Moore, B.D. and Seemann, J.R. (1998) Effects of short- and long-term treatments at elevated CO_2 on the expression of Rubisco genes and leaf carbohydrate accumulation in *Arabidopsis thaliana* (L.) Henyh. *Plant Physiol.* **116:** 715–723.

Cheng, W.H., Endo, A., Penney, J., *et al.* (2002) A unique short-chain dehydrogenase/reductase in Arabidopsis glucose signaling and abscisic acid biosynthesis and functions. *Plant Cell* **14:** 2723–2743.

Chinnusamy, V., Schumaker, K. and Zhu, J.K. (2004) Molecular genetic perspectives on cross-talk and specificity in abiotic stress signalling in plants. *J. Exp. Bot.* **55:** 225–236.

Claeyssen, E. and Rivoal, J. (2007) Isozymes of plant hexokinase: occurrence, properties and functions. *Phytochemistry* **68:** 709–731.

Contento, A.L., Kim, S.J., Bassham, D.C. (2004) Transcriptome profiling of the response of *Arabidopsis* suspension culture cells to Suc starvation. *Plant Physiol.* **135:** 2330–2347.

Couée, I., Sulmon, C., Gouesbet, G. and Amrani, A. (2006) Involvement of soluble sugars in reactive oxygen species balance and responses to oxidative stress in plants. *J. Exp. Bot.* **57:** 449–459.

Cox, P.M., Betts, R.A., Jones, C.D., Spall, S.A. and Totterdell, I.J. (2000) Acceleration of global warming due to carbon-cycle feedbacks in a coupled climate model. *Nature* **408:** 184–187.

Dai, N., Schaffer, A., Petreikov, M., Shahak, Y., Giller, Y., Ratner, K., Levine, A. and Granot, D. (1999) Overexpression of Arabidopsis hexokinase in tomato plants inhibits growth, reduces photosynthesis, and induces rapid senescence. *Plant Cell* **11:** 1253–1266.

De Smet, I., Signora, L., Beeckman, T., Inze, D., Foyer, C.H. and Zhang, H. (2003) An ABA-sensitive lateral root developmental checkpoint in Arabidopsis. *Plant J.* **33:** 543–555.

Delaney, T.P., Uknes, S., Vernooij, B., *et al.* (1994) A central role of salicylic acid in plant disease resistance. *Science* **266:** 1247–1250.

Diaz, C., Purdy, S., Christ, A., Morot-Gaudry, J.F., Wingler, A. and Masclaux-Daubresse, C. (2005) Characterization of markers to determine the extent and variability of leaf senescence in Arabidopsis. A metabolic profiling approach. *Plant Physiol.* **138:** 898–908.

Dijkwel, P.P., Huijser, C., Weisbeek, P.J., Chua, N.-H. and Smeekens, S. (1997) Sucrose control of phytochrome A signalling in *Arabidopsis*. *Plant Cell* **9:** 583–595.

Dreher, K. and Callis, J. (2007) Ubiquitin, hormones and biotic stress in plants. *Ann. Bot.* **99:** 787–822.

Driscoll, S.P., Prins, A., Olmos, E., Kunert, K.J. and Foyer, C.H. (2006) Specification of adaxial and abaxial stomata, epidermal structure and photosynthesis to CO_2 enrichment in maize leaves. *J. Exp. Bot.* **57:** 381–390.

Eastmond, P.J., van Dijken, A.J., Spielman, M., Kerr, A., Tissier, A.F., Dickinson, H.G., Jones, J.D., Smeekens, S.C. and Graham, I.A. (2002) Trehalose-6-phosphate synthase 1,which catalyses the first step in trehalose synthesis, is essential for Arabidopsis embryo maturation. *Plant J.* **29:** 225–235.

Ehness, R. and Roitsch, T. (1997) Co-ordinated induction of mRNAs for extracellular invertase and a glucose transporter in *Chenopodium rubrum* by cytokinins. *Plant J.* **11:** 539–548.

Ehness, R., Ecker, M., Godt, D.E. and Roitsch, T. (1997) Glucose and stress independently regulate source and sink metabolism and defense mechanisms via signal transduction pathways involving protein phosophorylation. *Plant Cell* **9:** 1825–1841.

Evans, J.R. (2004) Developmental constraints on photosynthesis: effects of light and nutrition. In: *Advances in Photosynthesis and Respiration: Photosynthesis and the Environment*, Vol. 5 (ed. N.R. Baker). Springer, Berlin, pp. 281–304.

Fellbrich, G., Romanski, A., Varet, A., *et al.* (2002) *NPP1*, a *Phytophthora*-associated trigger of plant defense in parsley and Arabidopsis. *Plant J.* **32:** 375–390.

Feller, U., Anders, I. and Mae, T. (2007) Rubiscolytics: fate of Rubisco after its enzymatic function in a cell is terminated. *J. Exp. Bot.* **59:** 1615–1624.

Feng, S., Martinez, C., Gusmaroli, G., *et al.* (2008) Coordinated regulation of *Arabidopsis thaliana* development by light and gibberellins. *Nature* **451:** 475–479.

Finkelstein, R.R. and Gibson, S.I. (2002) ABA and sugar interactions regulating development: cross-talk or voices in a crowd? *Curr. Opin. Plant Biol.* **5:** 26–32.

Finkelstein, R.R., Wang, M.L., Lynch, T.J., Rao, S. and Goodman, H.M. (1998) The Arabidopsis abscisic acid response locus *ABI4* encodes an APETALA2 domain protein. *Plant Cell* **10:** 1043–1054.

Foyer, C.H. and Noctor, G. (2000) Oxygen processing in photosynthesis: regulation and signalling. *New Phytol.* **146:** 359–388.

Foyer, C.H. and Noctor, G. (2003) Redox signalling through chloroplasts and mitochondria in photosynthetic cells. *Physiol. Plant.* **119:** 355–364.

Fujita, M., Fujita, Y., Noutoshi, Y., Takahashi, F., Narusaks, Y., Yamaguchi-Shinozaji, K. and Shinozaki, K. (2006) Cross talk between abiotic and biotic stress responses: a current view from the point of convergence in the stress signalling network. *Curr. Opin. Plant Biol.* **9:** 1–7.

Gazzarrini, S. and McCourt, P. (2001) Genetic interactions between ABA, ethylene and sugar signaling pathways. *Curr. Opin. Plant Biol.* **4:** 387–391.

Gibson, S.I. (2004) Sugar and phytohormone response pathways: navigating a signalling network. *J. Exp. Bot.* **55:** 253–264.

Gibson, S.I. (2005) Control of plant development and gene expression by sugar signalling. *Curr. Opin. Plant Biol.* **8:** 93–102.

Glazebrook, J. (1999) Genes controlling expression of defense responses in *Arabidopsis. Curr. Opin. Plant Biol.* **2:** 280–286.

Gong, M., Hay, S., Marshall, K.R., Munro, A.W. and Scrutton, N.S. (2007) DNA binding suppresses human AIF-M2 activity and provides a connection between redox chemistry, reactive oxygen species, and apoptosis. *J. Biol. Chem.* **282:** 30331–30340.

Graham, J.S. and Ryan, C.A. (1981) Accumulation of metallocarboxy-peptidase inhibitor in leaves of wounded potato plants. *Biochem. Biophys. Res. Comm.* **101:** 1164–1170.

Grant, J.J. and Loake, G.J. (2000) Role of reactive oxygen intermediates and cognate redox signalling in disease resistance. *Plant Physiol.* **124:** 21–29.

Gray, J.E., Holroyd, G.H., van der Lee, F.M., *et al.* (2000) The *HIC* signalling pathway links CO_2 perception to stomatal development. *Nature* **408:** 713–716.

Grime, J.P. (2001) *Plant Strategies: Vegetation Processes and Ecosystem Properties.* Wiley, Chichester.

Hammond-Kosack, K.E. and Jones, J.D.G. (1996) Resistance gene-dependent plant defense responses. *Plant Cell* **8:** 1773–1791.

Hartig, K. and Beck, E. (2006) Crosstalk between auxin, cytokinins, and sugars in the plant cell cycle. *Plant Biol.* **8:** 389–396.

Hayashi, Y., Yamada, K., Shimada, T., Matsushima, R., Nishizawa, N., Nishimura, M. and Hara-Nishimura, I. (2001) A proteinase-storing body that prepares for cell death or stresses in the epidermal cells of Arabidopsis. *Plant Cell Physiol.* **42:** 894–899.

Hendriks, J.H.M., Kolbe, A., Gibon, Y., Stitt, M. and Geigenberger, P. (2003) ADP-glucose pyrophosphorylase is activated by posttranslational redox-modification in response to light and to sugars in leaves of Arabidopsis and other plant species. *Plant Physiol.* **133:** 838–849.

Herbers, K. and Sonnewald, U. (1998) Altered gene expression brought about by inter- and intracellular formed hexoses and its possible implications for plant–pathogen interactions. *J. Plant Res.* **111:** 323–328.

Ho, S.L., Chao, Y.C., Tong, W.F. and Yu, S.M. (2001) Sugar co-ordinately and differentially regulates growth- and stress-related gene expression via a complex signal transduction network and multiple control mechanisms. *Plant Physiol.* **125:** 877–890.

Hollander-Czytko, H., Andersen, J.L. and Ryan, C.A. (1985) Vacuolar localization of wound-induced carboxy peptidase inhibitor in potato leaves. *Plant Physiol.* **78:** 76–79.

Hoth, S., Morgante, M., Sanchez, J., Hanafey, M., Tingey, S. and Chua, N. (2002) Genome-wide expression profiling in *Arabidopsis thaliana* reveals new targets of abscisic acid and largely impaired gene regulation in the abi1-1-1 mutant. *J. Cell Sci.* **115:** 4891–4900.

Huijser, C., Kortstee, A., Pego, J.V., Wisman, E., Weisbeek, P. and Smeekens, S.C.M. (2000) The *Arabidopsis SUCROSE UNCOUPLED-6* gene is identical to *ABSCISIC ACID INSENSITIVE-4*: involvement of abscisic acid in sugar responses. *Plant J.* **23:** 577–585.

Idso, S.B., Kimball, B.A., Pettit, G.R., Garner, L.C., Pettit, G.R. and Backhaus, R.A. (2000) Effects of atmospheric CO_2 enrichment on the growth and development of *Hymenocallis littoralis* (Amaryllidaceae) and the concentrations of several antineoplastic and antiviral constituents of its bulbs. *Am. J. Bot.* **87:** 769–773.

Jacob, J., Greitner, C. and Drake, B.G. (1995) Acclimation of photosynthesis in relation to Rubisco and non-structural carbohydrate contents and in situ carboxylase activity in *Scirpus olneyi* grown at elevated CO_2 in the field. *Plant Cell Environ.* **18:** 875–884.

Jang, J.C. and Sheen, J. (1994) Sugar sensing in higher plants. *Plant Cell* **6:** 1665–1679.

Jang, J.C. and Sheen, J. (1997) Sugar sensing in higher plants. *Trends Plant Sci.* **2:** 208–214.

Jang, J.C., León, P., Zhou, L. and Sheen, J. (1997) Hexokinase as a sugar sensor in higher plants. *Plant Cell* **9:** 5–19.

Jiang, C., Gao, X., Liao, L., Harberd, N.P. and Fu, X. (2007) Phosphate-starvation root architecture and anthocyanin-accumulation responses are modulated by the GA-DELLA signaling pathway in Arabidopsis. *Plant Physiol.* **145:** 1460–1470.

Jitla, D., Rogers, G., Seneweera, S., Basra, A., Oldfield, R. and Conroy, J. (1997) Accelerated early growth of rice at elevated CO_2. *Plant Physiol.* **115:** 15–22.

Johnston, C.A., Taylor, J.P., Gao, Y., Kimple, A.J., Grigston, J.C., Chen, J.G., Siderovski, D.P., Jones, A.M. and Willard, F.S. (2007) GTPase acceleration as the rate-limiting step in Arabidopsis G protein-coupled sugar signaling. *Proc. Natl Acad. Sci. USA* **104:** 7317–7322.

Kaliff, M., Staal, J., MyrenÅs, M. and Dixelius, C. (2007) ABA is required for *Leptosphaeria maculans* resistance via ABI1 and ABI4 dependent signalling. *Mol. Plant Microbe Interact.* **20:** 335–345.

Kim, S., Kang, J.-Y., Cho, D.-I., Park, J.H. and Kim, S.Y. (2004) ABF2, an ABRE-binding bZIP factor, is an essential component of glucose signaling and its overexpression affects multiple stress tolerance. *Plant J.* **40:** 75–87.

Kinsman, E., Lewis, C., Davies, M., Young, J., Francis, D., Vilhar B. and Ougham, H. (1997) Elevated CO_2 stimulates cells to divide in grass meristems: a differential effect in two natural populations of *Dactylis glomerata*. *Plant Cell Environ.* **20:** 1309–1316.

Klüsener, B., Young, J., Murata, Y., Allen, G., Mori, I., Hugouvieux, V. and Schroeder, J. (2002) Convergence of calcium signaling pathways of pathogenic elicitors and abscisic acid in Arabidopsis guard cells. *Plant Physiol.* **130:** 2152–2163.

Koiwa, H., Bressan, R.A. and Hasegawa, P.M. (1997) Regulation of protease inhibitors and plant defense. *Trends Plant Sci.* **2:** 379–384.

Kolbe, A., Tiessen, A., Schluepmann, H., Paul, M., Ulrich, S. and Geigenberger, P. (2005) Trehalose 6-phosphate regulates starch synthesis via posttranslational redox activation of ADP-glucose pyrophosphorylase. *Proc. Natl Acad. Sci. USA* **102:** 11118–11123.

Koussevitzky, S., Nott, A., Mockley, T.C., Hong, F., Sachetto-Martins, G., Surpin, M., Lim, J., Mittler, R. and Chory, J. (2007) Signals from chloroplasts converge to regulate nuclear gene expression. *Science* **316:** 715–719.

Kwak, J., Mori, I., Pei, Z., *et al.* (2003) NADPH oxidase AtrbohD and AtrbohF genes function in ROS-dependent ABA signaling in Arabidopsis. *EMBO J.* **22:** 2623–2633.

Lake, J.A., Woodward, F.I. and Quick, W.P. (2002) Long distance CO_2 signalling in plants. *J. Exp. Bot.* **53:** 183–193.

Laskowski, M. and Kato, I. (1980) Protein inhibitors of proteinases. *Annu. Rev. Biochem.* **49:** 685–693.

Leakey, A.D.B., Uribelarrea, M., Ainsworth, E.A., Naidu, S.L., Rogers, A., Ort, D.R. and Long, S.P. (2006) Photosynthesis, productivity, and yield of maize are not affected by open-air elevation of CO_2 concentration in the absence of drought. *Plant Physiol.* **140:** 779–790.

León, P. and Sheen, J. (2003) Sugar and hormone connections. *Trends Plant Sci.* **8:** 110–116.

Leple, J.C., Bonade-Bottino, M., Augustin, S., Pilate, G., Le Tan, V.D., Delplanque, A., Cornu, D. and Jouanin, L. (1995) Toxicity of *Chrysomila tremulae* (Coleoptera: Chrysomelidae) of transgenie poplars expressing a cysteine proteinase inhibitor. *Mol. Breed.* **1:** 319–328.

Leung, J. and Giraudat, J. (1998) Abscisic acid signal transduction. *Annu. Rev. Plant Physiol.* **49:** 199–222.

Lewis, C.E., Peratoner, G., Cairns, A.J., Causton, D.R. and Foyer, C.H. (1999) Acclimation of the summer annual species, *Lolium temulentum*, to CO_2 enrichment. *Planta* **210:** 104–114.

Lewis, C.E., Noctor, G., Causton, D. and Foyer, C.H. (2000) Regulation of assimilate partitioning in leaves. *Aust. J. Plant Physiol.* **27:** 507–519.

Liang, C., Brookhart, G., Feng, G.H., Reeck, G.R. and Kramer, K.J. (1991) Inhibition of digestive proteinases of stored grain coleopteran by oryzacystatin, a cysteine proteinase inhibitor from rice seed. *FEBS Lett.* **278:** 139–142.

Long, S.P., Ainsworth, E.A., Rogers, A. and Ort, D.R. (2004) Rising atmospheric carbon dioxide: plants FACE the future. *Annu. Rev. Plant Biol.* **55:** 591–628.

Loreti, E., Poggi, A., Novi, G., Alpi, A. and Perata, P. (2005) A genome-wide analysis of the effects of sucrose on gene expression in Arabidopsis seedling under anoxia. *Plant Physiol.* **137:** 1130–1138.

Martin, C. and Glover, B.J. (2007) Functional aspects of cell patterning in aerial epidermis. *Curr. Opin. Plant Biol.* **10:** 70–82.

Mason, H.S., DeWald, D.B., Creelman, R.A. and Mullet, J.E. (1992) Coregulation of soybean vegetative storage protein gene expression by methyl jasmonate and soluble sugars. *Plant Physiol.* **98:** 859–867.

Matros, A., Amme, S., Kettig, B., Buck-Sorlin, G.-H., Sonnewald, U. and Mock, H.-P. (2006) Growth at elevated CO_2 concentrations leads to modified profiles of secondary metabolites in tobacco cv. Samsun NN and to increased resistance against infection with *potato virus Y*. *Plant Cell Environ.* **29:** 126–137.

Mittler, R. (2002) Oxidative stress, antioxidants and stress tolerance. *Trends Plant Sci.* **7:** 405–410.

Mittler, R., Vanderauwera, S., Gollery, M. and Van Breusegem, F. (2004) Reactive oxygen gene network of plants. *Trends Plant Sci.* **9:** 490–498.

Moore, B.D., Cheng, S.-H., Sims, D. and Seemann, J.R. (1999) The biochemical and molecular basis for photosynthetic acclimation to elevated atmospheric CO_2. *Plant Cell Environ.* **22:** 567–582.

Moore, B., Zhou, L., Rolland, F., Hall, Q., Cheng, W.H., Liu, Y.-X., Hwang, I., Jones, T. and Sheen, J. (2003) Role of the Arabidopsis glucose sensor HXK1 in nutrient, light, and hormonal signalling. *Science* **300:** 332–336.

Morikami, A., Masaki, T., Tsukagoshi, H., Mitsui, N., Maeo, K.-I. and Nakamura, K. (2005) Transcription factors for sugar-inducible genes. *Plant Biotechnol.* **22:** 371–378.

Müller, J., Boller, T. and Wiemken, A. (1998) Trehalose affects sucrose synthase and invertase activities in soybean (*Glycine max* L. Merr.) roots. *J. Plant Physiol.* **153:** 255–257.

Nemani, R.R., Keeling, C.D., Hashimoto, H., Jolly, W.M., Piper, S.C., Tucker, C.J., Myneni, R.B. and Running, S.W. (2003) Climate-driven increases in global terrestrial net primary production from 1982–1999. *Science* **300:** 1560–1563.

Nie, G.Y., Hendrix, D.L., Webber, A.N., Kimball, B.A. and Long, S.P. (1995) Increased accumulation of carbohydrates and decreased photosynthetic gene transcript levels in wheat grown at an elevated CO_2 concentration in the field. *Plant Physiol.* **108:** 975–983.

Nishikawa, N., Kato, M., Hyodo, H., Ikoma, Y., Sugiura, M. and Yano, M. (2005) Effect of sucrose on ascorbate level and expression of genes involved in the ascorbate biosynthesis and recycling pathway in harvested broccoli florets. *J. Exp. Bot.* **56:** 65–72.

Noctor, G. and Foyer, C. (1998) Ascorbate and glutathione: keeping active oxygen under control. *Annu. Rev. Plant Physiol.* **49:** 249–279.

Ohto, M., Nakamura-Kito, K. and Nakamura, K. (1992) Induction of expression of genes coding for sporamin and β-amylase by polygalacturonic acid in leaf-petiole cuttings of sweet potato. *Plant Physiol.* **99:** 422–427.

Orozco-Cardenas, M. and Ryan, C.A. (1999) Hydrogen peroxide is generated systemically in plant leaves by wounding and systemin via the octadecanoid pathway. *Proc. Natl Acad. Sci. USA* **96:** 6553–6557.

Pastori, G.M. and Foyer, C.H. (2002) Common components, networks, and pathways of cross-tolerance to stress. The central role of 'redox' and abscisic acid-mediated controls. *Plant Physiol.* **129:** 460–468.

Paul, M.J. and Pellny, T.K. (2003) Carbon metabolite feedback regulation of leaf photosynthesis and development. *J. Exp. Bot.* **54:** 539–547.

Paul, M., Pellny, T. and Goddijn, O. (2001) Enhancing photosynthesis with sugar signals. *Trends Plant Sci.* **6:** 197–200.

Pego, J.V., Kortstee, A., Huijser, C. and Smeekens, S.C.M. (2000) Photosynthesis, sugars and the regulation of gene expression. *J. Exp. Bot.* **51:** 407–416.

Pellny, T. K., Van Aken, O., Dutilleul, C., Wolff, T., Groten, K., Bor, M., De Paepe, R., Reyss, A., Van Breusegem, F., Noctor, G. and Foyer, C. H. (2008) Mitochondrial respiratory pathways modulate nitrate sensing and nitrogen-dependent regulation of plant architecture in *Nicotiana sylvestris*. *Plant J.* **54:** 976–992.

Penna, S. (2003) Building stress tolerance through over-producing trehalose in transgenic plants. *Trends Plant Sci.* **8:** 355–357.

Perata, P., Matsukura, C., Vernieri, P. and Yamaguchi, J. (1997) Sugar repression of a gibberellin-dependent signalling pathway in barley embryos. *Plant Cell* **9:** 2197–2208.

Perfus-Barbeoch, L., Jones, A.M. and Assmann, S.M. (2004) Plant heterotrimeric G protein function: insights from Arabidopsis and rice mutants. *Curr. Opin. Plant Biol.* **7:** 719–731.

Polle, A. (2001) Dissecting the superoxide dismutase–ascorbate peroxidase–glutathione pathway in chloroplasts by metabolic modelling. Computer simulations as a step towards flux analysis. *Plant Physiol.* **126:** 445–462.

Potters, G., Pasternak, T.P., Guisez, Y., Palme, K.J. and Jansen, M.A.K. (2007) Stress-induced morphogenic responses: growing out of trouble? *Trends Plant Sci.* **12:** 98–105.

Prasad, T.K., Anderson, M.D., Martin, B.A. and Stewart, C.R. (1994) Evidence for chilling-induced oxidative stress in maize seedlings and a regulatory role for hydrogen peroxide. *Plant Cell* **6:** 65–74.

Price, J., Laxmi, A., St Martin, S.K. and Jang, J.C. (2004) Global transcription profiling reveals multiple sugar signal transduction mechanisms in *Arabidopsis*. *Plant Cell* **16:** 2128–2150.

Prins, A., Verrier, P., Kunert, K.J. and Foyer, C.H. (2008a) Chapter 3: CO_2 enrichment modulates both proteases and protease inhibitors in maize. In: *Photosynthesis. Energy from the Sun: 14th International Congress on Photosynthesis* (eds J.F. Allen, E. Gantt, J.H. Golbeck and B. Osmond). 1375-1378

Prins, A., Van Heerden, P.D.R., Olmos, E., Kunert, K.J. and Foyer, C.H. (2008b) Cysteine proteinases regulate chloroplast protein content and composition in tobacco leaves: a model for dynamic interactions with ribulose-1, 5-bisphosphate carboxylase/oxygenase (Rubisco) vesicular bodies. *J. Exp. Bot.* **59:** 1935–1950.

Quilliam, R.S., Swarbrick, P.J., Scholes, J.D. and Rolfe, S.A. (2006) Imaging photosynthesis in wounded leaves of *Arabidopsis thaliana*. *J. Exp. Bot.* **57:** 55–69.

Robert-Seilaniantz, A., Navarro, L., Bari, R. and Jones, J.D.G. (2007) Pathological hormone imbalances. *Curr. Opin. Plant Biol.* **10:** 372–379.

Robertson, E.J. and Leech, R.M. (1995) Significant changes in cell and chloroplast development in young wheat leaves (*Triticum aestivum* cv Hereward) grown in elevated CO_2. *Plant Physiol.* **107:** 63–71.

Robertson, E.J., Williams, M., Harwood, J.L., Lindsay, J.G., Leaver, C.J. and Leech, R.M. (1995) Mitochondria increase three-fold and mitochondrial proteins and lipid change dramatically in postmeristematic cells in young wheat leaves grown at elevated CO_2. *Plant Physiol.* **108:** 469–474.

Rock, C. (2000) Pathway to abscisic acid-regulated gene expression. *New Phytol.* **148:** 357–396.

Rogg, L.E. and Bartel, B. (2001) Auxin signaling: derepression through regulated proteolysis. *Dev. Cell* **1:** 595–604.

Roitsch, T. (1999) Source-sink regulation by sugar and stress. *Curr. Opin. Plant Biol.* **2:** 198–206.

Roitsch, T., Balibrea, M.E., Hofmann, M., Proels, R. and Sinha, A.K. (2003) Extracellular invertase: key metabolic enzyme and PR protein. *J. Exp. Bot.* **54:** 513–524.

Rolland, F., Moore, B. and Sheen, J. (2002) Sugar sensing and signalling in plants. *Plant Cell* **14:** 185–205.

Rolland, F., Baena-Gonzalez, E. and Sheen, J. (2006) Sugar sensing and signalling in plants: conserved and novel mechanisms. *Annu. Rev. Plant Biol.* **57:** 675–709.

Rook, F., Corke, F., Card, R., Munz, G., Smith, C. and Bevan, M.W. (2001) Impaired sucrose-induction mutants reveal the modulation of sugar-induced starch biosynthetic gene expression by abscisic acid signalling. *Plant J.* **26:** 421–433.

Ryu, J.Y., Song, J.Y., Lee, J.M., Jeong, S.W., Chow, W.S., Choi, S.B., Pogson, B.J. and Park, Y.I. (2004) Glucose-induced expression of carotenoid biosynthesis genes in the dark is mediated by cytosolic pH in the cyanobacterium *Synechocystis* sp. PCC6803. *J. Biol. Chem.* **279:** 25320–25325.

Sage, R.F. (2001) Environmental and evolutionary pre-conditions for the origin and diversification of the C_4 photosynthetic syndrome. *Plant Biol.* **3:** 202–213.

Sánchez, J.P., Duque, P. and Chua, N.-H. (2004) ABA activates ADPR cyclise and cADPR induces a subset of ABA-responsive genes in Arabidopsis. *Plant J.* **38:** 381.

Schluepmann, H., Pellny, T., van Dijken, A., Smeekens, S. and Paul, M. (2003) Trehalose 6-phosphate is indispensable for carbohydrate utilization and growth in *Arabidopsis thaliana*. *Proc. Natl Acad. Sci. USA* **100:** 6849–6854.

Sheen, J., Zhou, L. and Jang, J.C. (1999) Sugars as signalling molecules. *Curr. Opin. Plant Biol.* **2:** 410–418.

Signora, L., De Smet, I., Foyer, C.H. and Zhang, H. (2001) ABA plays a central role in mediating the regulatory effects of nitrate on root branching in *Arabidopsis*. *Plant J.* **28:** 655–662.

Sims, D.A., Luo, Y. and Seemann, J.R. (1998) Importance of leaf versus whole plant CO_2 environment for photosynthetic acclimation. *Plant Cell Environ.* **21:** 1189–1196.

Smeekens, S. (2000) Sugar-induced signal transduction in plants. *Annu. Rev. Plant Physiol.* **51:** 49–81.

Soares, A.S., Driscoll, S.P., Olmos, E., Harbinson, J., Arrabaça, M.C. and Foyer, C.H. (2008) Adaxial/abaxial specification in the regulation of photosynthesis with respect to light orientation and growth with CO_2 enrichment in the C_4 species *Paspalum dilatatum*. *New Phytol.* **177:** 186–198.

Söderman, E., Brocard, I., Lynch, T. and Finkelstein, R. (2000) Regulation and function of the Arabidopsis ABA-insensitive4 gene in seed and abscisic acid response signaling networks. *Plant Physiol.* **124:** 1752–1765.

Solomon, M., Belenghi, B., Delledonne, M., Menachem, E. and Levine, A. (1999) The involvement of cysteine proteases and protease inhibitor genes in the regulation of programmed cell death in plants. *Plant Cell* **11:** 431–443.

Spreitzer, R.J. and Salvucci, M.E. (2002) Rubisco: structure, regulatory interactions, and possibilities for a better enzyme. *Annu. Rev. Plant Biol.* **53:** 449–475.

Stitt, M. (1991) Rising CO_2 levels and their potential significance for carbon flow in photosynthetic cells. *Plant Cell Environ.* **14:** 741–762.

Sugiyama, T., Mizuno, M. and Hayashi, M. (1984) Partitioning of nitrogen among ribulose-1,5-bisphosphate carboxylase/oxygenase, phosphoenolpyruvate carboxylase, and pyruvate orthophosphate dikinase as related to biomass productivity in maize seedlings. *Plant Physiol.* **75:** 665–669.

Sulmon, C., Gouesbet, G. Couee, I. and El Amrani, A. (2004) Sugar-induced tolerance to atrazine in Arabidopsis seedlings: interacting effects of atrazine and soluble sugars on *psbA* mRNA and D1 protein levels. *Plant Sci.* **167:** 913–923.

Sun, J., Gibson, K.M., Kiirats, O., Okita, T.W. and Edwards, G.E. (2002) Interactions of nitrate and CO_2 enrichment on growth, carbohydrates, and Rubisco in Arabidopsis starch mutants. Significance of starch and hexose. *Plant Physiol.* **130:** 1573–1583.

Tajima, H. and Koizumi, N. (2006) Induction of BiP by sugar independent of a *cis*-element for the unfolded protein response in *Arabidopsis thaliana*. *Biochem. Biophys. Res. Commun.* **346:** 926–930.

Thevelein, J.M. and Hohmann, S. (1995) Trehalose synthase: guard to the gate of glycolysis in yeast? *Trends Biochem. Sci.* **20:** 3–10.

Thibaud, M.C., Gineste, S., Nussaume, L. and Robaglia, C. (2004) Sucrose increases pathogenesis-related PR-2 gene expression in *Arabidopsis thaliana* through an SA-dependent but NPR1-independent signaling pathway. *Plant Physiol. Biochem.* **42:** 81–88.

Thomann, A., Dieterle, M. and Genschik, P. (2005) Plant CULLIN-based E3s: phytohormones come first. *FEBS Lett.* **579:** 3239–3245.

Thomas, H., Ougham, H.J., Wagstaff, C. and Stead, A.D. (2003) Defining senescence and death. *J. Exp. Bot.* **54:** 1127–1132.

Torres, M. and Dangl, J. (2005) Functions of the respiratory burst oxidase in biotic interactions, abiotic stress and development. *Curr. Opin. Plant Biol.* **8:** 397–403.

Torres, M., Dangl, J. and Jones, J. (2002) Arabidopsis *gp91phox* homologues *AtrbohD* and *AtrbohF* are required for accumulation of reactive oxygen species in the plant defense response. *Proc. Natl Acad. Sci. USA* **99:** 517–522.

Tsukaya, H., Ohshima, T., Naito, S., Chino, M. and Komeda, Y. (1991) Sugar-dependent expression of the *CHS-A* gene for chalcone synthase from petunia in transgenic *Arabidopsis*. *Plant Physiol.* **97:** 1414–1421.

Tsugane, K., Kobayashi, K., Niwa, Y., Ohba, Y., Wada, K. and Kobayashi, H. (1999)

A recessive Arabidopsis mutant that grows photoautotrophically under salt stress shows enhanced active oxygen detoxification. *Plant Cell* **11:** 1195–1206.

Urbano, A., Lakshmanan, U., Choo, P-H., Kwan, J.C., Ng, P.Y., Guo, K., Dhakshinamoorthy, S. and Porter, A. (2005) AIF suppresses chemical stress-induced apoptosis and maintains the transformed state of tumor cells. *EMBO J.* **24:** 2815–2826.

Urwin, P.E., Atkinson, H.J., Waller, D.A. and McPherson, M.J. (1995) Engineered oryzacystatin-I expressed in transgenic hairy roots confers resistance to *Globodera pallida*. *Plant J.* **8:** 121–131.

Van Dijken, A.J.H., Schluepmann, H. and Smeekens, S.C.M. (2004) Arabidopsis trehalose-6-phosphate synthase 1 is essential for normal vegetative growth and transition to flowering. *Plant Physiol* **135:** 969–977.

Watling, J.R., Press, M.C. and Quick, W.P. (2000) Elevated CO_2 induces biochemical and ultrastructural changes in leaves of the C_4 cereal sorghum. *Plant Physiol.* **123:** 1143–1152.

Wingler, A., Fritzius, T., Wiemken, A., Boller, T. and Aeschbacher, R.A. (2000) Trehalose induces the ADP-glucose pyrophosphorylase gene, *ApL3*, and starch synthesis in Arabidopsis. *Plant Physiol.* **124:** 105–114.

Woodward, F.I. (2002) Potential impacts of global elevated CO_2 concentrations on plants. *Curr. Opin. Plant Biol.* **5:** 207–211.

Woodward, F.I., Lake, J.A. and Quick, W.P. (2002) Stomatal development and CO_2: ecological consequences. *New Phytol.* **153:** 477–484.

Xiao, W., Sheen, J. and Jang, J.C. (2000) The role of hexokinase in plant sugar signal transduction and growth and development. *Plant Mol. Biol.* **44:** 451–461.

Yanagisawa, S., Yoo, S.D. and Sheen, J. (2003) Differential regulation of EIN3 stability by glucose and ethylene signalling in plants. *Nature* **425:** 521–525.

Zhang, Y.M., Wong, T.Y., Chen, L.Y., Lin, C.S. and Liu, J.K. (2000) Induction of a futile Embden–Meyerhof–Parnas pathway in *Deinococcus radiodurans* by Mn: possible role of the pentose phosphate pathway in cell survival. *Appl. Environ. Microb.* **66:** 105–112.

Zhang, Y., Yang, C., Li, Y., Zheng, N., Chen, H., Zhao, Q., Gao, T., Guo, H. and Xie, Q. (2007) SDIR1 is a RING finger ligase that positively regulates stress-responsive abscisic acid signalling in Arabidopsis. *Plant Cell* **19:** 1912–1929.

Zhao, Y., Botella, M.A., Subramanian, L., Niu, X., Nielsen, S.S., Bressan, R.A. and Hasegawa, P.M. (1996) Two wound-inducible soybean cysteine proteinase inhibitors have greater insect digestive proteinase inhibitory activities than a constitutive homologue. *Plant Physiol.* **111:** 1299–1306.

12

Protein damage in the ageing process: Advances in quantitation and the importance of enzymatic defences

Mingzhan Xue, Antonysunil Adaikalakoteswari, Naila Rabbani and Paul J. Thornalley

1 Markers of protein glycation

1.1 Physiological glycating agents

Glycation is a major cause of spontaneous damage to cellular and extracellular proteins in physiological systems, accounting for 0.1–0.2% of lysine and arginine residues (Thornalley, 1999; Thornalley *et al.*, 2003). For some proteins with limited protein turnover, such as those in lens fibre cells, the extent of protein glycation may be up to 10-fold higher (Ahmed *et al.*, 2003). Glycation adducts occur in proteins of tissues and body fluids as glycated amino-acid residues, formed by glycation of proteins *in situ*.

Glycation adducts are formed in non-enzymatic, spontaneous reactions of monosaccharides with proteins. Other important glycating agents are reactive dicarbonyls such as glyoxal, methylglyoxal and 3-deoxyglucosone (3-DG). Dicarbonyls are formed endogenously by lipid peroxidation and the degradation of glycolytic intermediates and glycated proteins (*Figure 1*). Dicarbonyl compounds are potent glycating agents, 200–50 000-fold more reactive than glucose. The physiological concentrations of dicarbonyls are typically 10 000–50 000-fold lower than glucose yet, nevertheless, dicarbonyl compounds remain important precursors of advanced glycation end products (AGEs) in physiological systems. Glycation by glucose and other monosaccharides is directed to amino groups of lysine and N-terminal amino-acid residues of proteins, whereas glycation by dicarbonyl compounds is mainly but not exclusively directed to arginine residues (Thornalley, 2005).

1.2 Early-stage glycation adducts and advanced glycation end products

The initial glycation adducts formed by glycation by glucose and other monosaccharides are called early-stage glycation adducts. The predominant early glycation adduct

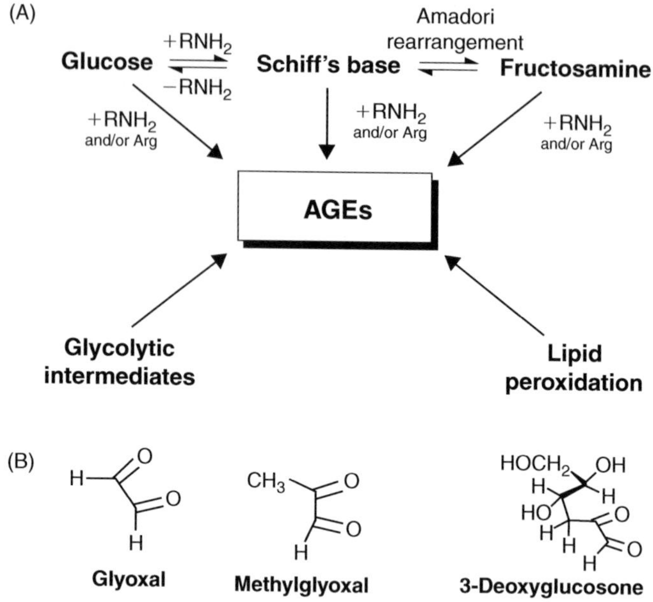

Figure 1. (A) Formation of early and advanced glycation adducts from glucose and glycolytic intermediates and products of lipid peroxidation. (B) Physiological reactive dicarbonyl glycating agents.

is N_ε-fructosyl-lysine (FL). Schiff's base and FL adducts degrade slowly in further, advanced reactions to form many different glycation adducts. These adducts are collectively called 'advanced glycation end products' (AGEs). Dicarbonyls react with proteins directly to also form AGEs. Important AGEs quantitatively are hydroimidazolones derived from arginine residues modified by glyoxal, methylglyoxal and 3-DG: N_δ-(5-hydro-4-imidazolon-2-yl)-ornithine (G-H1), N_δ-(5-hydro-5-methyl-4-imidazolon-2-yl)-ornithine (MG-H1) and N_δ-(5-hydro-5-(2,3,4-trihydroxy-butyl)-4-imidazolon-2-yl)-ornithine and related structural isomers (3DG-H). Other important and widely studied AGEs are N_ε-carboxymethyl-lysine (CML) and N_ε-carboxyethyl-lysine (CEL), and the protein cross-links, pentosidine and glucosepane (Ahmed *et al.*, 2002; Biemel *et al.*, 2002; Sell and Monnier, 1989; Thornalley *et al.*, 2003; Thorpe and Baynes, 2002). Further AGEs and related derivatives of emerging importance are N_ε-carboxymethyl-cysteine (CMC; Zeng and Davies, 2005), N_ω-carboxymethylarginine (CMA; Odani *et al.*, 2001) and ornithine (Sell and Monnier, 2004) – the latter formed as a degradation product of hydroimidazolones (*Figure 2*). Some examples of protein contents of AGE residues with relation to subject age are given in *Table 1*.

1.3 *Protein fluorescence as a marker of advanced glycation end products*

AGEs have been determined by 'total AGE fluorescence', measuring fluorescence with excitation and emission wavelengths of 350 nm and 450 nm, respectively (Monnier *et*

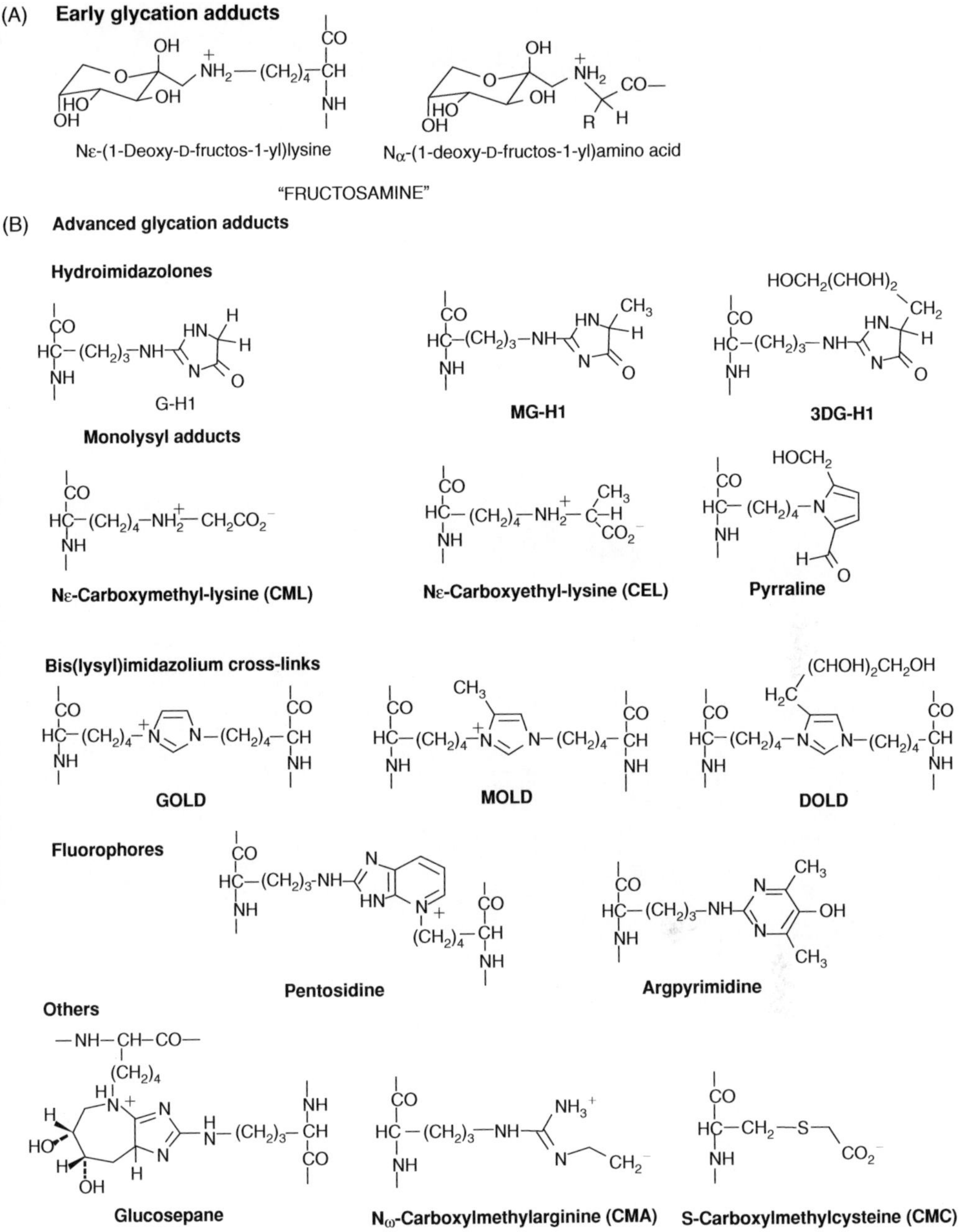

Figure 2. *Protein glycation adduct residues. (A) Early glycation adducts. (B) Advanced glycation end products. For the corresponding free adducts at physiological pH, the N-terminal amino group is protonated –NH_3^+ and the C-terminal carbonyl is a carboxylate –CO_2^- moiety.*

Table 1. *Glycation adduct residues in human proteins and dependence on donor age.*

Adduct	Protein	Content	Age dependence	Reference
CML residue (mmol mol^{-1} lys)	Lens	4.95 ± 0.50	CML (mmol mol^{-1} lys) = 0.079 × age (years) + 0.48	(Dunn *et al.*, 1989)
	Skin collagen	1.70 ± 0.31	CML (mmol mol^{-1} lys) = 0.02 × age (years)	(Dunn *et al.*, 1991)
	Cartilage collagen	4.24 ± 1.51	CML (mmol mol^{-1} lys) = 0.073 × age (years) – 1.17	(Verzijl *et al.*, 2000a)
CMhL (mmol mol^{-1} hlys)	Skin collagen	0–6	[CMhL] (mmol mol^{-1} hlys) = 0.06 × age (years)	(Dunn *et al.*, 1991)
CEL residue (mmol mol^{-1} lys)		3.78 ± 0.28	CEL (mmol mol^{-1} lys) = 0.0322 × age (years) + 1.28	(Ahmed *et al.*, 1997)
GOLD (mmol mol^{-1} lys)	Lens protein	0.11 ± 0.05	GOLD (mmol mol^{-1} lys) = 0.0023 × age (years) + 0.0033	(Brinkmann Frye *et al.*, 1998)
MOLD (mmol mol^{-1} lys)	Lens protein	0.32 ± 0.22	MOLD (mmol mol^{-1} lys) = 0.0068 × age (years)	(Brinkmann Frye *et al.*, 1998)
Pentosidine	Skin collagen	0–40	Pentosidine (μmol mol^{-1} lys) = 0.39 age (years)	(Dyer *et al.*, 1993)
	Cartilage collagen	0–120	Pentosidine (μmol mol^{-1} lys) = 1.74 age (years) – 21.8	(Verzijl *et al.*, 2000a)
	Lens protein	0.5–5.0	Pentosidine (μmol mol^{-1} lys) = 0.046 age (years) + 0.005	(Dyer *et al.*, 1991)
MG-H1 (mmol mol^{-1} arg)	Lens protein	15	MG-H1 (pmol mg^{-1} protein) = 70 × age (years) + 705	(Ahmed *et al.*, 2003)

Table 2. Fluorophores associated with protein damage by glycation, oxidation and ageing.

Fluorophore	Structure	Fluorescence characteristics: $\lambda_{excitation}$, $\lambda_{emission}$	Reference
Glycation			
Pentosidine	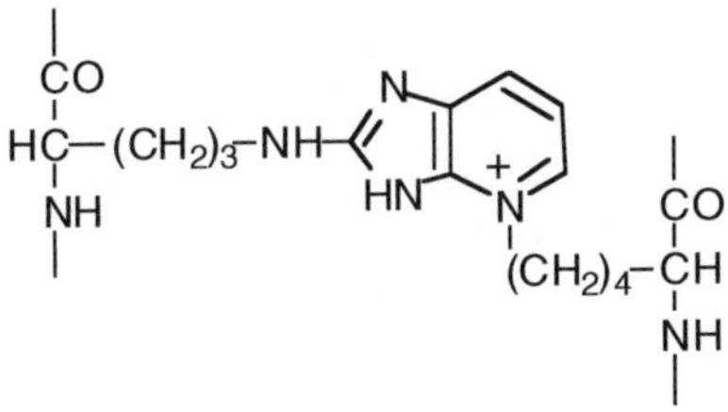	335 nm, 385 nm	(Sell and Monnier, 1989)
Argpyrimidine	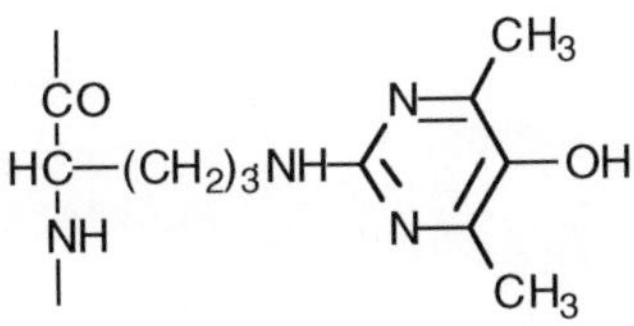	320 nm, 385 nm	(Shipanova *et al.*, 1997)
Vesperlysine A (LM-1)	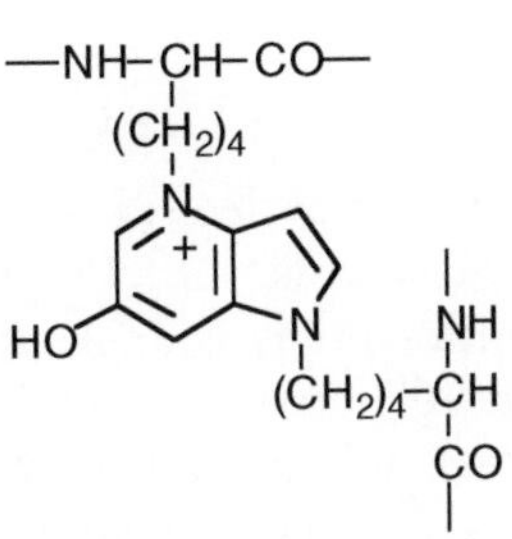	345 nm, 405 nm	(Tessier *et al.*, 1999b)
"AGE fluorescence"	Multiple	350 nm, 440 nm	(Monnier *et al.*, 1984)
Oxidation			
N-Formyl-kynurenine	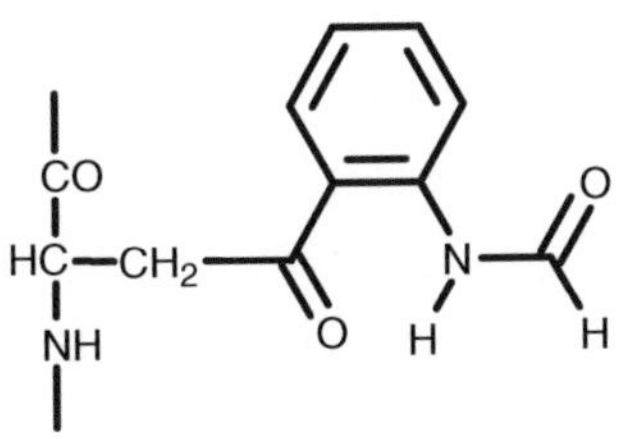	325 nm, 434 nm	(Fukunaga *et al.*, 1982)
Kynurenine	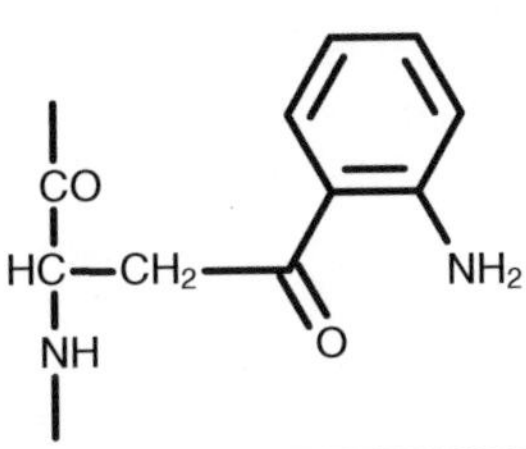	365 nm, 480 nm	(Fukunaga *et al.*, 1982)

continued

Table 2. *Fluorophores associated with protein damage by glycation, oxidation and ageing (continued).*

Fluorophore	Structure	Fluorescence characteristics: $\lambda_{excitation}$, $\lambda_{emission}$	Reference
Oxidation			
2-Amino-6-(3,5-diformyl-1,4-dihydro-4-methyl-pyridin-1-yl) hexanoic acid	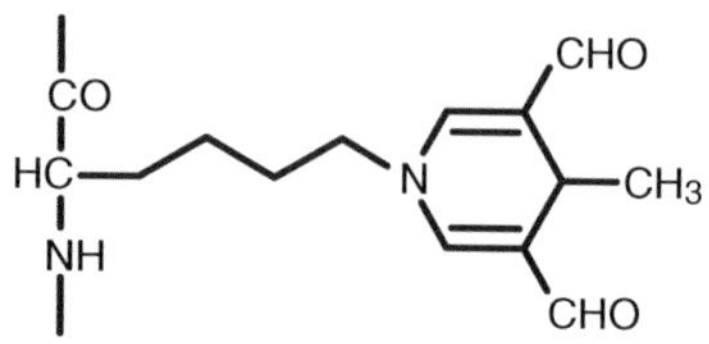	387 nm, 455 nm	(Yamada *et al.*, 2001)
"Fluorescent oxidation products"	Multiple fluorophores	360 nm, 430 nm	(Shimasaki, 1994)
Lipofuscin	Multiple fluorophores	340 – 390 nm, 430 – 490 nm	(Li-*et al.*, 2006; Sohal, 1981)

al., 1984), with recent modification of this method to detect low molecular mass AGE peptides or free adducts (Thomas *et al.*, 2004; Wrobel *et al.*, 1997). These methods are compromised by several analytical problems: (i) the major AGEs quantitatively are not fluorescent and are therefore not detected (Thornalley *et al.*, 2003); (ii) there is significant interference from the oxidative adduct and fluorophore, N-formylkynurenine (Buxton and Guilbault, 1974); and (iii) several different fluorophores contribute to the global measure of fluorescence, each with a different specific fluorescence, such that no quantitative calibration can be achieved (*Table 2*). Estimates of total AGE fluorescence as determined currently represent a qualitative measure of damage by both glycation and oxidation adduct fluorophores.

1.4 Enzymatic defence against glycation

Protein glycation in physiological systems is prevented and repaired by components of the enzymatic defence against glycation (Thornalley, 2003a). It involves enzymatic activities that suppress the formation of glycation adducts and repair sites of early glycation: glyoxalase 1 (Glo1), aldoketo reductases (AKRd) and aldehyde dehydrogenases detoxify reactive dicarbonyl glycating agents (Shinohara *et al.*, 1998; Suzuki *et al.*, 1998), and amadoriase and fructosamine 3-phosphokinase catalyse the removal of FL residues and free adducts (Delpierre *et al.*, 2000; Szwergold *et al.*, 2001; Takahashi *et al.*, 1997; *Figure 3*). The enzymatic defence against glycation suppresses damage to biological macromolecules but it is an imperfect defence; glycation adducts of proteins, nucleotides and basic phospholipids are formed in normal physiological states. The enzymatic defence against glycation is overwhelmed in some disease states and

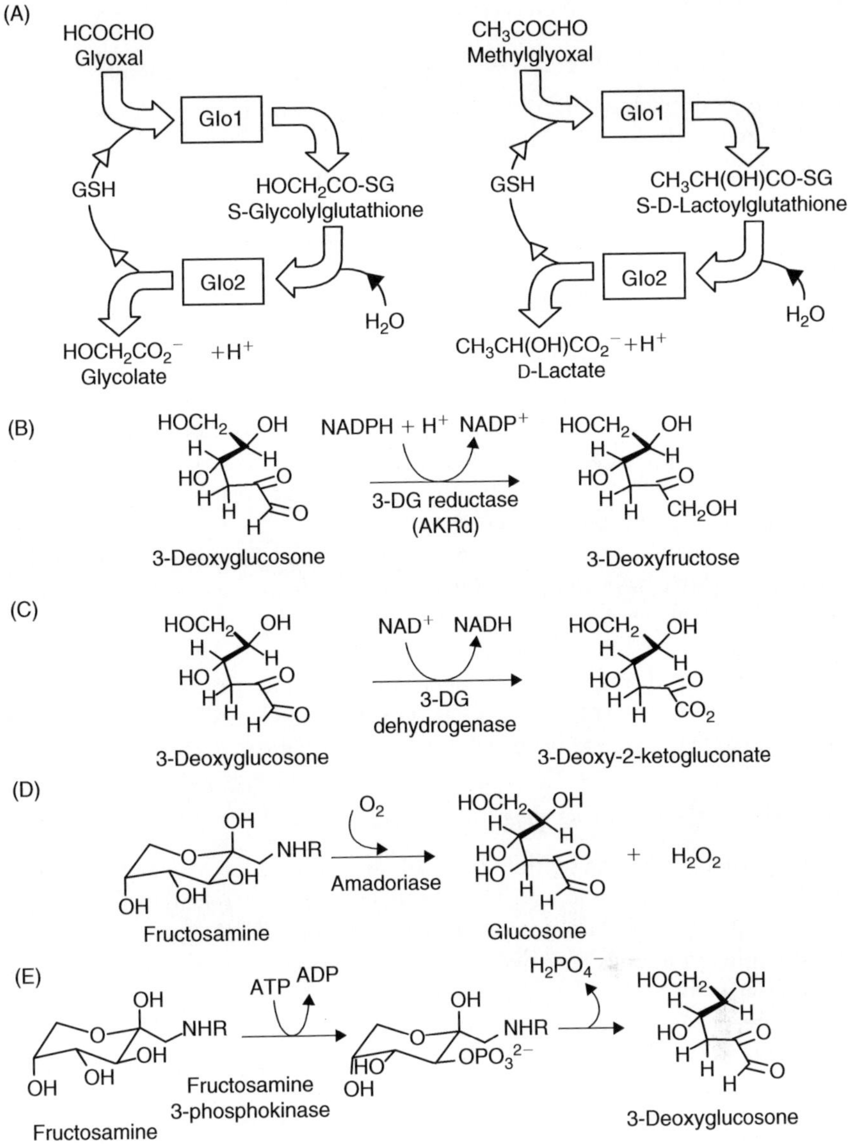

Figure 3. *The enzymatic defence against glycation. (A) Metabolism of glyoxal and methylglyoxal by the glyoxalase system. (B) Metabolism of 3-deoxyglucosone by 3-DG reductase, an aldoketo reductase (AKRd). (C) Metabolism of 3-deoxyglucosone by 3-DG dehydrogenase reductase. (D) and (E) Repair of early glycated proteins by amadoriase and fructosamine-3-phosphokinase.*

glycation adduct concentrations increase. The decline of expression of enzymes of the enzymatic defence against glycation may be key to increased protein glycation in the ageing phenotype (see later).

2 Markers of protein oxidation

2.1 *Physiological oxidizing agents*

The oxidation of proteins in physiological systems occurs by spontaneous autoxidation of cysteinyl thiols (Saez *et al.*, 1982) and by interaction of proteins with oxidizing agents such as hydrogen peroxide H_2O_2, hydroperoxides ROOH, hypochlorite ClO^-, peroxynitrite and other oxidizing reactive intermediates (hydroxyl radical $HO^{\bullet}$, carbonate radical anion $CO_3^{\bullet-}$, and others; Finkel and Holbrook, 2000). These processes may be catalysed by trace redox active metal ions such as iron (III) or ferric ion Fe^{3+}, and Cu (II) or cupric Cu^{2+} ion (Castellani *et al.*, 2004). Hydrogen peroxide in physiological systems originates mainly from the dismutation of superoxide $O_2^{\bullet-}$ formed by the inefficiency of mitochondrial respiration (Chance *et al.*, 1979), superoxide-forming enzymes such as phagocyte and vascular NADPH oxidases (Lassegue and Clempus, 2003), uncoupling of enzymatic reactions (for example, nitric oxide synthases; Du *et al.*, 1999), and autoxidation of reducing metabolites (Finkel and Holbrook, 2000). Hypochlorite is formed by the enzymatic reaction of myeloperoxidase in neutrophils (Naskalski *et al.*, 2002). Extracellular proteins are more susceptible to oxidation than cellular proteins because the non-enzymatic and enzymatic activities of antioxidants mostly have an intracellular location. However, one of the sources of the highest flux of hydrogen peroxide generation, the mitochondria (Chance *et al.*, 1979), is also intracellular. Although the cytoplasm in physiological systems is maintained under reducing conditions by the presence of high concentrations of low-molecular-mass peptide thiols (usually glutathione, GSH), there is nevertheless a residual low level of oxidative damage to both cellular and extracellular proteins.

2.2 *Oxidation of cysteine residues*

The thiol group of cysteine residues is usually the most easily oxidized site within proteins (*Figure 4*). Oxidation occurs to form cysteine sulphenic acid, cys-S-OH (Luo *et al.*, 2005), or sulphenylamide derivatives (van Montfort *et al.*, 2003). Cysteinyl sulphenic acid and sulphenyl-amide derivatives may play a role in the redox regulation of cysteine-dependent enzymes (Poole *et al.*, 2004). Effective oxidants are hydrogen peroxide, hypochlorite, peroxynitrite and other reactive oxygen species (ROS) and reactive nitrogen species (RNS). Exposure of rat heart tissue to hydrogen peroxide *ex vivo*, for example, formed cysteine sulphenic acid residues in the following proteins: myosin heavy and light chains, aconitase, acyl-CoA dehydrogenase, ATP synthase, aspartate aminotransferase, actin, tropomyosin, troponin T, ADP, ATP carrier protein T1, malate dehydrogenase, L-lactate dehydrogenase, cytochrome c oxidase and myoglobin (Kiley and Storz, 2004). Cysteine sulphenic acid residues are readily reversed by reduction with GSH, thioredoxin and glutaredoxin. Protein sulphenic acid formation is involved in the regulation of tyrosine phosphatases, peroxiredoxins, methionine sulphoxide reductases and other proteins (Poole *et al.*, 2004). GSH reacts with cysteine sulphenic acid residues to form a glutathionyl mixed disulphide – a process called protein thiolation. Protein thiolation is also considered to mediate redox regulation of enzymes and other proteins (Giustarini *et al.*, 2004). Disulphide and sulphenic acid residue formation are readily reversed by reduction with GSH. Cysteine residues may also be oxidized to cysteine sulphinic acid residues, cys-SO_2H. This is mainly an irreversible oxidation (Poole *et al.*,

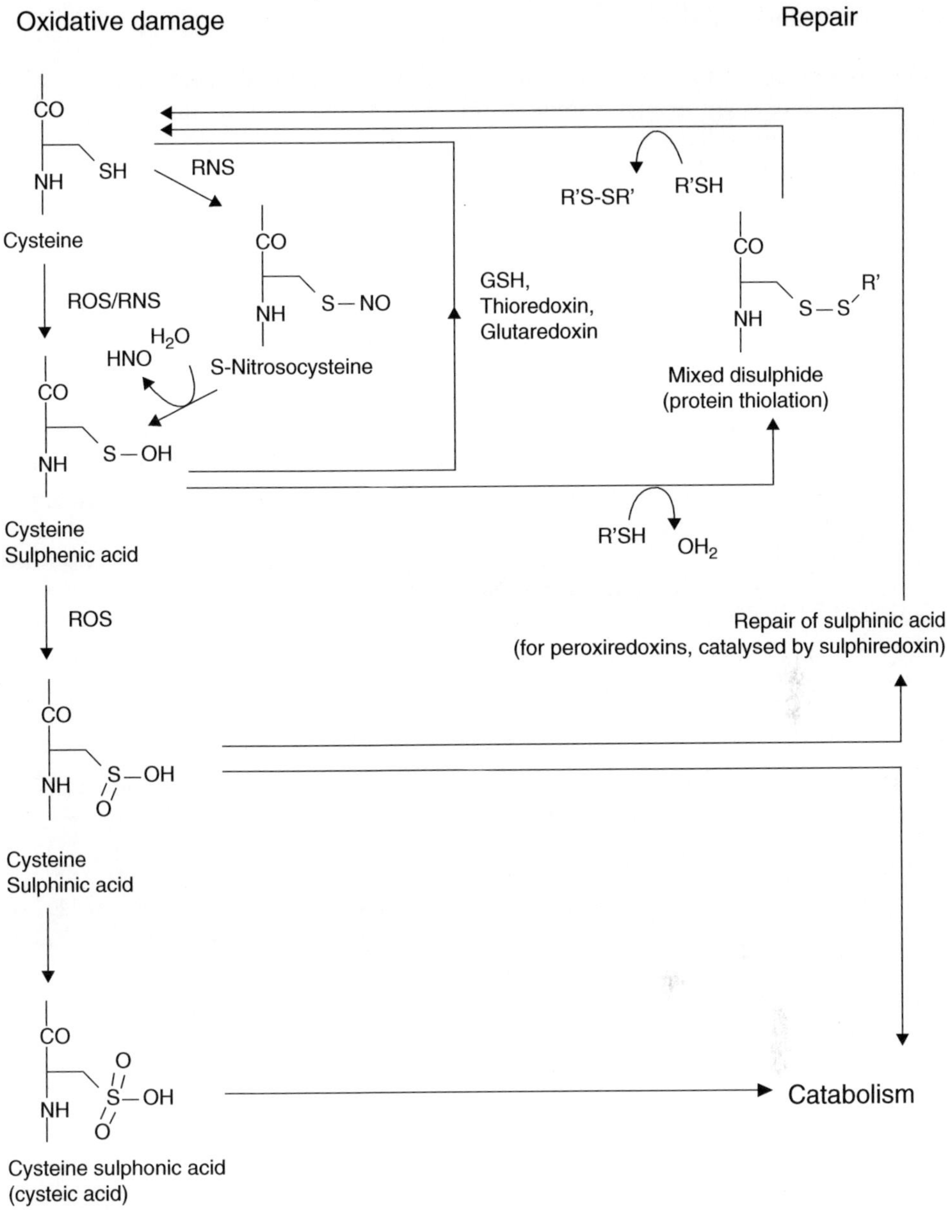

Figure 4. *Oxidation and repair of cysteine residues in proteins. For the corresponding free adducts at physiological pH, the N-terminal amino group is protonated* $-NH_3^+$ *and the C-terminal carbonyl is a carboxylate* $-CO_2^-$ *moiety.*

2004). Recent research has identified the enzyme sulphiredoxin that repairs sulphinic acid formation but this is specific for the antioxidant protein peroxiredoxins (Woo *et al.*, 2005). Further oxidation to a sulphonic acid or cysteic acid cys-SO_3H residues is possible but requires strong oxidants and is an irreversible oxidation (Poole *et al.*, 2004).

2.3 Oxidation of methionine residues

Methionine residues in proteins are susceptible to oxidation, leading to the formation of *R*- and *S*-epimers of methionine sulphoxide (MetSO) and methionine sulphone ($MetSO_2$) residues (*Figure 5*). MetSO residues are formed by exposure to hydrogen peroxide (Finch *et al.*, 1993), hypochlorite (Panzenbock and Stocker, 2005) and peroxynitrite – but in the latter case the yield of MetSO residues was decreased markedly by physiological concentrations of carbon dioxide (Tien *et al.*, 1999). Severe oxidizing conditions are thought to be required to form methionine sulphone (Chowdhury *et al.*, 1995). MetSO is a major quantitative marker of protein oxidation in cellular and extracellular proteins. Some examples of MetSO residue detection are given in *Table 3*. MetSO residue formation in physiological systems can be repaired, however, by enzymatic reduction catalysed by methionine sulphoxide reductase (Msr, EC 1.4.8.6; Weissbach *et al.*, 2005). The physiological co-factor providing reducing equivalents to sustain the reducing reaction is thioredoxin, which is itself reduced by NADPH catalysed by thioredoxin reductase. There are two forms of Msr: MsrA reduces the *S*-epimer of MetSO and MsrB reduces the *R*-epimer (Boschi-Muller *et al.*, 2005; Hansel *et al.*, 2005). Msr isozymes reduce MetSO residues in proteins and MetSO free adduct (Boschi-Muller *et al.*, 2005; Moskovitz *et al.*, 1996). Human Msr is expressed in all tissues, with highest expression in the heart, skeletal muscle, liver and cerebellum (Hansel *et al.*, 2003; Hu *et al.*, 2005; Jung *et al.*, 2002; Kuschel *et al.*, 1999). mRNA splice variants of MsrA targets the enzyme to mitochondria and to nuclear and cytosolic subcellular locations (Hansel *et al.*, 2002, 2005; Vougier *et al.*, 2003).

Table 3. *Detection of protein oxidation adduct residues and free adducts in biomedical research: methionine sulphoxide*

Adduct	Sample	Species	Comment	Reference
MetSO residue	Plasma protein	Human	Normal healthy controls	(Ahmed *et al.*, 2005a)
	Haemoglobin	Human	Normal healthy controls and patients with diabetes	(Ahmed *et al.*, 2005a)
	Skin collagen	Human	Increase with ageing mmol MetSO mol Met^{-1} = 1.6 × age (years) + 20.7	(Wells-Knecht *et al.*, 1997)
	Lens protein	Bovine	Hydrogen peroxide oxidation. αA-crystallin: met-1 oxidized. αB-crystallin: met-1 and met-68 oxidized	(Finley *et al.*, 1998)
	Cardiac mitochondrial proteins	Human	Multiple mitochondrial proteins modified identified by proteomics	(Taylor *et al.*, 2003)
	Glomeruli, retina, nerve, skeletal muscle, plasma	Rat	Normal control and streptozotocin-induced diabetic rats	(Thornalley *et al.*, 2003)
MetSO free adduct	Plasma	Human	Normal healthy controls	(Ahmed *et al.*, 2005a)
	Urine	Human	Normal healthy controls	(Ahmed *et al.*, 2005a)

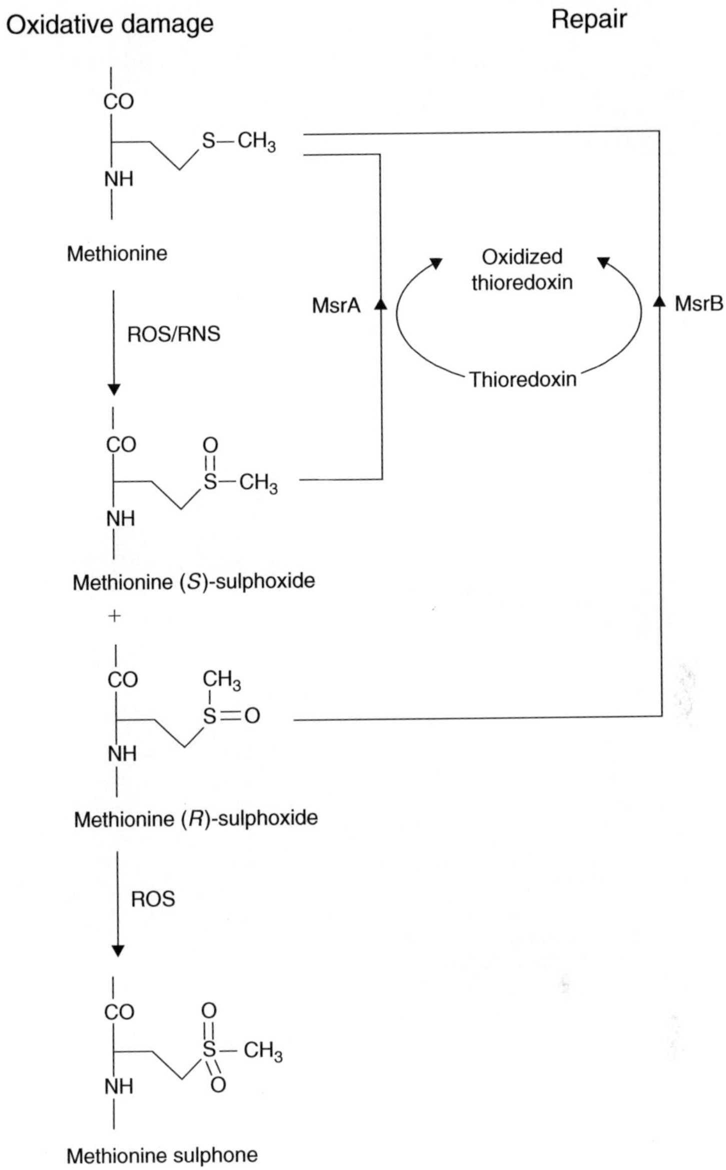

Figure 5. *Oxidation and repair of methionine residues in proteins.*

2.4 Oxidation of tyrosine residues

Tyrosine residues that are adjacent in proteins may be oxidized to form a cross-linked dimer, 2,2′-dihydroxy-5,5′-bis(β-carboxy-β-aminoethyl)diphenyl or dityrosine (McCormick *et al.*, 1998; *Figure 6A*). When formed from tyrosine residues in proteins, this represents a non-sulphhydryl cross-link. Dityrosine is formed by exposure of proteins to hydrogen peroxide – including peroxidase catalysed reactions, hypochlorite

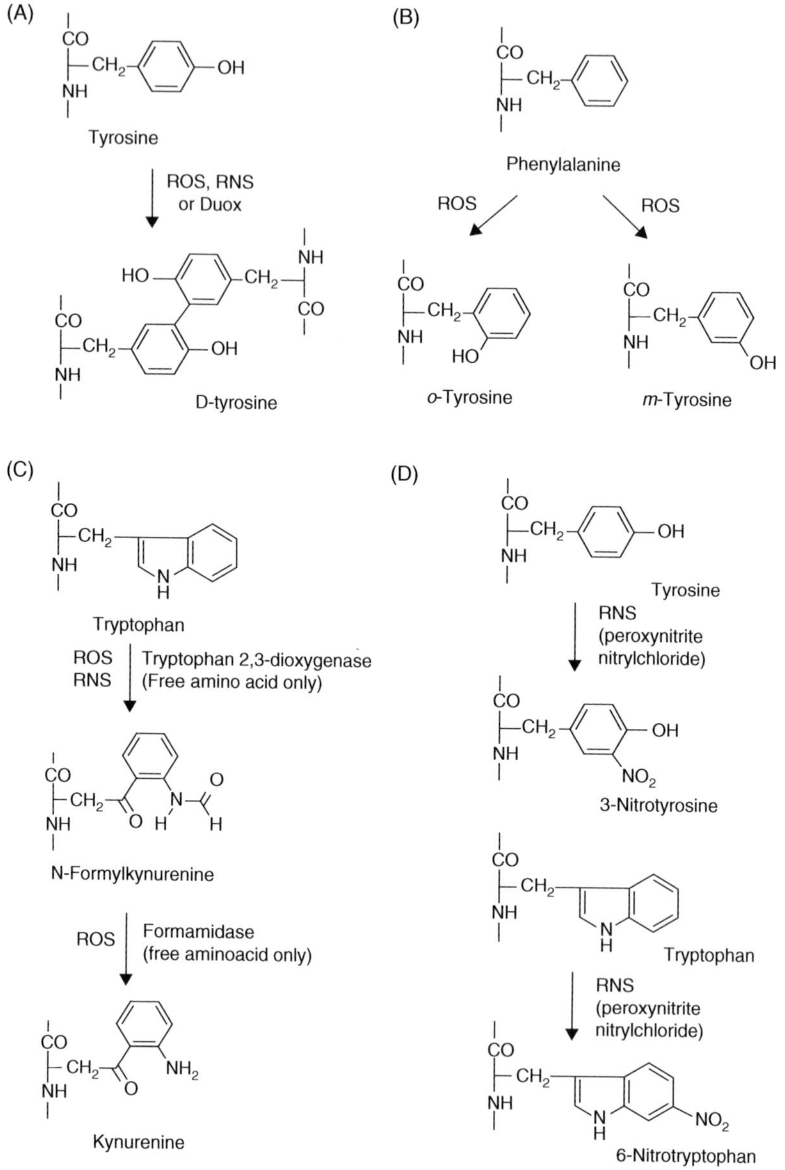

Figure 6. *(A) Oxidative dimerization of tyrosine residues in proteins forming dityrosine. (B) Hydroxylation of phenylalanine residues in proteins forming* o- *and* m-*tyrosine. (C) Oxidation of tryptophan residues in proteins to N-formylkynurenine and kynurenine. (D) Nitration of tyrosine and tryptophan residues in proteins to 3-nitrotyrosine and 6-nitrotryptophan residues, respectively.*

and peroxynitrite – the efficacy of formation depending on the proximity of two tyrosine residue side-chain moieties for cross-link formation (Aeschbach *et al.*, 1976; Krishnan *et al.*, 2003; Ushijima *et al.*, 1984). Dityrosine formation stimulated by

Table 4. Detection of protein oxidation adduct residues and free adducts in biomedical research: dityrosine and o*-tyrosine.*

Type of adduct	Sample	Species	Comment	Reference
Dityrosine residues	Carotid intimal protein	Human	Increased levels in atherosclerotic plaques	(Fu *et al.*, 1998a)
	Lens protein	Human		(Souza *et al.*, 2000)
	Heart and muscle	Rat	Cytosolic and mitochondrial protein in control	(Leeuwenburgh *et al.*, 1999a)
	Heart, skeletal muscle, brain and liver	Mouse	Decreased formation in dietary restriction	(Leeuwenburgh *et al.*, 1997b)
	Cuticle	*C. elegans*	Duox shown as precursor of cuticle protein dityrosine	(Edens *et al.*, 2001)
Dityrosine free adduct	Urine	Human	Normal healthy controls	(Ahmed *et al.*, 2005a)
o-Tyrosine residues	Skin collagen	Human	Increase with ageing *o*-Tyr (μmol mol^{-1} Phe) = 0.88 × age (years) + 1.2	(Wells-Knecht *et al.*, 1997)

peroxynitrite was favoured at low steady-state concentrations of peroxynitrite (Pfeiffer *et al.*, 2000). Dityrosine residue formation in cellular proteins was suggested as a marker of protein oxidation (Giulivi and Davies, 2001). Dityrosine residues are formed enzymatically by the dual oxidase/peroxidase activity of the phagocyte oxidase subunit gp91*phox*. Dityrosyl residues are formed in cuticle collagen and other proteins to stabilize the cuticular extracellular matrix (Edens *et al.*, 2001). Some examples of dityrosine residue detection are given in *Table 4*.

2.5 Oxidation of phenylalanine residues

Phenylalanine residues in proteins may be oxidized in iron (III)- and copper (II)-catalysed oxidations with hydrogen peroxide leading to the formation of *o*- and *m*-tyrosine (Leeuwenburgh *et al.*, 1997a, 1997b; *Figure 6B*). *o*- and *m*-tyrosine residue concentrations were increased in mitochondrial proteins of heart muscle after exercise (Leeuwenburgh *et al.*, 1999a). Antioxidants had little effect on the levels of *o*-tyrosine in urine of ageing and exercised rats (Leeuwenburgh *et al.*, 1999b). Some examples of *o*- and *m*-tyrosine residue detection are given in *Table 4*.

2.6 Oxidation of tryptophan residues

Tryptophan (trp) and tryptophan residues are susceptible to oxidation by hydrogen peroxide, hypochlorite and peroxynitrite under physiological conditions (Kato *et al.*, 1997; Simat and Steinhart, 1998). The major products of trp oxidation were oxindolylalanine, 3a-hydroxy-1,2,3,3a,8,8a-hexahydropyrrolo[2,3-b]indole-2-carboxylic acid,

Table 5. Detection of protein oxidation adduct residues and free adducts in biomedical research: N-formylkynurenine.

Adduct	Sample	Species	Comment	Reference
NFK residues	Plasma protein	Human	Normal controls	(Ahmed *et al.*, 2005a)
	Lens protein	Bovine	αA-crystallins oxidized by hydrogen peroxide: W9 and W60 modified	(Finley *et al.*, 1998)
	Collagen type IV	Human	Peroxynitrite oxidation	(Kato *et al.*, 1997)
	Cardiac mitochondria	Human	Multiple mitochondrial proteins modified identified by proteomics, especially complex I subunit B17.2	(Castegna *et al.*, 2003; Taylor *et al.*, 2003)
	Cerebrospinal fluid protein	Human	Healthy controls: 3.17 ± 0.50 mmol mol^{-1} trp; increased in Alzheimer's disease	(Ahmed *et al.*, 2004)
NFK free adduct	Urine	Human	Normal controls	(Ahmed *et al.*, 2005a)
	Cerebrospinal fluid	Human	Healthy controls: 391 ± 196 nM	(Ahmed *et al.*, 2004)

N-formyl-kynurenine (NFK), dioxindolylalanine, kynurenine (Kyn) and 5-hydroxytryptophan. The rate of oxidation was slower for trp residues in proteins than for free trp where oxindolylalanine and NFK were the main degradation products (Simat and Steinhart, 1998; *Figure 6C*). NFK may be formed by oxidation of trp with peroxidase compound III (Ximenes *et al.*, 2001) and model iron (III) complexes (Itakura *et al.*, 1994). NFK is fluorescent with excitation and emission wavelength maxima of 325 nm and 434 nm, respectively. Kynurenine is also fluorescent but it is a weaker fluorophore than NFK with excitation and emission wavelength maxima of 365 nm and 480 nm, respectively (Fukunaga *et al.*, 1982). Some examples of NFK residue detection are given in *Table 5*.

Other oxidized amino acids adducts studied as markers of protein oxidation are: 3,4-dihydroxyphenylalanine (dopa), hydroxyleucine, hydroxyvaline and others (Fu *et al.*, 1998a, 1998b).

2.7 Protein carbonyls, advanced protein oxidation products and non-tryptophan fluorescence

'Protein carbonyls' have been employed as a measure of protein oxidation (Berlett and Stadtman, 1997). Total protein carbonyls have been determined by derivatization with 2,4-dinitrophenyl-hydrazine and the related hydrazones formed detected by characteristic absorbance at 360–390 nm (Levine *et al.*, 1994), immunodetection (Nakamura and Goto, 1996) or by reduction with tritiated borohydride (Yan and Sohal, 2000). Protein carbonyls are considered to be mainly 2-aminoadipic semialdehyde (AASA), formed from oxidative deamination of lysine, and glutamic semialdehyde (GSA), formed by oxidation of proline and arginine residues (Requena *et al.*, 2001). AASA may oxidize further to 2-aminoadipic acid and this has been detected in human skin collagen (Sell *et al.*, 2007). These analytes have been detected discretely after reduction to 6-hydroxy-2-aminocaproic acid (HACA) and 5-hydroxy-2-aminovaleric acid (HAVA; Requena *et al.*, 2001). The GSA residue content of rat heart mitochondria was 20-fold higher than AASA (Pamplona *et al.*, 2002b). There is doubt whether the measure of protein carbonyls is reporting oxidative damage: for a recent validation, this putative measure of protein oxidation was unresponsive in models of oxidative stress (Kadiiska *et al.*, 2005).

'Advanced oxidation protein products' (AOPPs) is a further indirect measure of protein oxidation where molecular species contributing to the assessment have been incompletely defined. It is a measure of the ability of protein oxidation products to oxidize iodide to iodine (WitkoSarsat *et al.*, 1996) – thought to be related to N-chloramine derivatives formed by oxidation of protein with hypochlorite generated by myeloperoxidase (Capeillere-Blandin *et al.*, 2004). Although this marker was found to increase in skeletal muscle of ageing rats (Cakatay *et al.*, 2003), it is not clear if this has significance for oxidative damage of proteins.

Non-tryptophan protein fluorescence has been linked with protein damage due to its association with formation of AGEs (Monnier *et al.*, 1984) and formation of the ageing pigment lipofuscin (Sohal, 1981). Formation of AGEs often involves production of trace amounts of the intense fluorophores pentosidine (Sell and Monnier, 1989) and vesperlysine (fluorophore LM-1) (Tessier *et al.*, 1999), and the weaker fluorophore argpyrimidine (Shipanova *et al.*, 1997). Protein oxidation may lead to the oxidation of tryptophan to form the fluorophores NFK and kynurenine (Fukunaga *et al.*, 1982). Interaction of proteins with lipid peroxidation products, such as malondialdehyde, also leads to the formation of fluorophores such as 3,5-diformyl-1,4-dihydropyridin-4-yl-pyridinium derivatives (Yamada *et al.*, 2001) and related cross-links (Itakura *et al.*, 1996). These and other compounds (retinoid derivatives in retinal tissue, for example; Eldred and Lasky, 1993; Sparrow *et al.*, 1999) are thought to contribute to the fluorescence of lipofuscin. The mixture of fluorophores is complex (Li *et al.*, 2006) and includes fluorescence of endogenous fluorescent metabolites – reduced nicotinamide adenine dinucleotides (NADH and NADPH), flavin adenine dinucleotide (FAD) and porphyrins (Na *et al.*, 2000).

3 Markers of protein nitration

Nitration of proteins in physiological systems occurs via the interaction of proteins with reactive nitrogen species such as the nitronium cation, NO_2^+, nitryl chloride, NO_2Cl, and related species. Susceptible sites for nitration are tyrosine and tryptophan residues, giving rise to the formation of 3-nitrotyrosine (3-NT) and 6-nitrotryptophan (Lerman *et al.*, 2002; Ryberg and Caidahl, 2007; *Figure 6D*). 3-NT residues are present in proteins of tissues, extracellular matrix, plasma and other body fluids formed by nitration of tyrosine residues. The cellular compartmentalization of 3-NT residue-containing proteins does not usually change compared to the unmodified protein – although albumin containing 3-NT residues showed preferential transport across the capillary endothelium (Predescu *et al.*, 2002). The 3-NT residue content of proteins has been markedly overestimated by measurement with immunoassay and chromatographic

Table 6. *Estimates of 3-nitrotyrosine residue content of plasma protein, haemoglobin and lipoproteins by the LC-MS/MS method and comparison with other reports of tandem mass spectrometric analysis.*

Adduct	Protein	Species	3-NT content / concentration	Reference
3-NT residue ($\mu mol\ mol^{-1}$)	Plasma protein	Human	0.6 ± 0.3 (n = 12)	(Ahmed *et al.*, 2005b)
	Plasma protein	Rat	0.67 ± 0.29 (n = 13)	P.J. Thornalley, unpublished
	Hemoglobin	Human	24 ± 2 (n = 12)	(Ahmed *et al.*, 2005b)
	Cerebrospinal fluid protein	Human	8.9 ± 4.7 (n = 18)	(Ahmed *et al.*, 2004)
3-NT free adduct (nM)	Plasma	Human	0.74 ± 0.30 (n = 12)	(Soderling *et al.*, 2003)
		Rat	2.05 ± 0.37 (n = 5)	(Rabbani *et al.*, 2007)
	Cerebrospinal fluid protein	Human	0.40 ± 0.28 (n = 18)	(Ahmed *et al.*, 2004)
Urinary 3-NT free adduct (nmol mmol creatinine^{-1})	Urine	Human	21 ± 10 (n = 12)	(Ahmed *et al.*, 2005b)

electrochemical and selective ion-monitoring mass spectrometry (gas chromatography with mass spectrometric detection, GC-MS, and liquid chromatography with mass spectrometric detection, LC-MS). Tandem mass spectrometric detection is required (GC-MS/MS or LC-MS/MS) or sensitive immunoassay methods that corroborate with these methods (Rabbani and Thornalley, 2008). Some examples of 3-NT residue detection are given in *Table 6*.

4 Advances in quantitation of protein damage

The assessment of protein damage – measurement of multiple trace amounts (fmol–pmol) of glycated, oxidized and nitrated amino acids in the presence of 10^3–10^6-fold excess of related unmodified amino acid – is a challenge of the analytical methodology employed. Physiological samples are also labile; glycation, oxidation and nitration adduct analytes may be formed readily by inappropriate sample processing. Immunoassays have been employed for the measurement of protein glycation, oxidation and nitration adduct analytes but have often given unreliable and overestimated values of analyte content. The reasons for this are: (i) poor antigen specificity, (ii) introduction of analyte into the assay matrix in blocking proteins (to decrease non-specific binding of the antibody), (iii) formation of analyte during sample processing, and (iv) lack of validation of matrix effects – particular interferences. Fluorescence-based assays are appropriate but are restricted to the minority of protein glycation, oxidation and nitration adduct analytes that are fluorescent. Fluorophores should also be resolved by a chromatographic step if quantitative measurement is desired. High performance liquid chromatography (HPLC) with electrochemical detection has been used for estimation of 3-NT residues but has led to marked overestimation. We have discussed these issues in relation to detection of glycation adducts (Ahmed and Thornalley, 2007), oxidation adducts (Thornalley, 2006a) and nitration adducts elsewhere (Rabbani and Thornalley, 2008). The gold standard reference method for analysis of protein glycation, oxidation and nitration adducts is liquid chromatography with tandem mass spectrometric detection (LC-MS/MS) with isotopic dilution analysis – as we have described (Thornalley, 2006b; Thornalley *et al.*, 2003). Protein glycation, oxidation and nitration adduct residues are determined after exhaustive enzymatic hydrolysis of proteins. Protein glycation, oxidation and nitration free adducts are determined in ultrafiltrates of physiological fluids (using typically a 10–12 kDa cut-off microspin filter). Many research teams are now employing this technology. Other techniques – particularly immunoassay – are more practicable for high sample throughput but the protocol and use in a particular sample matrix should be validated to the reference LC-MS/MS method before use. Future developments involve application of ultra-high-performance liquid chromatography with tandem mass spectrometric detection (UPLC-MS/MS) and robotic automation of enzymatic hydrolysis procedures.

5 Proteasomal proteolysis of glycated, oxidized and nitrated proteins

5.1 Proteasomal proteolysis of damaged proteins

Protein glycation, oxidation and nitration were originally viewed as post-translational modifications that accumulated mostly on extracellular proteins. For example, AGE residues were thought to be formed slowly in proteins throughout life and the protein residue content of AGE residues found considered to represent a life-long accumulation of these glycation adducts. Whilst this applies for chemically stable AGEs formed on long-lived proteins, it is now appreciated that FL and some AGEs (hydroimidazolones, for example) have relatively short chemical half-lives under physiological conditions (2–6 weeks) and they may also be formed on cellular and short-lived extracellular proteins. Similar arguments apply to the formation of protein oxidation and nitration adducts. Protein damage by glycation, oxidation and nitration is implicated in protein misfolding. Misfolded proteins are degraded by the proteasome to ensure the high quality of intracellular proteins (Goldberg, 2003); the median half-life of cellular proteins was 32 h (Gerner *et al.*, 2002). For example, mild oxidation of protein leads to proteasomal proteolysis (Grune *et al.*, 1996); incorporation of oxidized amino acids into proteins targets the modified protein for cellular proteolysis (Dunlop *et al.*, 2008); nitrated proteins are also degraded by cellular proteolysis (Greenacre *et al.*, 1999; Grune *et al.*, 2001); and changes in glycated proteins during differentiation of embryonic stem cells suggest they too are targeted for proteasomal degradation (Hernebring *et al.*, 2006). The turnover of proteins damaged by glycation, oxidation and nitration due to degradation by cellular proteolysis releases glycated, oxidized and nitrated amino acids called glycation, oxidation and nitration free adducts (Thornalley *et al.*, 2003). They are also formed by digestion of damaged glycated proteins in food, and absorbed from the gastrointestinal tract (Ahmed and Thornalley, 2007). Damaged proteins are degraded preferentially by the 20S proteasome in a ubiquitin-independent process (Grune *et al.*, 1996; Hernebring *et al.*, 2006). The mechanism of presentation of glycated, oxidized and nitrated proteins to the proteasome has not been fully characterized but heat-shock proteins 70, 90 and others are likely involved (Bassi *et al.*, 2005; Whittier *et al.*, 2004).

5.2 Protein glycation, oxidation and nitration free adducts

Protein glycation, oxidation and nitration free adducts are exported from cells, leak into plasma and are excreted in the urine. Indeed, glycation, oxidation and nitration free adducts are present in all body fluids – cerebrospinal fluid, synovial fluid and others (Ahmed *et al.*, 2004, 2006). Glycation, oxidation and nitration free adducts are the major form in which glycation adducts are excreted from the body in urine (Ahmed *et al.*, 2005; Kirschbaum, 2001; Thornalley *et al.*, 2003). Major quantitative glycation adducts, FL, hydroimidazolones, CML and CEL, have high renal clearances – 35–93 ml min^{-1} in normal human subjects but this declines in uremia (Thornalley *et al.*, 2003). MetSO and 3-NT both had low renal clearances (<10 ml min^{-1}; Thornalley *et al.*, 2003). This reflects the efficient renal reduction of MetSO by MetSO reductase (Weissbach *et al.*, 2002) and renal re-uptake and/or metabolism of 3-NT (Ohshima *et al.*, 1990). Loss of renal function leads to profound increase in protein glycation, oxidation and nitration free adducts in plasma. This was found experimentally after bilateral nephrectomy (Rabbani *et al.*, 2007) and clinically in patients with end-stage renal disease (Agalou *et al.*, 2005). The plasma concentrations of protein glycation, oxidation and nitration free adducts increased less markedly in experimental bilateral ureteral ligation than in bilateral nephrectomy, demonstrating the ability of the kidney to remove these damaged amino acids from the circulation (Rabbani *et al.*, 2007; *Figure 7*).

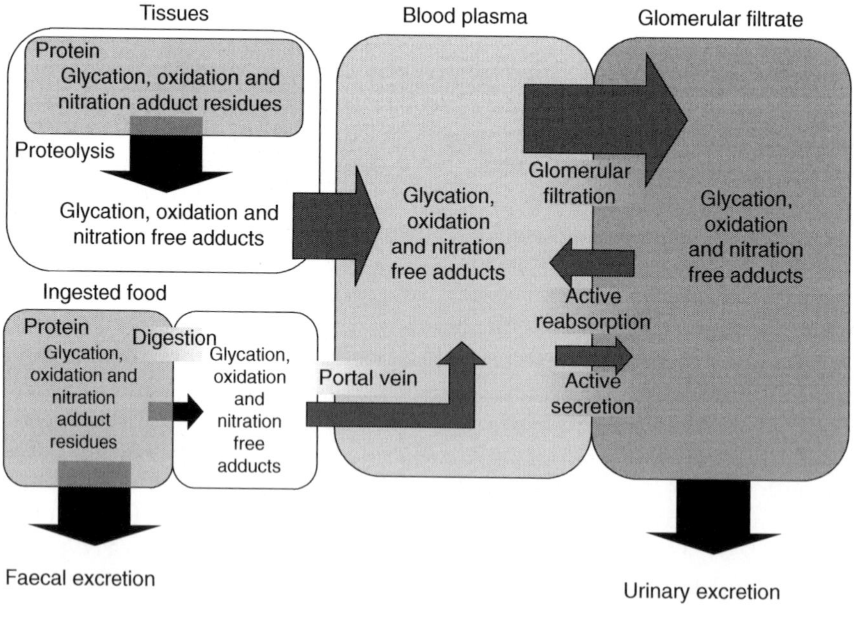

Figure 7. *Biodistribution scheme illustrating flows of formation and removal of protein oxidation free adducts.*

NFK free adduct is found in plasma, urine and other physiological fluids *in vivo* (Buxton and Guilbault, 1974). NFK free adduct is also formed enzymatically by tryptophan 2,3-dioxygenase (TDO; EC 1.13.11.17; Knox and Mehler, 1950). It is usually found together with kynurenine formamidase (KFA; EC 3.5.1.9) which converts NFK to kynurenine (Pabarcus and Casida, 2002; Seifert, 1993). This is part of the kynurenine pathway, which forms kynurenic acid and also 3-hydroxykynurenine and quinolinic acid – intermediates involved in NAD^+ synthesis (Heyes *et al.*, 1997; Schwarcz and Pellicciari, 2002; *Figure 5B*). TDO expression was induced via tumour necrosis factor-α and interferon-γ dependent pathways in response to endotoxin (Fujigaki *et al.*, 2001). Kynurenine pathway metabolites were also increased in uremia (Pawlak *et al.*, 2001). TDO and KFA act on tryptophan and NFK free adduct, respectively, and not tryptophan and NFK residues.

Free 3-NT is also present in plasma, urine, cerebrospinal fluid, synovial fluid and presumably interstitial fluid and other body fluids. Free 3-NT is transported across cell membranes by the L-aromatic amino-acid transporter and dopamine transporters (Blanchard-Fillion *et al.*, 2006). Free 3-NT is metabolized in rats and human subjects, leading to the formation of the major urinary metabolite 3-nitro-4-hydroxyphenylacetic acid (3-NHA) – with minor formation of 3-nitro-4-hydroxyphenyllactic acid (Ohshima *et al.*, 1990; Tabrizi-Fard *et al.*, 1999). 3-NHA is not a good marker of flux of formation of 3-NT, however, as it may be formed by 3-NT-independent pathways (Pannala *et al.*, 2006).

6 Accumulation of protein damage and the ageing phenotype

6.1 Studies in human subjects

Evidence of increased protein damage in ageing of human subjects has been sought by analysis of markers of damage in proteins of long biological half-life – typically proteins of lens fibre cells (with limited proteasomal degradation; Bloemendal *et al.*, 2004; Pereira *et al.*, 2003) and skin and articular collagen (half-lives of 14.8 and 115 years, respectively; Verzijl *et al.*, 2000b).

The AGE residue content of human lens proteins increased with donor age (Ahmed *et al.*, 1997, 2003; Brinkmann Frye *et al.*, 1998; Dunn *et al.*, 1989; Dyer *et al.*, 1991; Sell and Monnier, 2004). Increased pentosidine residue content of skeletal muscle (Haus *et al.*, 2007) and CML residue content of vimentin and a proteasome subunit was also found in fibroblasts of elderly human donors (Gonzalez-Dosal *et al.*, 2006; Kueper *et al.*, 2007). The change in oxidative markers was less clear: there was a small increase in dityrosine and no increase in *o*-tyrosine with increasing age in lens protein (Wells-Knecht *et al.*, 1993).

AGE residue content of human skin and articular collagen increased with donor age (Brinkmann Frye *et al.*, 1998; Dunn *et al.*, 1991; Sell *et al.*, 1996, 2007; Verzijl *et al.*, 2000a, 2000b). Protein oxidation markers, methionine sulphoxide and *o*-tyrosine, were increased in skin collagen with age (Wells-Knecht *et al.*, 1997). A study of human subjects and seven other mammalian species showed an inverse correlation between lifespan and rate of accumulation of increase of pentosidine residue content in skin collagen (Sell *et al.*, 1996).

6.2 *Studies in rodents*

In rats, fluorescence measurements reflective of protein damage were increased with age in rat-tail tendon collagen and glomerular basement membrane of Wistar rats (Odetti *et al.*, 1994; Verbeke *et al.*, 1997). Similar fluorescence increases with age were found in skin and aortic collagen of Sprague Dawley rats (Novelli *et al.*, 1998). The CML and pentosidine residue content of Brown Norway rats increased significantly with age in skin collagen (Cefalu *et al.*, 1995). Immunoblotting of rat skeletal muscle with anti-AGE antibody 6D12 that recognizes CML and CEL residues showed increases of immunoreactivity of 33- and 36-month-old rats, with a particular increase in immunoreactivity of β-enolase with ageing (Snow *et al.*, 2007); hepatic mitochondrial glutamate dehydrogenase showed similar increased 6D12 antibody immunoreactivity in aged Wistar rats (Hamelin *et al.*, 2007). Ageing in Wistar rats increased the CML residue content but not the CEL residues or protein carbonyl contents of rat-heart mitochondrial protein (Pamplona *et al.*, 2002a). In mice, pentosidine residue content of tendon, ear and skin collagen increased with age, increasing faster in the short-lived DBA/2 strain than the long-lived C57BL/6 strain (Sell and Monnier, 1997).

For protein oxidation, there was an age-related increase in dityrosine residue content but no increased *o*-tyrosine residue content of rat heart and skeletal muscle protein of ageing mice (Leeuwenburgh *et al.*, 1997b). Urinary excretion of dityrosine free adduct was increased in aged rats (Leeuwenburgh *et al.*, 1999b). Increased protein nitration is thought to play a role in ageing (Drew and Leeuwenburgh, 2002), although there was no increase in 3-NT residues of heart, skeletal muscle and liver protein extracts of old (24 months old) compared to young (9 months old) female Long-Evans/Wistar hybrid rats (Leeuwenburgh *et al.*, 1998). Several other studies have found evidence by immunoblotting and proteomics analysis for increased nitration of specific proteins in skeletal muscle, heart and brain (Gokulrangan *et al.*, 2007; Kanski *et al.*, 2005a, 2005b).

6.3 *Studies in flies,* Drosophila melanogaster, *and nematode,* Caenorhabditis elegans

The accumulation of AGE residues in protein extracts of young and old *Drosophila melanogaster* was assessed by 'AGE fluorescence' (excitation wavelength 365 nm, emission wavelength of 440 nm). Young flies (10 days old) had 44% less fluorescence than senescent flies (75 days old). Fluorescence was decreased by raising the flies on a medium containing aminoguanidine (20–40 mM). This is a scavenger of dicarbonyl precursors of AGEs but is also a scavenger of pyruvate and glucose at this high concentration (Thornalley, 2003b) and hence could have imposed a chemical caloric restriction. Reduction of fluorescence in flies failed to increase their mean lifespan, however, probably because high concentrations of aminoguanidine were toxic (Oudes *et al.*, 1998). Chemical derivatization of protein carbonyls on mitochondrial matrix proteins of the flight muscles of flies suggested aconitase was the predominant protein that exhibited an age-associated increased damage. There was a concomitant 50% loss in aconitase activity (Das *et al.*, 2001). A comparison among five different species of dipteran flies, namely, *Drosophila melanogaster* (fruit fly), *Musca domestica* (house fly), *Sarcophaga bullata* (flesh fly), *Calliphora vicina* (blow fly) and *Phaenecia sericata* (a species of blow fly) showed the average lifespan was inversely correlated with the level of protein carbonyls (Sohal *et al.*, 1995).

In the nematode *Caenorhabditis elegans*, a short- (mev-1) lifespan mutant accumulated protein carbonyls at a faster rate than the wild type, whereas a long-lifespan mutant (age-1) exhibited no increase except at the end of an extended lifespan (Adachi *et al.*, 1998). Recently we have shown that there is an accumulation of methylglyoxal-derived AGE residues in mitochondrial proteins of *C. elegans* with age, and prevention of this by over-expression of Glo1 was associated with life extension (Morcos *et al.*, 2008). Autofluorescent (excitation 340 nm, emission 430 nm) of wild-type animals showed slow accumulation during the reproductive phase and more rapidly during the post-reproductive period. Mutations that lower efficacy of insulin signalling and extend lifespan (daf-2(e1370) insulin receptor and age-1(hx546) PI3-kinase) prevented accumulation of autofluorescence; conversely, elimination of the insulin-inhibited DAF-16/FOXO transcription factor causes an increased autofluorescence. Mutations that increased mitochondrial ROS formation did not affect autofluorescence. Dietary restriction decreased autofluorescence. Genetically identical siblings that age poorly (as judged by decrepit locomotory capacity) had increased autofluorescence compared with same-aged siblings that aged more healthily (Gerstbrein *et al.*, 2005). There was an increase in protein oxidation and nitration in *C. elegans* with ageing, as judged by LC-MS/MS detection of MetSO and 3-NT residues. Over-expression of Glo1 prevented this increase – an effect mediated by prevention of glycation of mitochondrial proteins and related increase in mitochondrial formation of ROS (Morcos *et al.*, 2008).

6.4 Studies in mammalian cell senescence

There have been few studies of quantitative changes in protein glycation in cell senescence. Studies with anti-AGE antibodies suggested that protein AGE residue content increased during replicative senescence of WI-38 human embryonic fibroblasts (Ahmed *et al.*, 2007). Aminoguanidine (2 mM and 4 mM) delayed the senescence of human diploid fibroblasts *in vitro* (Wang *et al.*, 2007). This is a higher concentration than required for inhibition of dicarbonyl glycation and is expected to also scavenge pyruvate (Thornalley, 2003b). The effect, therefore, may be mediated by a chemically induced decrease in cell respiration by removal of pyruvate from anaerobic glycolysis.

In the BJ fibroblast model of proliferative and non-dividing cell senescence, protein carbonyls and lipofuscin fluorescence increased as cells entered senescence (Sitte *et al.*, 2000a, 2000b). A similar effect was found in MCR5 human fibroblasts during proliferative senescence (Sitte *et al.*, 1998).

7 Caloric restriction and protein damage

Caloric restriction (CR) is a proven intervention that increases lifespan in rodents, flies and nematodes (Pletcher *et al.*, 2005; Walker *et al.*, 2005; Wang *et al.*, 2004; Weindruch and Walford, 1982). It may do so in human and non-human primates too but conclusive data are not yet available (Weindruch, 2006). CR (30%) over a 7-year study period decreased FL residue content of skin collagen of rhesus, but not squirrel monkeys. The CML and pentosidine residue contents of skin collagen were not affected in both squirrel and rhesus monkeys. These results suggest that caloric restriction, when maintained long-term in non-human primates, tends to decrease early glycation – and probably glycaemic status – but not levels of these AGE residues (Sell *et al.*, 2003). CR of rhesus monkeys decreased oxidative damage in skeletal muscle, as judged by measurement of protein carbonyls and immunoassay of 3-NT residues (Zainal *et al.*, 2000). CR (40%) decreased FL content of plasma protein in Sprague Dawley rats (Davis *et al.*, 1994). CR (40%) of Brown-Norway rats from 4–25 months of age produced decreased levels of glycated haemoglobin, plasma proteins (18–33%), and CML (25%) and pentosidine (50%) residue content and fluorescence (15%) of skin collagen (Cefalu *et al.*, 1995). In mice, the age-related increases of pentosidine residue content of tendon, ear and skin collagen were suppressed by CR (20–40%) in the short-lived DBA/2 strain but not in the long-lived C57BL/6 strain (Sell and Monnier, 1997). In Fischer 344 rats, the rate of glycation of collagen was found to be tissue specific but unrelated to the type of carbohydrate (cornstarch, glucose, sucrose or fructose). Restricting the amount of calories consumed resulted in lower serum glucose concentrations, glycated haemoglobin levels and pentosidine concentrations in tail tendon collagen (Lingelbach *et al.*, 2000). Markers of skin collagen glycation predicted early deaths in *ad libitum* and caloric restricted C57BL/6NNia mice (Sell *et al.*, 2000). It is not only long-lived extracellular proteins that are protected from damage in dietary restriction. AGE residue content of liver and kidney protein extracts was decreased by dietary restriction (Teillet *et al.*, 2000, 2002). After 4 months of CR (40%) in Wistar rats, the CEL, CML, malondialdehyde-lysine and GSA residue contents of rat-heart mitochondrial protein were decreased with respect to controls (Hamelin *et al.*, 2007). Long-term CR (1 year, 40%) decreased CEL and malondialdehyde-lysine residue content of rat-heart mitochondrial protein without change in CML residue and protein carbonyl contents (Pamplona *et al.*, 2002a). For protein oxidation, CR prevented the age-related increase in dityrosine residue content of heart and skeletal muscle but did not decrease *o*-tyrosine residue content in ageing mice (Leeuwenburgh *et al.*, 1997b). CR suppressed the age-related increases in lipofuscin in mice (Moore *et al.*, 1995).

8 Protein damage in the ageing phenotype: Increased formation of damaging agents, decreased defences or both?

Studies of damage to the mammalian proteome in the last decade have revealed the susceptibility of proteins, extracellular and cellular, long-lived and short-lived, to spontaneous damage by glycation, oxidation and nitration. Studies of cellular proteolysis have also revealed the dynamic nature of protein glycation, oxidation and nitration. There is a continual flux of formation of glycating, oxidizing and nitration agents, a related flux of formation of protein glycation, oxidation and nitration adduct residues, and targeting of damaged proteins for proteolysis with formation and excretion of protein glycation, oxidation and nitration free adducts. If the renal clearance of these free adducts is impaired or lost, the concentration of glycation, oxidation and nitration free adducts increases profoundly (Rabbani *et al.*, 2007). Recovery from oxidative stress and replacement of proteins damaged irreversibly involves the proteasomal system (Shang and Taylor, 1995). Decreasing damage to the proteome by glycation and oxidation involves the activation of the 20S proteasome (Hernebring *et al.*, 2006). It might be expected, therefore, that age-related loss of proteasomal activity is linked to accumulation of damaged proteins and increased risk of irreversible tissue damage and cell senescence. This has indeed been found (Louie *et al.*, 2002; Sitte *et al.*, 2000a, 2000b).

Damage to proteins by glycation, oxidation and nitration is not random. There are neighbouring group interactions that activate particular amino-acid residues for modification. For example, a lysine residue in the peptide backbone of an α-helix that has basic and acidic amino-acid residues 3–4 residues from it in either direction, is activated for glycation by glucose (Venkatraman *et al.*, 2001). Where reactivity towards glycation, oxidation and nitration coincides with functional activity of an amino acid residue, the functional consequence of damage is likely to be profound; for example, in dicarbonyl glycation of arginine residues in RGD and GFOGER integrin-binding domains of type-IV collagen (Dobler *et al.*, 2006) and oxidation of active-site cysteinyl thiols in protein tyrosine phosphatases (Denu and Tanner, 1998). Damage of short-lived cellular and extracellular proteins is not without consequence: glycation of mitochondrial proteins leads to impairment of mitochondrial electronic transport and increased formation of ROS (Sell and Monnier, 2004). Glycation of the co-repressor protein sin3A in endothelial cells gives rise to increased angiopitein-2 production and increased apoptosis (Yao *et al.*, 2007).

The life-long accumulation of protein damage on short-lived proteins and short-lived glycation and oxidation adducts on proteins reveals a critical feature of protein damage in ageing. Accumulation of protein damage in ageing is often associated with a decline in cellular protective and protein reparative activities. This defence is critical in countering the effects of endogenous and exogenous glycating, oxidizing and nitrating agents (*Figures 8 and 9*). Decline of the enzymatic defences against protein

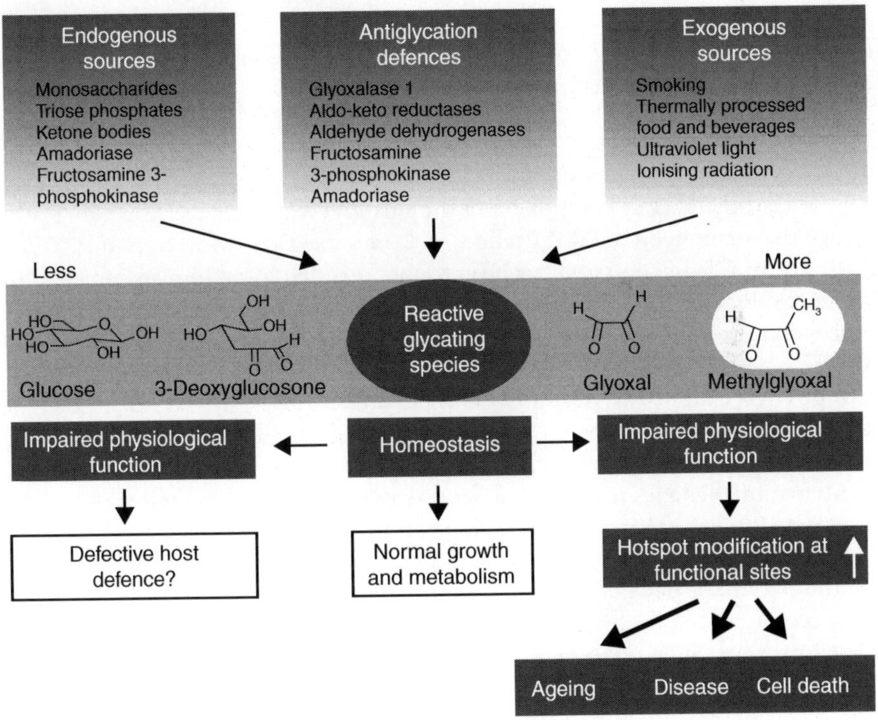

Figure 8. *The sources of and cellular responses to glycating agents, countered by the enzymatic defence against glycation.*

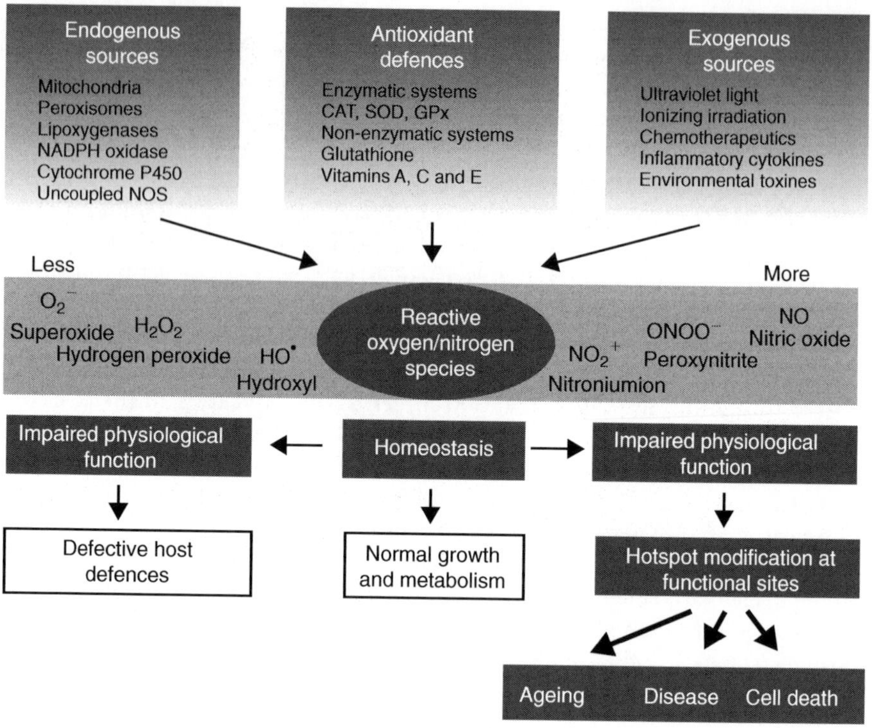

Figure 9. *The sources of and cellular responses to reactive oxygen and nitrogen species, countered by the enzymatic defence against glycation.*

damage may also lead to increased formation of damaging agents in the ageing process. The recent study of a key component of the anti-glycation defence, Glo1, in *C. elegans* revealed that expression of Glo1 declines with increased age (Morcos *et al.*, 2008). Similar decline of Glo1 expression has been found in rodents and human subjects (Kuhla *et al.*, 2006; Piec *et al.*, 2005; Sharmaluthra and Kale, 1994). The downstream consequence of decline in Glo1 expression was shown by achieving life extension with over-expression and life-shortening by silencing of Glo1. The consequent increase in dicarbonyl glycation with decline in Glo1 expression was critical for mitochondrial proteins. Decline of this protective enzymatic system produced increased steady-state levels of dicarbonyl glycation damage to short-lived mitochondrial proteins, and potentiated further and made more diverse damage by enhancing mitochondrial ROS formation with increased oxidative and nitration damage (increased MetSO and 3-NT residues) and ageing. We would expect, therefore, that genes of proteins associated with the enzymatic defence against glycation, oxidation and nitration may afford life-extension (or at least healthier ageing) with engineered or pharmacologically induced over-expression. Such genes have been called 'vitagenes' (Weindruch and Walford, 1982). Indeed, some genes for enzymes of the anti-glycation and antioxidant defences are in this class: Glo1 (Morcos *et al.*, 2008), MSR (Moskovitz *et al.*, 2001; Petropoulos and Friguet, 2005) and catalase (Schriner and Linford, 2006). These and others have been collated in the GenAge database (de Magalhaes and Toussaint, 2004).

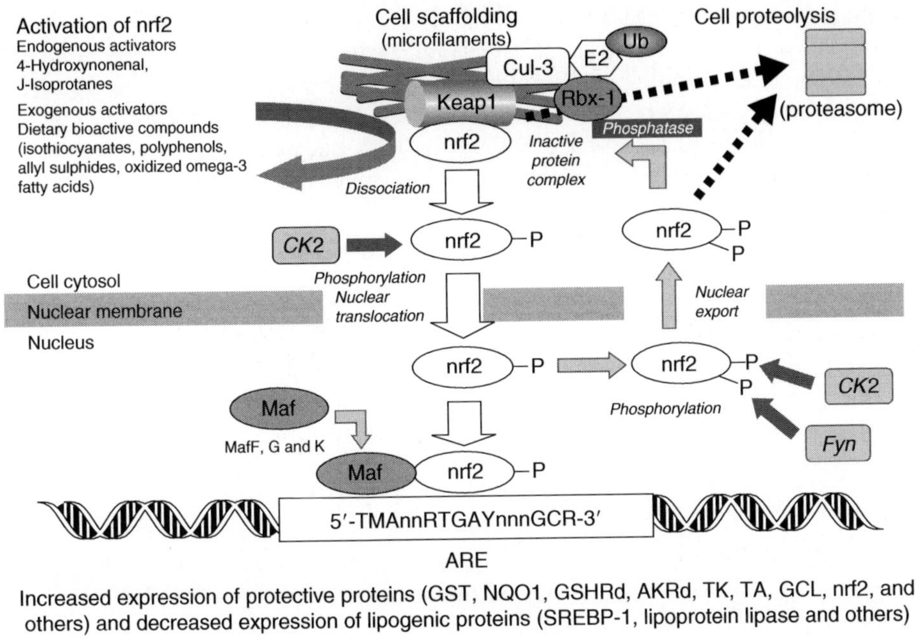

Figure 10. *Anti-ageing responses induced by activators of nrf2 (AKRd, aldo-keto reductase; CK2, casein kinase 2; Cul3, Cullin-3; E2, ubiquitin activating protein; Fyn, 59-kDa src family-related protein tyrosine kinase; GCL, γ-glutamylcysteine ligase; GSHRd glutathione reductase; GSTs, glutathione transferases; Keap-1, Kelch-like ECH-associated protein 1; NQO1, quinone reductase; –P, phosphorylation; SREBP1, sterol response element binding protein-1; TK, transketolase; TA, transaldolase; Ub, ubiquitin).*

Studies of the ageing process in multiple model systems indicates that genes with expression products associated with prevention of damage to the proteome (and genome), and detoxification and removal of waste products and xenobiotics, are linked to healthy ageing (Schriner and Linford, 2006). Many of these genes have antioxidant response elements (AREs) in their promoters and hence have expression regulated by transcription factor NF-E2-related factor-2 (nrf2; *Figure 10*). Such genes are also involved in countering glycation, oxidation and nitration – for example, aldoketo reductases and glutathione reductase (Kwak *et al.*, 2003). A strategy for healthy ageing is to provide for increased expression of vitagenes and this may been achieved in part by inducing expression of ARE-linked genes with pharmacological activators of nrf2. Life extension by CR was lost in *C. elegans* by ablation of the nrf2 gene homolog SKN-1 (Bishop and Guarente, 2007), which suggested nrf2 and related gene signalling has a critical role in achieving increased longevity. Recent research suggests, however, that the effect of nrf2 signalling in CR-induced life extension in rodents is due to cancer prevention effects of increased ARE-linked gene expression (Pearson *et al.*, 2008). Development of tumours is a major cause of mortality in experimental rodent models of CR-induced life extension (Weindruch, 1992). Candidate nrf2 activators are dietary bioactive compounds such as glucosinolate-derived isothiocyanates, polyphenols, allyl sulphides, oxidized omega-3 fatty acids and others (Ben Dor *et al.*, 2005; Chen *et al.*, 2004; Gao *et al.*, 2007; Mann *et al.*, 2007; Tanigawa *et al.*, 2007; Thornalley and IARC Workgroup, 2004). Studies of the polyphenol in grape juice, resveratrol, were encouraging initially (Kang *et al.*, 2002; Wood *et al.*, 2004) and disappointing in recent attempted validations (Bass *et al.*, 2007). It remains a key challenge in ageing research to evaluate life extension effects of dietary bioactives that may have multiple health benefits – prevention of cancer, cardiovascular disease and stroke (Thornalley and IARC Workgroup, 2004) and healthy ageing.

Acknowledgements

The authors would like to thank the Biosciences and Biotechnology Research Council (UK), the Wellcome Trust, British Heart Foundation and European Union 7th Framework Programme for support for our research.

References

Adachi, H., Fujiwara, Y. and Ishii, N. (1998) Effects of oxygen on protein carbonyl and aging in *Caenorhabditis elegans* mutants with long (*age-1*) and short (*mev-1*) life spans. *J. Gerontol. Ser. A Biol. Sci. Med. Sci.* **53:** B240–B244.

Aeschbach, R., Amado, R. and Neukom, H. (1976) Formation of dityrosine crosslinks in proteins by oxidation of tyrosine residues. *Biochim. Biophys. Acta* **439:** 292–301.

Agalou, S., Ahmed, N., Babaei-Jadidi, R., Dawnay, A. and Thornalley, P.J. (2005) Profound mishandling of protein glycation degradation products in uremia and dialysis. *J. Am. Soc. Nephrol.* **16:** 1471–1485.

Ahmed, E.K., Picot, C.R., Bulteau, A.L. and Friguet, B. (2007) Protein oxidative modifications and replicative senescence of WI-38 human embryonic fibroblasts. *Ann. N.Y. Acad. Sci.* **1119:** 88–96.

Ahmed, M.U., Brinkmann Frye, E., Degenhardt, T.P., Thorpe, S.R. and Baynes, J.W. (1997) N^{ε}-(Carboxyethyl)lysine, a product of chemical modification of proteins by methylglyoxal, increases with age in human lens proteins. *Biochem. J.* **324:** 565–570.

Ahmed, N. and Thornalley, P.J. (2007) Advanced glycation endproducts: what is their relevance to diabetic complications? *Diab. Obes. Metab.* **9:** 233–245.

Ahmed, N., Argirov, O.K., Minhas, H.S., Cordeiro, C.A. and Thornalley, P.J. (2002) Assay of advanced glycation endproducts (AGEs): surveying AGEs by chromatographic assay with derivatisation by aminoquinolyl-*N*-hydroxysuccimidyl-carbamate and application to N^{ε}-carboxymethyl-lysine- and N^{ε}-(1-carboxyethyl) lysine-modified albumin. *Biochem. J.* **364:** 1–14.

Ahmed, N., Thornalley, P.J., Dawczynski, J., Franke, S., Strobel, J., Stein, G. and Haik, G.M. Jr (2003) Methylglyoxal-derived hydroimidazolone advanced glycation endproducts of human lens proteins. *Invest. Ophthalmol. Vis. Sci.* **44:** 5287–5292.

Ahmed, N., Ahmed, U., Thornalley, P.J., Hager, K., Fleischer, G.A. and Munch, G. (2004) Protein glycation, oxidation and nitration marker residues and free adducts of cerebrospinal fluid in Alzheimer's disease and link to cognitive impairment. *J. Neurochem.* **92:** 255–263.

Ahmed, N., Babaei-Jadidi, R., Howell, S.K., Beisswenger, P.J. and Thornalley, P.J. (2005a) Degradation products of proteins damaged by glycation, oxidation and nitration in clinical type 1 diabetes. *Diabetologia* **48:** 1590–1603.

Ahmed, N., Babaei-Jadidi, R., Howell, S.K., Beisswenger, P.J. and Thornalley, P.J. (2005b) Degradation products of proteins damaged by glycation, oxidation and nitration in clinical type 1 diabetes. *Diabetologia* **48:** 1590–1603

Ahmed, N., Ahmed, U., Thornalley, P.J., Watts, R., Tarr, J., Haigh, R. and Winyard, P. (2006) Profound increase in proteolytic products of glycated and oxidised proteins in synovial fluid and plasma in osteoarthritis and rheumatoid arthritis, corrected by tnf-α antibody therapy in rheumatoid arthritis. Rheumatology **45(Suppl 1):** i53.

Bass, T.M., Weinkove, D., Houthoofd, K., Gems, D. and Partridge, L. (2007) Effects of resveratrol on lifespan in *Drosophila melanogaster* and *Caenorhabditis elegans*. *Mech. Ageing Dev.* **128:** 546–552.

Bassi, A.M., Ledda, S., De Pascale, M.C., Penco, S., Rossi, S., Odetti, P. and Cottalasso, D. (2005) Antioxidant status in J774A.1 macrophage cell line during chronic exposure to glycated serum. *Biochem. Cell Biol. Biochim. Biol. Cell.* **83:** 176–187.

Ben Dor, A., Steiner, M., Gheber, L., Danilenko, M., Dubi, N., Linnewiel, K., Zick, A., Sharoni, Y. and Levy, J. (2005) Carotenoids activate the antioxidant response element transcription system. *Mol. Cancer Ther.* **4:** 177–186.

Berlett, B.S. and Stadtman, E.R. (1997) Protein oxidation in aging, disease and oxidative stress. *J. Biol. Chem.* **272:** 20313–20316.

Biemel, K.M., Friedl, D.A. and Lederer, M.O. (2002) Identification and quantification of major Maillard cross-links in human serum albumin and lens protein – evidence for glucosepane as the dominant compound. *J. Biol. Chem.* **277:** 24907–24915.

Bishop, N.A. and Guarente, L. (2007) Two neurons mediate diet-restriction-induced longevity in *C. elegans*. *Nature* **447:** 545–549.

Blanchard-Fillion, B., Prou, D., Polydoro, M., *et al.* (2006) Metabolism of 3-nitrotyrosine induces apoptotic death in dopaminergic cells. *J. Neurosci.* **26:** 6124–6130.

Bloemendal, H., de Jong, W., Jaenicke, R., Lubsen, N.H., Slingsby, C. and Tardieu, A. (2004) Ageing and vision: structure, stability and function of lens crystallins. *Prog. Biophys. Mol. Biol.* **86:** 407–485.

Boschi-Muller, S., Olry, A., Antoine, M. and Branlant, G. (2005) The enzymology and biochemistry of methionine sulphoxide reductases. *Biochim. Biophys. Acta Proteins Proteom.* **1703:** 231–238.

Brinkmann Frye, E., Degenhardt, T.P., Thorpe, S.R. and Baynes, J.W. (1998) Role of the Maillard reaction in aging of tissue proteins. Advanced glycation end product-dependent increase in imidazolium cross-links in human lens proteins. *J. Biol. Chem.* **273:** 18714–18719.

Buxton, T. and Guilbault, G.G. (1974) Fluorometric analysis for N′-formylkynurenine in plasma and urine. *Clin. Chem.* **20:** 765–768.

Cakatay, U., Telci, A., Kayali, R., Tekeli, F., Akcay, T. and Sivas, A. (2003) Relation of aging with oxidative protein damage parameters in the rat skeletal muscle. *Clin. Biochem.* **36:** 51–55.

Capeillere-Blandin, C., Gausson, V., Descamps-Latscha, B. and Witko-Sarsat, V. (2004) Biochemical and spectrophotometric significance of advanced oxidized protein products. *Biochim. Biophys. Acta Mol. Basis Dis.* **1689:** 91–102.

Castegna, A., Thongboonkerd, V., Klein, J.B., Lynn, B., Markesbery, W.R. and Butterfield, D.A. (2003) Proteomic identification of nitrated proteins in Alzheimer's disease brain. *J Neurochem* **85**: 1394–1401.

Castellani, R.J., Honda, K., Zhu, X.W., Cash, A.D., Nunomura, A., Perry, G. and Smith, M.A. (2004) Contribution of redox-active iron and copper to oxidative damage in Alzheimer disease. *Ageing Res. Rev.* **3:** 319–326.

Cefalu, W.T., Bell-Farrow, A.D., Wang, Z.Q., Sonntag, W.E., Fu, M.-X., Baynes, J.W. and Thorpe, S.R. (1995) Calorific restriction decrease age-dependent accumulation of the glycoxidation products, N^{ε}-carboxymethyl-lysine and pentosidine, in rat skin collagen. *J. Gerontol.* **50A:** B337–B341.

Chance, B., Sies, H. and Boveris, A. (1979) Hydroperoxide metabolism in mammalian organs. *Physiol. Rev.* **59:** 527–605.

Chen, C., Pung, D., Leong, V., Hebbar, V., Shen, G., Nair, S., Li, W. and Kong, A.N.T. (2004) Induction of detoxifying enzymes by garlic organosulfur compounds through transcription factor Nrf2: effect of chemical structure and stress signals. *Free Radic. Biol. Med.* **37:** 1578–1590.

Chowdhury, S.K., Eshraghi, J., Wolfe, H., Forde, D., Hlavac, A.G. and Johnston, D. (1995) Mass-spectrometric identification of amino-acid transformations during oxidation of peptides and proteins – modifications of methionine and tyrosine. *Analyt. Chem.* **67:** 390–398.

Das, N., Levine, R.L., Orr, W.C. and Sohal, R.S. (2001) Selectivity of protein oxidative damage during aging in *Drosophila melanogaster*. *Biochem. J.* **360:** 209–216.

Davis, L.J., Hakim, G., Masiello, P., Novelli, M., Bergamini, E. and Licastro, F. (1994) Plasma protein's glycation is decreased in Sprague Dawley rats under caloric restriction. *Arch. Gerontol. Geriatr.* **19:** 295–301.

de Magalhaes, J.P. and Toussaint, O. (2004) GenAge: a genomic and proteomic network map of human ageing. *FEBS Lett.* **571:** 243–247.

Delpierre, G., Rider, M.H., Collard, F., Stroobant, V., Vanstapel, F., Santos, H. and Van Schaftingen, E. (2000) Identification, cloning, and heterologous expression of a mammalian fructosamine-3-kinase. *Diabetes* **49:** 1627–1634.

Denu, J.M. and Tanner, K.G. (1998) Specific and reversible inactivation of protein tyrosine phosphatases by hydrogen peroxide: evidence for a sulfenic acid intermediate and implications for redox regulation. *Biochemistry* **37:** 5633–5642.

Dobler, D., Ahmed, N., Song, L.J., Eboigbodin, K.E. and Thornalley, P.J. (2006) Increased dicarbonyl metabolism in endothelial cells in hyperglycemia induces anoikis and impairs angiogenesis by RGD and GFOGER motif modification. *Diabetes* **55:** 1961–1969.

Drew, B. and Leeuwenburgh, C. (2002) Aging and the role of reactive nitrogen species. *Ann. N.Y. Acad. Sci.* **959:** 66–81.

Du, X., Stockklauser-Farber, K. and Rosen, P. (1999) Generation of reactive oxygen intermediates, activation of NF-κβ and induction of apoptosis in human endothelial cells by glucose: role of nitric oxide synthase? *Free Radic. Biol. Med.* **27:** 752–763.

Dunlop, R.A., Dean, R.T. and Rodgers, K.J. (2008) The impact of specific oxidized amino acids on protein turnover in J774 cells. *Biochem. J.* **410:** 131–140.

Dunn, J.A., Patrick, J.S., Thorpe, S.R. and Baynes, J.W. (1989) Oxidation of glycated proteins: age-dependent accumulation of *N*ε-(carboxymethyl)lysine in lens proteins. *Biochemistry* **28:** 9464–9468.

Dunn, J.A., McCance, D.R., Thorpe, S.R., Lyons, T.J. and Baynes, J.W. (1991) Age-dependent accumulation of N^{ε}-carboxymethyl-lysine and N^{ε}-carboxymethyl-hydroxylysine in human skin collagen. *Biochemistry* **30:** 1205–1210.

Dyer, D.G., Blackledge, J.A., Thorpe, S.R. and Baynes, J.W. (1991) Formation of pentosidine during nonenzymatic browning of protein by glucose: identification of glucose and other carbohydrates as possible precursors of pentosidine *in vivo*. *J. Biol. Chem.* **266:** 11654–11660.

Dyer, D.G., Dunn, J.A., Thorpe, S.R., Bailie, K.E., Lyons, T.J., McCance, D.R. and Baynes, J.W. (1993) Accumulation of Maillard reaction products in skin collagen in diabetes and aging. *J Clin Invest* 91: 2463–2469.

Edens, W.A., Sharling, L., Cheng, G.J., *et al.* (2001) Tyrosine cross-linking of extracellular matrix is catalyzed by Duox, a multidomain oxidase/peroxidase with homology to the phagocyte oxidase subunit gp91 phox. *J. Cell Biol.* **154:** 879–891.

Eldred, G.E. and Lasky, M.R. (1993) Retinal age pigments generated by self-assembling lysosomotropic detergents. *Nature* **361:** 724–726.

Finch, J.W., Crouch, R.K., Knapp, D.R. and Schey, K.L. (1993) Mass-spectrometric identification of modifications to human serum-albumin treated with hydrogen-peroxide. *Arch. Biochem. Biophys.* **305:** 595–599.

Finkel, T. and Holbrook, N.J. (2000) Oxidants, oxidative stress and the biology of ageing. *Nature* **408:** 239–247.

Finley, E.L., Dillon, J., Crouch, R.K. and Schey, K.L. (1998) Identification of tryptophan oxidation products in bovine alpha-crystallin. *Protein Science* **7:** 2391–2397.

Fu, S., Davies, M.J., Stocker, R. and Dean, R.T. (1998a) Evidence for roles of radicals in protein oxidation in advanced human atherosclerotic plaque. *Biochem. J.* **333:** 519–525.

Fu, S., Fu, M.-X., Baynes, J.W., Thorpe, S.R. and Dean, R.T. (1998b) Presence of dopa and amino acid hydroperoxides in proteins modified with advanced glycation end products (AGEs): amino acid oxidation products as a possible source of oxidative stress by AGE proteins. *Biochem. J.* **330:** 233–239.

Fujigaki, S., Saito, K., Sekikawa, K., Tone, S., Takikawa, O., Fujii, H., Wada, H., Noha, A. and Seishima, M. (2001) Lipopolysaccharude induction of indoleamine 2,3-dioxygenase is mediated dominantly by an IFN-gamma dependent mechanism. *Eur. J. Immunol.* **31:** 2313–2318.

Fukunaga, Y., Katsuragi, Y., Izumi, T. and Sakiyama, F. (1982) Fluorescence characteristics of kynurenine and N′-formylkynurenine. Their use as reporters of the environment of tryptophan 62 in hen egg-white lysozyme. *J. Biochem.* **92:** 129–141.

Gao, L., Wang, J.K., Sekhar, K.R., *et al.* (2007) Novel n-3 fatty acid oxidation products activate *Nrf2* by destabilizing the association between *Keap1* and *Cullin3*. *J. Biol. Chem.* **282:** 2529–2537.

Gerner, C., Vejda, S., Gelbmann, D., Bayer, E., Gotzmann, J., Schulte-Hermann, R. and Mikulitis, W. (2002) Concomitant determination of absolute values of cellular protein amounts, synthesis rates, and turnover rates by quantitative proteome profiling. *Mol. Cell Proteom.* **1:** 528–537.

Gerstbrein, B., Stamatas, G., Kollias, N. and Driscoll, M. (2005) *In vivo* spectrofluorimetry reveals endogenous biomarkers that report healthspan and dietary restriction in *Caenorhabditis elegans*. *Aging Cell* **4:** 127–137.

Giulivi, C. and Davies, K.J.A. (2001) Mechanism of the formation and proteolytic release of H_2O_2 induced dityrosine and tyrosine oxidation products in hemoglobin and red blood cells. *J. Biol. Chem.* **276:** 24129–24136.

Giustarini, D., Rossi, R., Milzani, A., Colombo, R. and Dalle-Donne, I. (2004) S-glutathionylation: from redox regulation of protein functions to human diseases. *J. Cell. Mol. Med.* **8:** 201–212.

Gokulrangan, G., Zaidi, A., Michaelis, M.L. and Schoneich, C. (2007) Proteomic analysis of protein nitration in rat cerebellum: effect of biological aging. *J. Neurochem.* **100:** 1494–1504.

Goldberg, A.L. (2003) Protein degradation and protection against misfolded or damaged proteins. *Nature* **426:** 895–899.

Gonzalez-Dosal, R., Sorensen, M.D., Clark, B.F.C., Rattan, S.I.S. and Kristensen, P. (2006) Phage-displayed antibodies for the detection of glycated proteasome in aging cells. *Ann. N.Y. Acad. Sci.* **1067:** 474–478.

Greenacre, S.A.B., Evans, P., Halliwell, B. and Brain, S.D. (1999) Formation and loss of nitrated proteins in peroxynitrite-treated rat skin *in vivo*. *Biochem. Biophys. Res. Commun.* **262:** 781–786.

Grune, T., Reinheckel, T. and Davies, K.J.A. (1996) Degradation of oxidized proteins in K562 human hematopoietic cells by proteasome. *J. Biol. Chem.* **271:** 15504–15509.

Grune, T., Klotz, L.O., Gieche, J., Rudeck, M. and Sies, H. (2001) Protein oxidation and proteolysis by the nonradical oxidants singlet oxygen or peroxynitrite. *Free Radic. Biol. Med.* **30:** 1243–1253.

Hamelin, M., Mary, J., Vostry, M., Friguet, B. and Bakala, H. (2007) Glycation damage targets glutamate dehydrogenase in the rat liver mitochondrial matrix during aging. *Eur. J. Biochem.* **274:** 5949–5961.

Hansel, A., Kuschel, L., Hehl, S., Lemke, C., Agricola, H.J., Hoshi, T. and Heinemann, S.H. (2002) Mitochondrial targeting of the human peptide methionine sulfoxide reductase (MSRA), an enzyme-involved in the repair of oxidized proteins. *FASEB J.* **16:** U393–U406.

Hansel, A., Jung, S., Hoshi, T. and Heinemann, S.H. (2003) A second human methionine sulfoxide reductase (hMSRB2) reducing methionine-R-sulfoxide displays a tissue expression pattern distinct from hMSRB1. *Redox Rep.* **8:** 384–388.

Hansel, A., Heinemann, S.H. and Hoshi, T. (2005) Heterogeneity and function of mammalian MSRs: enzymes for repair, protection and regulation. *Biochim. Biophys. Acta Proteins Proteom.* **1703:** 239–247.

Haus, J.M., Carrithers, J.A., Trappe, S.W. and Trappe, T.A. (2007) Collagen, cross-linking, and advanced glycation end products in aging human skeletal muscle. *J. Appl. Physiol.* **103:** 2068–2076.

Hernebring, M., Brolen, G., Aguilaniu, H., Semb, H. and Nystrom, T. (2006) Elimination of damaged proteins during differentiation of embryonic stem cells. *Proc. Natl Acad. Sci. USA* **103:** 7700–7705.

Heyes, M.P., Chen, C.Y., Major, E.O. and Saito, K. (1997) Different kynurenine pathway enzymes limit quinolinic acid formation by various human cell types. *Biochem. J.* **326:** 351–356.

Hu, P., Yu, L., Zhang, M., Zheng, L., Lan, F., Fu, Q. and Zhao, S. (2005) Cloning, tissue expression pattern characterization and chromosome localization of human peptide methionine sulfoxide reductase cDNA. *Chin. Sci. Bull.* **45:** 2251–2257.

Itakura, K., Uchida, K. and Kawakishi, S. (1994) Selective formation of oxindole-type and formylkynurenine-type products from tryptophan and its peptides treated with a superoxide-generating system in the presence of iron(III) EDTA – a possible involvement with iron oxygen complex. *Chem. Res. Toxicol.* **7:** 185–190.

Itakura, K., Uchida, K. and Osawa, T. (1996) A novel fluorescent malondialdehyde-lysine adduct. *Chem. Phys. Lipids* **84:** 75–79.

Jung, S., Hansel, A., Kasperczyk, H., Hoshi, T. and Heinemann, S.H. (2002) Activity, tissue distribution and site-directed mutagenesis of a human peptide methionine sulfoxide reductase of type B: hCBS1. *FEBS Lett.* **527:** 91–94.

Kadiiska, M.B., Gladen, B.C., Baird, D.D., *et al.* (2005) Biomarkers of oxidative stress study II. Are oxidation products of lipids, proteins, and DNA markers of CCl4 poisoning? *Free Radic. Biol. Med.* **38:** 698–710.

Kang, H.L., Benzer, S. and Min, K.T. (2002) Life extension in Drosophila by feeding a drug. *Proc. Natl Acad. Sci. USA* **99:** 838–843.

Kanski, J., Behring, A., Pelling, J. and Schoneich, C. (2005a) Proteomic identification of 3-nitrotyrosine-containing rat cardiac proteins: effects of biological aging. *Am. J. Physiol. Heart Circ. Physiol.* **288:** H371–H381.

Kanski, J., Hong, S.J. and Schoneich, C. (2005b) Proteomic analysis of protein nitration in aging skeletal muscle and identification of nitrotyrosine-containing sequences *in vivo* by nanoelectrospray ionization tandem mass spectrometry. *J. Biol. Chem.* **280:** 24261–24266.

Kato, Y., Kawakishi, S., Aoki, T., Itakura, K. and Osawa, T. (1997) Oxidative modification of tryptophan residues exposed to peroxynitrite. *Biochem. Biophys. Res. Commun.* **234:** 82–84.

Kiley, P.J. and Storz, G. (2004) Exploiting thiol modifications. *PLoS Biol.* **2:** e400.

Kirschbaum, B. (2001) Total urine antioxidant capacity. *Clin. Chim. Acta* **305:** 167–173.

Knox, W.E. and Mehler, A.H. (1950) The conversion of tryptophan to kynurenine in liver. *J. Biol. Chem.* **187:** 419–430.

Krishnan, S., Chi, E.Y., Wood, S.J., *et al.* (2003) Oxidative dimer formation is the critical rate-limiting step for Parkinson's disease alpha-synuclein fibrillogenesis. *Biochemistry* **42:** 829–837.

Kueper, T., Grune, T., Prahl, S., *et al.* (2007) Vimentin is the specific target in skin glycation: structural prerequisites, functional consequences, and role in skin aging. *J. Biol. Chem.* **282:** 23427–23436.

Kuhla, B., Boeck, K., Luth, H.J., Schmidt, A., Weigle, B., Schmitz, M., Ogunlade, V., Munch, G. and Arendt, T. (2006) Age-dependent changes of glyoxalase I expression in human brain. *Neurobiol. Aging* **27:** 815–822.

Kuschel, L., Hansel, A., Schoherr, R., Weissbach, H., Brot, N., Hoshi, T. and Heinemann, S.H. (1999) Molecular cloning and functional expression of a human peptide methionine sulfoxide reductase (hMsrA). *FEBS Lett.* **456:** 17–21.

Kwak, M.K., Wakabayashi, N., Itoh, K., Motohashi, H., Yamamoto, M. and Kensler, T.W. (2003) Modulation of gene expression by cancer chemopreventive dithiolethiones through the *Keap1-Nrf2* pathway. Identification of novel gene clusters for cell survival. *J. Biol. Chem.* **278:** 8135–8145.

Lassegue, B. and Clempus, R.E. (2003) Vascular NAD(P)H oxidases: specific features, expression, and regulation. *Am. J. Physiol. Regul. Integr. Comp. Physiol.* **285:** R277–R297.

Leeuwenburgh, C., Rasmussen, L.M., Hsu, F.F., Mueller, D.M., Pennathur, S. and Heinecke, J.W. (1997a) Mass spectrometric quantification of markers for protein oxidation by tyrosyl radical, copper and hydroxyl radical in low density lipoprotein isolated from human atherosclerotic plaques. *J. Biol. Chem.* **272:** 3520–3526.

Leeuwenburgh, C., Wagner, P., Holloszy, J.O., Sohal, R.S. and Heinecke, J.W. (1997b) Caloric restriction attenuates dityrosine cross-linking of cardiac and skeletal muscle proteins in aging mice. *Arch. Biochem. Biophys.* **346:** 74–80.

Leeuwenburgh, C., Hansen, P., Shaish, A., Holloszy, J.O. and Heinecke, J.W. (1998) Markers of protein oxidation by hydroxyl radical and reactive nitrogen species in tissues of aging rats. *Am. J. Physiol. Regul. Integr. Comp. Physiol.* **274:** R453–R461.

Leeuwenburgh, C., Hansen, P.A., Holloszy, J.O. and Heinecke, J.W. (1999a) Hydroxyl radical generation during exercise increases mitochondrial protein oxidation and levels of urinary dityrosine. *Free Radic. Biol. Med.* **27:** 186–192.

Leeuwenburgh, C., Hansen, P.A., Holloszy, J.O. and Heinecke, J.W. (1999b) Oxidized amino acids in the urine of aging rats: potential markers for assessing oxidative stress *in vivo*. *Am. J. Physiol. Regul. Integr. Comp. Physiol.* **276:** R128–R135.

Lerman, I., Harrison, B.C., Freeman, K., Hewett, T.E., Allen, D.L., Robbins, J. and Leinwand, L.A. (2002) Genetic variability in forced and voluntary endurance exercise performance in seven inbred mouse strains. *J. Appl. Physiol.* **92:** 2245–2255.

Levine, R.L., Williams, J.A., Stadtman, E.R. and Shacter, E. (1994) Carbonyl assays for determination of oxidatively modified proteins. *Methods Enzymol.* **233:** 346–357.

Li, G., Liao, Y., Wang, X., Sheng, S. and Yin, D. (2006) *In situ* estimation of the entire color and spectra of age pigment-like materials: application of a front-surface 3D-fluorescence technique. *Exp. Gerontol.* **41:** 328–336.

Lingelbach, L.B., Mitchell, A.E., Rucker, R.B. and McDonald, R.B. (2000) Accumulation of advanced glycation endproducts in aging male Fischer 344 rats during long-term feeding of various dietary carbohydrates. *J. Nutrit.* **130:** 1247–1255.

Louie, J.L., Kapphalm, R.J. and Ferrington, D.A. (2002) Proteasome function and protein oxidation in the aged retina. *Exp. Eye Res.* **75:** 271–284.

Luo, D.Y., Smith, S.W. and Anderson, B.D. (2005) Kinetics and mechanism of the reaction of cysteine and hydrogen peroxide in aqueous solution. *J. Pharmaceut. Sci.* **94:** 304–316.

Mann, G.E., Rowlands, D.J., Li, F.Y.L., de Winter, P. and Siow, R.C.M. (2007) Activation of endothelial nitric oxide synthase by dietary isoflavones: Role of NO in *Nrf2*-mediated antioxidant gene expression. *Cardiovasc. Res.* **75:** 261–274.

McCormick, M.L., Gaut, J.P., Lin, T.S., Britigan, B.L., Buettner, G.R. and Heinecke, J.W. (1998) Electron paramagnetic resonance detection of free tyrosyl radical generated by myeloperoxidase, lactoperoxidase, and horseradish peroxidase. *J. Biol. Chem.* **273:** 32030–32037.

Monnier, V.M., Kohn, R.R. and Cerami, A. (1984) Accelerated AGE-related browning of human collagen in diabetes-mellitus. *Proc. Natl Acad. Sci. USA* **81:** 583–587.

Moore, W.A.L., Davey, V.A., Weindruch, R., Walford, R. and Ivy, G.O. (1995) The effect of caloric restriction on lipofuscin accumulation in mouse brain with age. *Gerontology* **41:** 173–183.

Morcos, M., Du, X., Pfisterer, F., *et al.* (2008) Glyoxalase-1 prevents mitochondrial protein modification and enhances lifespan in *Caenorhabditis elegans*. *Aging Cell* **7:** 260–269.

Moskovitz, J., Weissbach, H. and Brot, N. (1996) Cloning and expression of a mammalian gene involved in the reduction of methionine sulfoxide residues in proteins. *Proc. Natl Acad. Sci. USA* **93:** 2095–2099.

Moskovitz, J., Bar-Noy, S., Williams, W.M., Berlett, B.S. and Stadtman, E.R. (2001) Methionine sulfoxide reductase (MsrA) is a regulator of antioxidant defense and lifespan in mammals. *Proc. Natl Acad. Sci. USA* **98:** 12920–12925.

Na, R.H., Stender, I.M., Ma, L.X. and Wulf, H.C. (2000) Autofluorescence spectrum of skin: component bands and body site variations. *Skin Res. Technol.* **6:** 112–117.

Nakamura, A. and Goto, S. (1996) Analysis of protein carbonyls with 2,4-dinitrophenyl hydrazine and its antibodies by immunoblot in two-dimensional gel electrophoresis. *J. Biochem.* **119:** 768–774.

Naskalski, J.W., Marcinkiewicz, J. and Drozdz, R. (2002) Myeloperoxidase-mediated protein oxidation: its possible biological functions. *Clin. Chem. Lab. Med.* **40:** 463–468.

Novelli, M., Masiello, P., Bombara, M. and Bergamini, E. (1998) Protein glycation in the aging male Sprague-Dawley rat: effects of antiaging diet restriction. *J. Gerontol.* **53A:** B94–B101.

Odani, H., Iijima, K., Nakata, M., *et al.* (2001) Identification of Nω-carboxymethylarginine, a new advanced glycation endproduct in serum proteins of diabetic patients: possibly a new marker of aging and diabetes. *Biochem. Biophys. Res. Comm.* **285:** 1232–1236.

Odetti, P., Pronzato, M.A., Noberasco, G., Cosso, L., Traverso, N., Cottalasso, D. and Marinari, U.M. (1994) Relationships between glycation and oxidation related fluorescences in rat collagen during aging. *Lab. Invest.* **70:** 61–67.

Ohshima, H., Friesen, M., Brouet, I. and Bartsch, H. (1990) Nitrotyrosine as a new marker for endogenous nitrosation and nitration of proteins. *Food Chem. Toxicol.* **28:** 647–652.

Oudes, A.J., Herr, C.M., Olsen, Y. and Fleming, J.E. (1998) Age-dependent accumulation of advanced glycation end-products in adult *Drosophila melanogaster*. *Mech. Ageing Dev.* **100:** 221–229.

Pabarcus, M.K. and Casida, J.E. (2002) Kynurenine formamidase: determination of primary structure and modeling-based prediction of tertiary structure and catalytic triad. *Biochim. Biophys. Acta* **1596:** 201–211.

Pamplona, R., Portero-Otin, M., Bellmunt, M.J., Gredilla, R. and Barja, G. (2002a) Aging increases N^{ε}-carboxymethyl-lysine and caloric restriction decreases N^{ε}-carboxyethyl-lysine and N^{ε}-malondialdehyde-lysine in rat heart mitochondrial proteins. *Free Radic. Res.* **36:** 47–54.

Pamplona, R., Portero-Otin, M., Requena, J., Gredilla, R. and Barja, G. (2002b) Oxidative, glycoxidative and lipoxidative damage to rat heart mitochondrial proteins is lower after 4 months of caloric restriction than in age-matched controls. *Mech. Ageing Dev.* **123:** 1437–1446.

Pannala, A.S., Mani, A.R., Rice-Evans, C.A. and Moore, K.P. (2006) pH-dependent nitration of para-hydroxyphenylacetic acid in the stomach. *Free Radic. Biol. Med.* **41:** 896–901.

Panzenbock, U. and Stocker, R. (2005) Formation of methionine sulfoxide-containing specific forms of oxidized high-density lipoproteins. *Biochim. Biophys. Acta Proteins Proteom.* **1703:** 171–181.

Pawlak, D., Tankiewicz, A. and Buczko, W. (2001) Kynurenine and its metabolites in the rat with experimental renal insufficiency. *J. Physiol. Pharmacol.* **52:** 755–766.

Pearson, K.J., Lewis, K.N., Price, N.L., *et al.* (2008) Nrf2 mediates cancer protection but not prolongevity induced by caloric restriction. *Proc. Natl Acad. Sci. USA* **105:** 2325–2330.

Pereira, P., Shang, F., Hobbs, M., Girao, H. and Taylor, A. (2003) Lens fibers have a fully functional ubiquitin-proteasome pathway. *Exp. Eye Res.* **76:** 623–631.

Petropoulos, I. and Friguet, B. (2005) Protein maintenance in aging and replicative senescence: a role for the peptide methionine sulfoxide reductases. *Biochim. Biophys. Acta Proteins Proteom.* **1703:** 261–266.

Pfeiffer, S., Schmidt, K. and Mayer, B. (2000) Dityrosine formation out competes tyrosine nitration at low steady-state concentrations of peroxynitrite. Implications for tyrosine modification by nitric oxide/superoxide *in vivo*. *J. Biol. Chem.* **275:** 6346–6352.

Piec, I., Listrat, A., Alliot, J., Chambon, C., Taylor, R.G. and Bechet, D. (2005) Differential proteome analysis of aging in rat skeletal muscle. *FASEB J.* **19:** 1143–1145.

Pletcher, S.D., Libert, S. and Skorupa, D. (2005) Flies and their golden apples: the effect of dietary restriction on Drosophila aging and age-dependent gene expression. *Ageing Res. Rev.* **4:** 451–480.

Poole, L.B., Karplus, P.A. and Claiborne, A. (2004) Protein sulfenic acids in redox signaling. *Annu. Rev. Pharmacol. Toxicol.* **44:** 325–347.

Predescu, D., Predescu, S. and Malik, A.B. (2002) Transport of nitrated albumin across continuous vascular endothelium. *Proc. Natl Acad. Sci. USA* **99:** 13932–13937.

Rabbani, N. and Thornalley, P.J. (2008) Assay of 3-nitrotyrosine on tissues and body fluids by liquid chromatography with tandem mass spectrometric detection. *Methods in Enzymol.* **440:** 337–359.

Rabbani, N., Sebekova, K., Sebekova, K., Jr., Heidland, A. and Thornalley, P.J. (2007) Protein glycation, oxidation and nitration free adduct accumulation after bilateral nephrectomy and ureteral ligation. *Kidney Internatl* **72:** 1113–1121.

Requena, J.R., Chao, C.-C., Levine, R.L. and Stadtman, E.R. (2001) Glutamic and aminoadipic semialdehydes are the main carbonyl products of metal-catalyzed oxidation of proteins. *Proc. Natl Acad. Sci. USA* **98:** 68–74.

Ryberg, H. and Caidahl, K. (2007) Chromatographic and mass spectrometric methods for quantitative determination of 3-nitrotyrosine in biological samples and their application to human samples. *J. Chromatogr. B* **851:** 160–171.

Saez, G.T., Thornalley, P.J., Hill, H.A.O., Hems, R. and Bannister, J.V. (1982) The production of free radicals by the autoxidation of cysteine and their effect on isolated hepatocytes. *Biochim. Biophys. Acta* **719:** 24–31.

Schriner, S. and Linford, N. (2006) Extension of mouse lifespan by overexpression of catalase. *Age* **28:** 209–218.

Schwarcz, R. and Pellicciari, R. (2002) Manipulation of brain kynurenines: glial targets, neuronal effects and clinical opportunities. *J. Pharmacol. Exp. Ther.* **303:** 1–10.

Seifert, J. (1993) Assay of tryptophan 2,3-dioxygenase using liver slices and high performance liquid chromatography. *J. Chromatogr.* **614:** 227–231.

Sell, D.R. and Monnier, V.M. (1989) Structure elucidation of a senescence crosslink from human extracellular matrix. Implication of pentoses in the aging process. *J. Biol. Chem.* **264:** 21597–21602.

Sell, D.R. and Monnier, V.M. (1997) Age-related association of tail tendon break time with tissue pentosidine in DBA/2 vs C57BL/6 mice: the effect of dietary restriction. *J. Gerontol.* **52A:** B277–B284.

Sell, D.R. and Monnier, V.M. (2004) Conversion of arginine into ornithine by advanced glycation in senescent human collagen and lens crystallins. *J. Biol. Chem.* **279:** 54173–54184.

Sell, D.R., Lane, M.A., Johnson, W.A., *et al.* (1996) Longevity and the genetic determination of collagen glycoxidation kinetics in mammalian senescence. *Proc. Natl Acad. Sci. USA* **93:** 485–490.

Sell, D.R., Kleinman, N.R. and Monnier, V.M. (2000) Longitudinal determination of skin collagen glycation and glycoxidation rates predicts early death in C57BL/6NNIA mice. *FASEB J.* **14:** 145–156.

Sell, D.R., Lane, M.A., Obrenovich, M.E., Mattison, J.A., Handy, A., Ingram, D.K., Cutler, R.G., Roth, G.S. and Monnier, V.M. (2003) The effect of caloric restriction on glycation and glycoxidation in skin collagen of nonhuman primates. *J. Gerontol. A Biol. Sci. Med. Sci.* **58:** B508–B516.

Sell, D.R., Strauch, C.M., Shen, W. and Monnier, V.M. (2007) 2-Aminoadipic acid is a marker of protein carbonyl oxidation in the aging human skin: effects of diabetes, renal failure and sepsis. *Biochem. J.* **404:** 269–277.

Shang, F. and Taylor, A. (1995) Oxidative stress and recovery from oxidative stress are associated with altered ubiquitin conjugating and proteolytic activities in bovine lens epithelial cells. *Biochem. J.* **307:** 297–303.

Sharmaluthra, R. and Kale, R.K. (1994) Age-related changes in the activity of the glyoxalase system. *Mech. Ageing Dev.* **73:** 39–45.

Shinohara, M., Thornalley, P.J., Giardino, I., Beisswenger, P.J., Thorpe, S.R., Onorato, J. and Brownlee, M. (1998) Overexpression of glyoxalase I in bovine endothelial cells inhibits intracellular advanced glycation endproduct formation and prevents hyperglycaemia-induced increases in macromolecular endocytosis. *J. Clin. Invest.* **101:** 1142–1147.

Shimasaki, H. (1994) Assay of Fluorescent Lipid-Peroxidation Products. *Methods in Enzymol* **233**: 338–346.

Shipanova, I.N., Glomb, M.A. and Nagaraj, R.H. (1997) Protein modification by methylglyoxal: chemical nature and synthetic mechanism of a major fluorescent adduct. *Arch. Biochem. Biophys.* **344:** 29–36.

Simat, T.J. and Steinhart, H. (1998) Oxidation of free tryptophan and tryptophan residues in peptides and proteins. *J. Agric. Food Chem.* **46:** 490–498.

Sitte, N., Merker, K. and Grune, T. (1998) Proteasome-dependent degradation of oxidized proteins in MRC-5 fibroblasts. *FEBS Lett.* **440:** 399–402.

Sitte, N., Merker, K., von Zglinicki, T., Davies, K.J.A. and Grune, T. (2000a) Protein oxidation and degradation during cellular senescence of human BJ fibroblasts. II. Aging of non-dividing cells. *FASEB J.* **14:** 2503–2510.

Sitte, N., Merker, K., von Zglinicki, T., Grune, T. and Davies, K.J.A. (2000b) Protein oxidation and degradation during cellular senescence of human BJ fibroblasts. I. Effects of proliferative senescence. *FASEB J.* **14:** 2495–2502.

Snow, L.M., Fugere, N.A. and Thompson, L.V. (2007) Advanced glycation end-product accumulation and associated protein modification in type II skeletal muscle with aging. *J. Gerontol. A Biol. Sci. Med. Sci.* **62:** 1204–1210.

Soderling, A.S., Ryberg, H., Gabrielsson, A., Larstad, M., Toren, K., Niari, S. and Caidahl, K. (2003) A derivatization assay using gaschromatography/negative chemical ionization tandem mass spectrometry to quantify 3-nitrotyrosine in human plasma. *J. Mass Spectr.* **38:** 1187–1196.

Sohal, R.S. (1981) *Age Pigments*. Elsevier/North Holland Biomedical Press, Amsterdam.

Sohal, R.S., Sohal, B.H. and Orr, W.C. (1995) Mitochondrial superoxide and hydrogen peroxide generation, protein oxidative damage, and longevity in different species of flies. *Free Radic. Biol. Med.* **19:** 499–504.

Souza, J.M., Giasson, B.I., Chen, Q., Lee, V.M.Y. and Ischiropoulos, H. (2000) Dityrosine cross-linking promotes formation of stable á-synuclein polymers-implication of nitrative and oxidative stress in the pathogenesis of neurodegenerative synucleinopathies. *J. Biol. Chem.* **275:** 18344–18349.

Sparrow, J.R., Parish, C.A., Hashimoto, M. and Nakanishi, K. (1999) A2E, a lipofuscin fluorophore, in human retinal pigmented epithelial cells in culture. *Invest. Ophthalmol. Vis. Sci.* **40:** 2988–2995.

Suzuki, K., Koh, Y.H., Mizuno, H., Hamaoko, R. and Taniguchi, N. (1998) Overexpression of aldehyde reductase protects PC12 cells from the cytotoxicity of methylglyoxal or 3-deoxyglucosone. *J. Biochem.* **123:** 353–357.

Szwergold, B.S., Howell, S. and Beisswenger, P.J. (2001) Human fructosamine-3-kinase. Purification, sequencing, substrate specificity, and evidence of activity *in vivo*. *Diabetes* **50:** 2139–2147.

Tabrizi-Fard, M.A., Maurer, T.S. and Fung, H.L. (1999) *In vivo* disposition of 3-nitro-l-tyrosine in rats: implications on tracking systemic peroxynitrite exposure. *Drug Metab. Dispos.* **27:** 429–431.

Takahashi, M., Pischetsrieder, M. and Monnier, V.M. (1997) Isolation, purification, and characterization of amadoriase isoenzymes (fructosyl amine-oxygen oxidoreductase EC 1.5.3) from *Aspergillus sp*. *J. Biol. Chem.* **272:** 3437–3443.

Tanigawa, S., Fujii, M. and Hou, D.X. (2007) Action of Nrf2 and Keap1 in ARE-mediated NQO1 expression by quercetin. *Free Radic. Biol. Med.* **42:** 1690–1703.

Taylor, S.W., Fahy, E., Murray, J., Capaldi, R.A. and Ghosh, S.S. (2003) Oxidative post-translational modification of tryptophan residues in cardiac mitochondrial proteins. *J. Biol. Chem.* **278:** 19587–19590.

Teillet, L., Verbeke, P., Gouraud, S., Bakala, H., Borot-Laloi, C., Heudes, D., Bruneval, P. and Corman, B. (2000) Food restriction prevents advanced glycation end product accumulation and retards kidney aging in lean rats. *J. Am. Soc. Nephrol.* **11:** 1488–1497.

Teillet, L., Ribiere, P., Gouraud, S., Bakala, H. and Corman, B. (2002) Cellular signaling, AGE accumulation and gene expression in hepatocytes of lean aging rats fed *ad libitum* or food-restricted. *Mech. Ageing Dev.* **123:** 427–439.

Tessier, F., Obrenovich, M. and Monnier, V.M. (1999a) Structure and mechanism of formation of human lens fluorophore LM-1. *J. Biol. Chem.* **274:** 20796–20804.

Tessier, F., Obrenovich, M. and Monnier, V.M. (1999b) Structure and Mechanism of Formation of Human Lens Fluorophore LM-1. Relationship to vesperlysine a and the advanced maillard reaction in aging, diabetes, and cataractogenesis. *J. Biol. Chem.* **274**: 20796-20804.

Thomas, M.C., Tsalamandris, C., MacIsaac, R., Medley, T., Kingwell, B., Cooper, M.E. and Jerums, G. (2004) Low-molecular-weight AGEs are associated with GFR and anemia in patients with type 2 diabetes. *Kidney Int.* **66:** 1167–1172.

Thornalley, P.J. (1999) Clinical significance of glycation. *Clin. Lab.* **45:** 263–273.

Thornalley, P.J. (2003a) The enzymatic defence against glycation in health, disease and therapeutics: a symposium to examine the concept. *Biochem. Soc. Trans.* **31:** 1343–1348.

Thornalley, P.J. (2003b) Use of aminoguanidine (Pimagedine) to prevent the formation of advanced glycation endproducts. *Arch. Biochem. Biophys.* **419:** 31–40.

Thornalley, P.J. (2005) Dicarbonyl intermediates in the Maillard reaction. *Ann. N.Y. Acad. Sci.* **1043:** 111–117.

Thornalley, P.J. (2006a) Protein oxidation marker residues and oxidised amino acids in disease – the damage and the debris of protein oxidation. In: *Protein Oxidation and Disease* (eds S.G. Pandalai and J. Pietzsch).Research Signpost, Kerala, pp. 143–178.

Thornalley, P.J. (2006b) Quantitative screening of protein glycation, oxidation, and nitration adducts by LC-MS/MS: protein damage in diabetes, uremia, cirrhosis, and Alzheimer's disease. In: *Redox Proteomics* (eds I. Dalle-Donne, A. Scaloni and D.A. Butterfield). Wiley, Hoboken, NJ, pp. 681–728.

Thornalley, P.J. and IARC Workgroup (2004) *Cruciferous Vegetables, Isothiocyanates and Indoles – IARC Handbook on Chemoprevention of Cancer.* IARC Press, Lyon.

Thornalley, P.J., Battah, S., Ahmed, N., Karachalias, N., Agalou, S., Babaei-Jadidi, R. and Dawnay, A. (2003) Quantitative screening of advanced glycation endproducts in cellular and extracellular proteins by tandem mass spectrometry. *Biochem. J.* **375:** 581–592.

Thorpe, S.R. and Baynes, J.W. (2002) CML: a brief history. In: *Maillard Reaction in Food Chemistry and Medical Science: Update for the Postgenomic Era.* Elsevier Science, Amsterdam, pp. 91–99.

Tien, M., Berlett, B.S., Levine, R.L., Chock, P.B. and Stadtman, E.R. (1999) Peroxynitrite-mediated modification of proteins at physiological carbon dioxide concentration: pH dependence of carbonyl formation, tyrosine nitration, and methionine oxidation. *Proc. Natl Acad. Sci.* **96:** 7809–7814.

Ushijima, Y., Nakano, M. and Goto, T. (1984) Production and Identification of bityrosine in horseradish peroxidase-H_2O_2-tyrosine system. *Biochem. Biophys. Res. Commun.* **125:** 916–918.

van Montfort, R.L.M., Congreve, M., Tisi, D., Carr, R. and Jhoti, H. (2003) Oxidation state of the active-site cysteine in protein tyrosine phosphatase 1B. *Nature* **423:** 773–777.

Venkatraman, J., Aggarwal, K. and Balaram, P. (2001) Helical peptide models for protein glycation: proximity effects in catalysis of the Amadori rearrangement. *Chem. Biol.* **8:** 611–625.

Verbeke, P., Perichon, M., BorotLaloi, C., Schaeverbeke, J. and Bakala, H. (1997) Accumulation of advanced glycation endproducts in the rat nephron: link with circulating AGEs during aging. *J. Histochem. Cytochem.* **45:** 1059–1068.

Verzijl, N., DeGroot, J., Oldehinkel, E., *et al.* (2000a) Age-related accumulation of Maillard reaction products in human articular cartilage collagen. *Biochem. J.* **350:** 381–387.

Verzijl, N., DeGroot, J., Thorpe, S.R., *et al.* (2000b) Effect of collagen turnover on the accumulation of advanced glycation endproducts. *J. Biol. Chem.* **275:** 39027–39031.

Vougier, S., Mary, J. and Friguet, B. (2003) Subcellular localization of methionine sulphoxide reductase A (MsrA): evidence for mitochondrial and cytosolic isoforms in rat liver cells. *Biochem. J.* **373:** 531–537.

Walker, G., Houthoofd, K., Vanfleteren, J.R. and Gems, D. (2005) Dietary restriction in *C. elegans*: from rate-of-living effects to nutrient sensing pathways. *Mech. Ageing Dev.* **126:** 929–937.

Wang, C., Weindruch, R., Fernandez, J.R., Coffey, C.S., Patel, P. and Allison, D.B. (2004) Caloric restriction and body weight independently affect longevity in Wistar rats. *Int. J. Obesity* **28:** 357–362.

Wang, P.C., Zhang, H., Zhang, Z.Y. and Tong, T.J. (2007) Aminoguanidine delays the replicative senescence of human diploid fibroblasts. *Chin. Med. J.* **120:** 2028–2035.

Weindruch, R. (1992) Effect of caloric restriction on age-associated cancers. *Exp. Gerontol.* **27:** 575–581.

Weindruch, R. (2006) Will dietary restriction work in primates? *Biogerontology* **7:** 169–171.

Weindruch, R. and Walford, R.L. (1982) Dietary restriction in mice beginning at 1 year of age – effect on life-span and spontaneous cancer incidence. *Science* **215:** 1415–1418.

Weissbach, H., Etienne, F., Hoshi, T., Heinemann, S.H., Lowther, W.T., Matthews, B., St John, G., Nathan, C. and Brot, N. (2002) Peptide methionine sulfoxide reductase: structure, mechanism of action, and biological function. *Arch. Biochem. Biophys.* **397:** 172–178.

Weissbach, H., Resnick, L. and Brot, N. (2005) Methionine sulfoxide reductases: history and cellular role in protecting against oxidative damage. *Biochim. Biophys. Acta* **1703:** 203–212.

Wells-Knecht, M.C., Huggins, T.G., Dyer, D.G., Thorpe, S.R. and Baynes, J.W. (1993) Oxidized amino-acids in lens protein with age – measurement of o-tyrosine and dityrosine in the aging human lens. *J. Biol. Chem.* **268:** 12348–12352.

Wells-Knecht, M.C., Lyons, T.J., McCance, D.R., Thorpe, S.R. and Baynes, J.W. (1997) Age-dependent increases in ortho-tyrosine and methionine sulfoxide in human skin collagen is not accelerated in diabetes. *J. Clin. Invest.* **100:** 839–846.

Whittier, J.E., Xiong, Y., Rechsteiner, M.C. and Squier, T.C. (2004) Hsp90 enhances degradation of oxidized calmodulin by the 20 S proteasome. *J. Biol. Chem.* **279:** 46135–46142.

WitkoSarsat, V., Friedlander, M., CapeillereBlandin, C., NguyenKhoa, T., Nguyen, N.T., Zingraff, J., Jungers, P. and DescampsLatscha, B. (1996) Advanced oxidation protein products as a novel marker of oxidative stress in uremia. *Kidney Int.* **49:** 1304–1313.

Woo, H.A., Jeong, W., Chang, T.S., Park, K.J., Park, S.J., Yang, J.S. and Rhee, S.G. (2005) Reduction of cysteine sulfinic acid by sulfiredoxin is specific to 2-cys peroxiredoxins. *J. Biol. Chem.* **280:** 3125–3128.

Wood, J.G., Rogina, B., Lavu, S., Howitz, K., Helfand, S.L., Tatar, M. and Sinclair,

D. (2004) Sirtuin activators mimic caloric restriction and delay ageing in metazoans. *Nature* **430:** 686–689.

Wrobel, K., Wrobel, K., Garay-Sevilla, M., Nava, L.E. and Malacara, J.M. (1997) Novel analytical approach to monitoring advanced glycosylation end products in human serum with on-line spectrophotometric and spectrofluorometric detection in a flow system. *Clin. Chem.* **43:** 1563–1569.

Ximenes, V.F., Catalani, L.H. and Campa, A. (2001) Oxidation of melatonin and tryptophan by an HRP cycle involving compound III. *Biochem. Biophys. Res. Commun.* **287:** 130–134.

Yamada, S., Kumazawa, S., Ishii, T., *et al.* (2001) Immunochemical detection of a lipofuscin-like fluorophore derived from malondialdehyde and lysine. *J. Lipid Res.* **42:** 1187–1196.

Yan, L.J. and Sohal, R.S. (2000) Analysis of oxidative modification of proteins. *Curr. Protocols Protein Sci.* **144:** 105–128.

Yao, D., Taguchi, T., Matsumura, T., *et al.* (2007) High glucose increases angiopoietin-2 transcription in microvascular endothelial cells through methylglyoxal modification of mSin3A. *J. Biol. Chem.* **282:** 31038–31045.

Zainal, T.A., Oberley, T.D., Allison, D.B., Szweda, L.I. and Weindruch, R. (2000) Caloric restriction of rhesus monkeys lowers oxidative damage in skeletal muscle. *FASEB J.* **14:** 1825–1836.

Zeng, J. and Davies, M.J. (2005) Evidence for the formation of adducts and S-(carboxymethyl)cysteine on reaction of α-dicarbonyl compounds with thiol groups on amino acids, peptides, and proteins. *Chem. Res. Toxicol.* **18:** 1232–1241.

Index

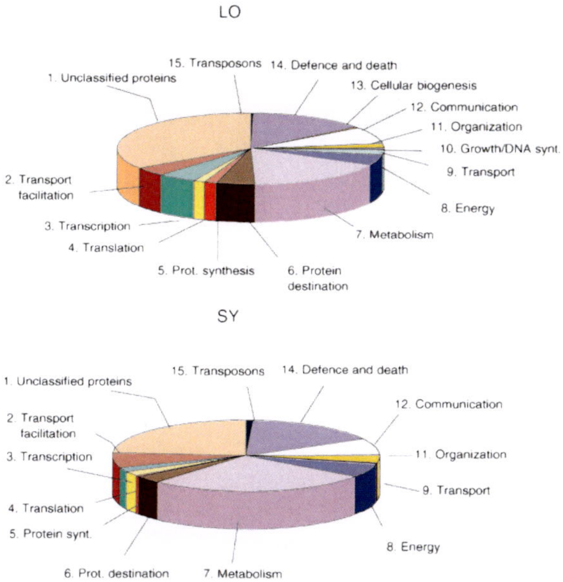

Categories	% in LO	% in SY
1. Unclassified proteins	34.5	23.8
2. Transport facilitation	4.0	6.8
3. Transcription	4.7	2.0
4. Translation	1.4	2.0
5. Protein synthesis	1.4	0.7
6. Protein destination	4.7	3.4
7. Metabolism	16.9	26.5
8. Energy	5.0	6.1
9. Cellular transport/ Transport mechanisms	2.2	0.7
10. Cell growth, cell division and DNA synthesis	0.7	0.0
11. Cellular organization	2.2	3.4
12. Cellular communication/ Signal transduction	7.6	7.5
13. Cellular biogenesis	0.7	0.0
14. Cell rescue, defence, cell death and ageing	13.7	16.3
15. Transposons	0.4	0.7

Chapter 5, Figure 1. *Functional categorization of suppressed substructive EST libraries. mRNA for these libraries was isolated from leaves directly exposed to 40 min of excess light (2200 ± 200 µmol m*$^{-2}$ *s*$^{-1}$*), 'local' (LO), and leaves undergoing systemic acquired acclimation, 'systemic' (SY), after 40 min. A list of all sequenced ESTs from these libraries is presented in Mühlenbock* et al., *(2008). Comparative distribution in percentage of different gene categories is presented in* Table 1.

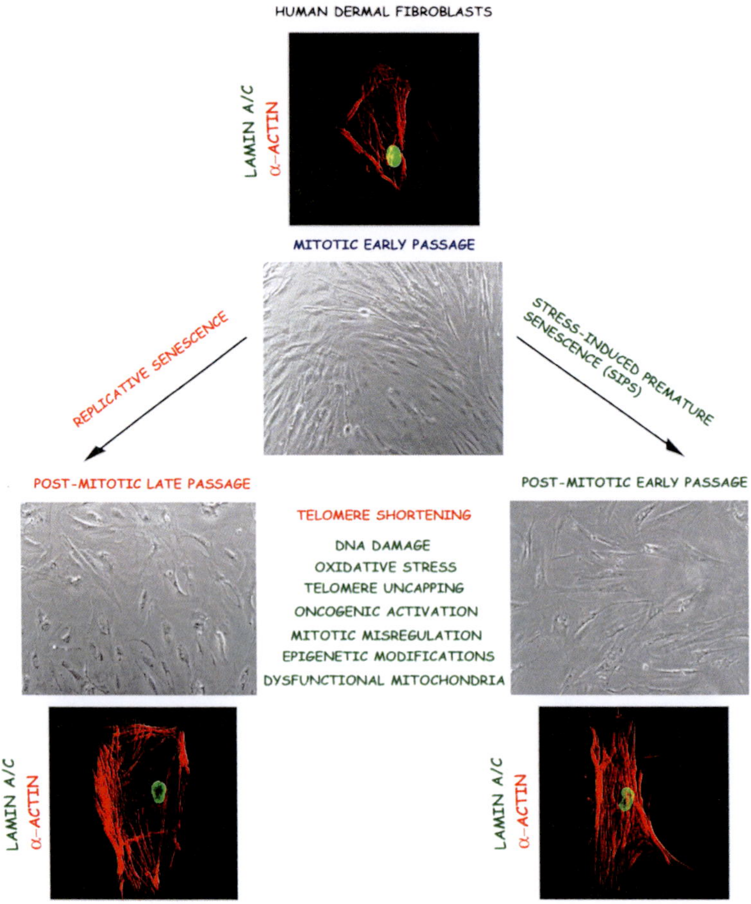

Chapter 8, Figure 1. *Human cells have two modes of regulating permanent growth arrest through either replicative senescence or stress-induced premature senescence (SIPS). A number of mechanisms leading to either replicative (marked in red) or premature (marked in green) senescence are outlined. Early mitotic passages of human dermal fibroblasts (marked in blue) were aged through serial passage (Pekovic et al., 2007) until they reached a post-mitotic stage through replicative senescence (marked in red). Alternatively, early mitotic fibroblasts were induced to senesce prematurely by two treatments with an oxidizing agent hydrogen peroxide (H_2O_2) using 300μM final concentration for 1 hr after which they became post-mitotic within the next 7 days (marked in green). Phase contrast images of cells were collected by a Nicon Diaphot 300 fluorescence microscope. Green/red colour merged images were collected by Zeiss confocal microscope following immuno-staining of cells with Sigma anti-lamin A rabbit antibody (in green) and anti-α-actin mouse antibody (in red). Senescent cultures display flat appearance and an increased cell size. As a result they have a lower number of cells per unit area devoid of cell-cell contacts as opposed to spindle-shaped morphology of early passage cultures in which any given cell is encircled by a number of neighbouring cells. Senescent cells display increased levels of senescence-associated β-galactosidase (SA-β-gal) activity, α-actin stress fibres in the cytoplasm and dysmorphic nuclei.*

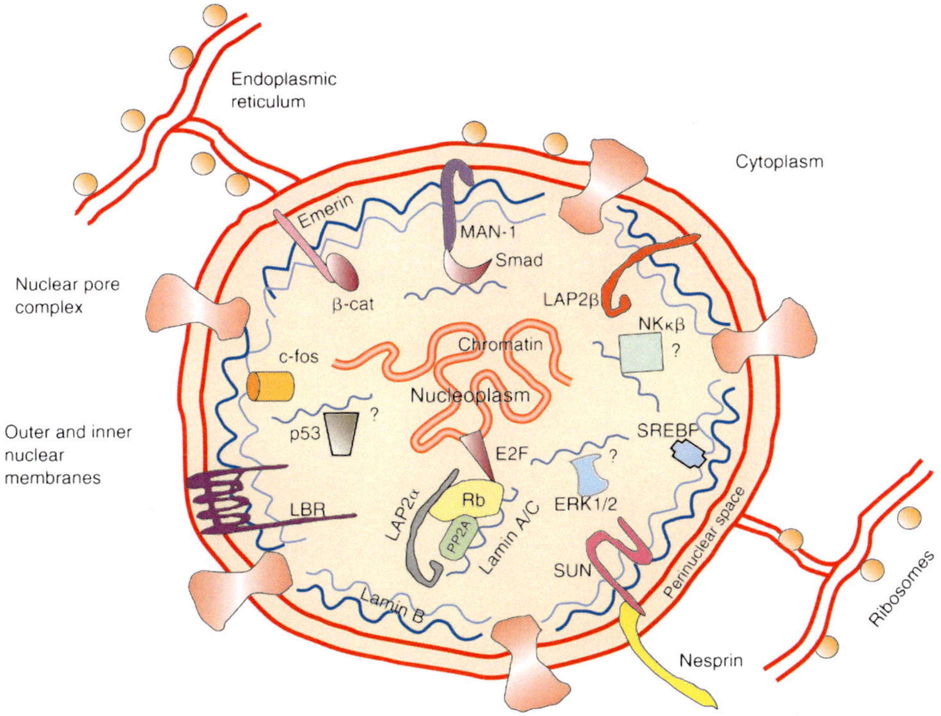

Chapter 8, Figure 2. *A schematic diagram depicting an overview of the major structural components of the nuclear envelope (NE) and their interactions and organization in the nucleus. These include the outer and inner nuclear membranes (ONM and INM; the former of which is continuous with the endoplasmic reticulum lumen lined with ribosomes), the nuclear pore complexes (NPCs), and the nuclear lamina (NL) composed of A-type and B-type lamin oligomers and INM and ONM proteins. Some selected INM proteins include: lamin B receptor (LBR), lamina-associated protein 2β (LAP2β), MAN1, emerin and SUN1. SUN1 acts as a tether for Nesprin 2 at the ONM through interactions which span the perinuclear space and link cytoskeleton to the nuclear membrane. Chromatin domains are also anchored to the nuclear lamina. A-type lamins and their binding partners (emerin, MAN1, LAP2α, SUN1) form functional complexes at the nuclear periphery and in the nucleoplasm, which have a role in gene regulation and signalling pathways. Not depicted are complexes involving B-type lamin binding partners such as LAP2β and LBR. Some selected nucleoplasmic partners include LAP2α, PP2A, Rb, E2F, β-catenin (β-cat), c-fos, Smads and SREBP1. Question marks indicate suggested but not yet proven A-type lamin interactions with p53, NF-κβ and ERK1/2.*

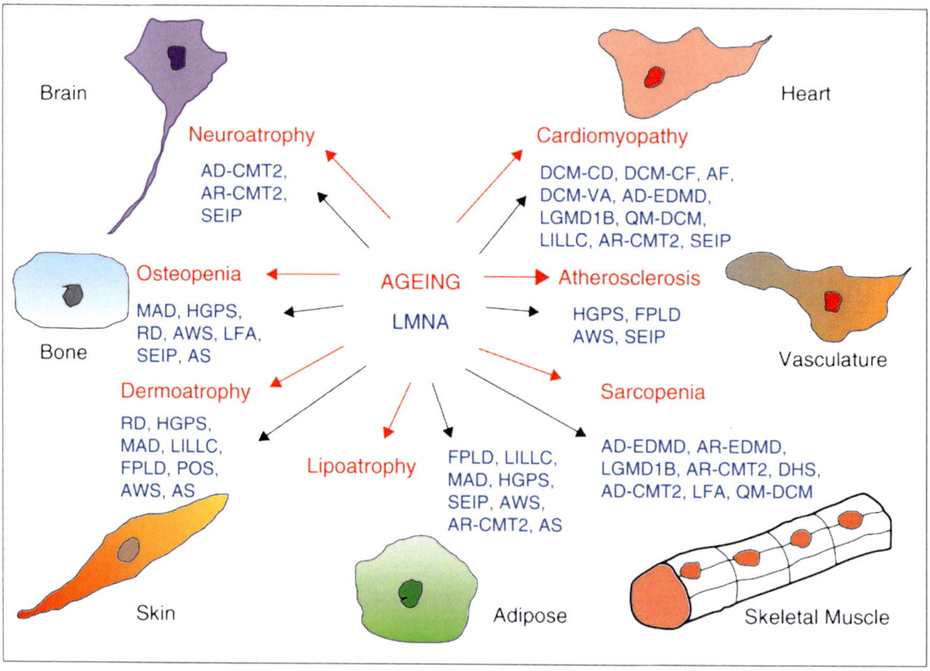

Chapter 8, Figure 3. *A schematic diagram drawing parallels between tissues affected in a number of age-associated degenerative disorders (labelled in red) and human diseases associated with mutations in the LMNA gene (labelled in blue). The full names of LMNA-associated diseases in the diagram are explained in the text (see Sections 5 and 6).*

Chapter 9, Figure 2. *Representative picture of non-transgenic and glyoxalase transgenic tobacco plants showing their relative tolerance to salt stress. Upper panel shows the relative growth of wild type (WT), Gly I, Gly II and Gly I + II transgenic plants in the continuous presence of 200 mM NaCl. Lower panel shows the growth of seedlings in the media containing 200 mM NaCl. Note the better growth and survival of the double transgenic plants.*

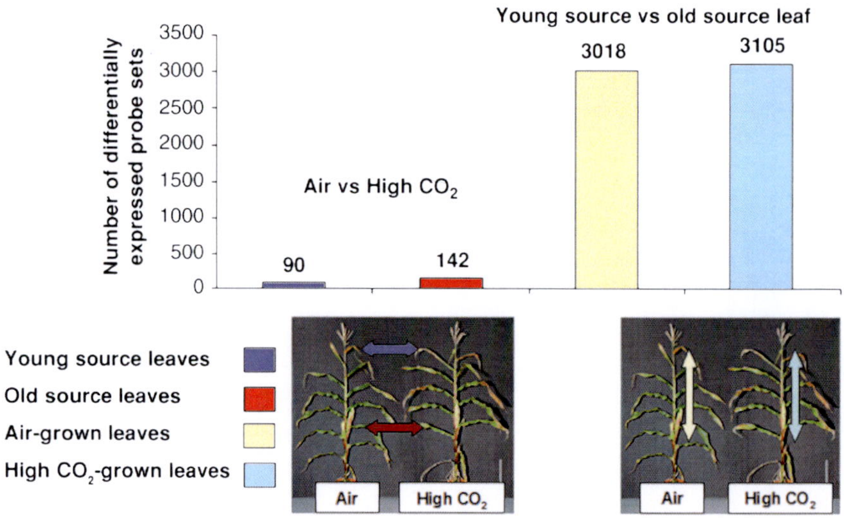

Chapter 11, Figure 2. *A comparison of the effect of age (young verses old source leaves; as illustrated in the lower figure on the right) and growth CO_2 availability (air: 350 μL L^{-1} CO_2 vs high: 700 μL L^{-1} CO_2; as illustrated in the lower figure on the left). Young source leaves, rank 12; mature source leaves, rank 3.*